Polysaccharide Materials: Performance by Design

ACS SYMPOSIUM SERIES **1017**

Polysaccharide Materials: Performance by Design

Kevin J. Edgar, Editor
Virginia Tech

Thomas Heinze, Editor
Center of Excellence for Polysaccharide Science

Charles M. Buchanan, Editor
Eastman Chemical Company

Sponsored by the
ACS Division of Cellulose and Renewable Materials

American Chemical Society, Washington DC

Library of Congress Cataloging-in-Publication Data

Polysaccharide materials : performance by design / [edited by] Kevin J. Edgar, Thomas Heinze, Charles M. Buchanan.
  p. cm. -- (ACS symposium series ; 1017)
  Includes bibliographical references and index.
  ISBN 978-0-8412-6986-6
  1. Carbohydrate drugs. 2. Polysaccharides. 3. Polymeric drug delivery systems. I. Edgar, Kevin J. II. Heinze, Thomas. III. Buchanan, Charles M.

  RS431.C23P65 2009
  615'.6--dc22

                                    2009034961

The paper used in this publication meets the minimum requirements of American National Standard for Information Sciences—Permanence of Paper for Printed Library Materials, ANSI Z39.48−1984.

Foreword

The ACS Symposium Series was first published in 1974 to provide a mechanism for publishing symposia quickly in book form. The purpose of the series is to publish timely, comprehensive books developed from the ACS sponsored symposia based on current scientific research. Occasionally, books are developed from symposia sponsored by other organizations when the topic is of keen interest to the chemistry audience.

Before agreeing to publish a book, the proposed table of contents is reviewed for appropriate and comprehensive coverage and for interest to the audience. Some papers may be excluded to better focus the book; others may be added to provide comprehensiveness. When appropriate, overview or introductory chapters are added. Drafts of chapters are peer-reviewed prior to final acceptance or rejection, and manuscripts are prepared in camera-ready format.

As a rule, only original research papers and original review papers are included in the volumes. Verbatim reproductions of previous published papers are not accepted.

ACS Books Department

Contents

Indexes

Preface

Events of the last few decades have led to remarkable growth in the importance of polysaccharides to the global economy. Polysaccharides are natural polymers found in all living organisms, with variety and complexity of structure and function that are unrivaled among natural polymers. Concerns about depletion of the global petroleum supply, and a consequent severe rise in petroleum prices, have enhanced interest in polysaccharide science. Much of this interest to date has been directed at improving the processes by which polysaccharides can be converted to biofuels such as ethanol and n-butanol. The recent instability in petroleum prices has underscored the importance of this research, which promises to create the basis for an industry producing renewable fuels that do not contribute to global warming.

Natural polysaccharides are also of growing and intense interest as sources of materials. Some of this interest is driven by the same price, climate change, security, and availability factors that propel the current interest in biofuels production. Petroleum is, after all, also the source of most of the materials used to create the objects we use in all aspects of our daily lives. Another key factor is our developing ability to control polysaccharide microstructure, determine structure-property relationships, and rationally design high-performance materials from polysaccharides. The ability to understand, control, and utilize the structures of these complex polymers, and their derivatives, promises to provide a new realm of renewable and abundant building blocks for modern materials scientists.

This book is based on a recent symposium on these subjects, entitled "Synthesis and Structure/Property Relationships of Polysaccharides", the Anselme Payen Award symposium of the ACS Cellulose and Renewable Materials Division in Chicago, Illinois, March 25-27, 2007, in honor of Charles Buchanan. Perusal of this volume will take the reader through a series of chapters covering recent progress in the important area of polysaccharides in drug delivery, where the vital roles that polysaccharides and derivatives in areas such as targeted delivery, drug bioavailability, and patient compliance are elucidated. The following chapters provide an enlightening view of the frontiers in the synthesis of polysaccharide derivatives of controlled microstructure, and the use of these polysaccharides and derivatives in advanced materials such as nanostructured composites. The volume concludes with two chapters; first, one on the utilization of complex polysaccharides as chemical feedstocks (biorefinery concept), which may be converted to small molecule building blocks in schemes analogous in concept, but rather different in chemistry, from those employed in the petroleum refinery. The final chapter discusses the all-important topic of the analytical chemistry of polysaccharides, specifically solid state NMR spectroscopy of polysaccharides. The challenges of analysis underpin all other studies of these complex materials.

We wish to thank the sponsors of the 2007 Payen Award Symposium, without whom this volume would not have been possible:

Eastman Chemical Company	Novamont Industries
Borregaard ChemCell	Rayonier Specialty Fibers
CyDex Pharmaceuticals	Weyerhaeuser

We thank the international group of reviewers, who gave willingly and generously of their time to help improve the quality of these chapters. We offer sincere appreciation to Jessica Rucker of ACS, whose patience and knowledge made the editorial process smooth and relatively painless. The editors are indebted to their respective institutions for financial and logistic support of this endeavor. Finally, thanks to all our authors, whose skill and insight created the science, and whose communication skills show through in every page of contributions; it is truly their book.

Kevin J. Edgar
> Macromolecules and Interfaces Inst. & Dept. of Wood Science & Forest Products, Virginia Tech
> 230 Cheatham Hall, Mail Code 0323
> Blacksburg, VA 24061, USA

Charles M. Buchanan
> Research Laboratories
> Eastman Chemical Company
> Kingsport, TN 37662, USA

Thomas Heinze
> Center of Excellence for Polysaccharide Science
> Friedrich Schiller University of Jena
> Humboldtstrasse 10
> D-07743 Jena, Germany

Introduction

Polysaccharide Chemistry: Frontiers and Challenges

Kevin J. Edgar

**Macromolecules and Interfaces Institute, and
Department of Wood Science and Forest Products
Virginia Tech, 230 Cheatham Hall, Blacksburg, VA 24061**

Polysaccharide chemistry is at a watershed moment. We have seen once-vigorous areas of science and industry decline, such as textile and paper chemistry. Yet, simultaneously, vast new frontiers are opening before us, because of the increasing demand for and decreasing supply of petroleum, and the growing appreciation for the enormous flexibility and information content of polysaccharides. The emerging biorefinery-based economy offers almost unlimited potential for contributions from polysaccharide chemists, in order to solve difficult problems associated with the conversion of biomass into biofuels and biomaterials. It would be difficult to imagine a field with more potential for global impact. At the same time, polysaccharide scientists are challenged to solve long standing scientific problems that could aid in meeting these critical global challenges, including the determination of fine structure, controlled synthesis, and elucidation of detailed structure-property relationships for polysaccharides and derivatives. Here I present a viewpoint on the state of the art, and the major challenges ahead for polysaccharide chemistry.

Introduction

It is not an exaggeration to say that we stand at a watershed moment in the history of polysaccharide science. On the one hand, every practitioner in the field is aware of the disruptions over the last couple of decades that have been caused by the decline of historical segments of polysaccharide science and industry, including paper, textiles, and pulp manufacture. On the other hand, the scientific advances in polysaccharide chemistry have been staggering. The importance of these natural polymers in the chemistry of life itself has been underlined again and again, by discoveries of roles for polysaccharides in fields like blood coagulation [1], pathogen recognition (*e.g.*, malaria [2] and Lyme disease [3]), and communication [4,5]. These scientific advances have been accompanied by massive societal shifts that have turned the attentions of the world towards polysaccharides. The declining supply of petroleum, the increases in petroleum demand that are resulting from economic growth in Asia, South America, and Africa, and the resultant price increases for fuel and petroleum-based materials have rocked modern society[6]. Global attention is now focused on ways to reduce dependence on petroleum, including conservation, solar and wind-based energy, and enhanced nuclear energy production. Biomass-derived biofuels must be an important part of the solution, and plant-derived biomass is roughly 2/3 polysaccharides [7]. Therefore the chemistry of depolymerization of polysaccharides to monomers [8], and fermentation of those monomers to fuels like ethanol[9] and n-butanol[10], receives unprecedented attention. At the same time, we must not forget our dependence on petroleum-based chemicals and materials. Considerable attention must also be focused on the development of polysaccharide-based materials to carry part of the load now borne by petroleum-based materials. The focus of this book and of this introductory chapter is on the development of well-characterized polysaccharides and derivatives, the in-depth understanding of their structure-property relationships, and the subsequent intelligent design of polysaccharides and derivatives for high-performance applications. In this introductory chapter, I will highlight some of the current frontiers, and some of the challenges which must be met if we are to utilize polysaccharides to their fullest potential in high-value applications, in many cases replacing petroleum-based materials.

Challenges

Polysaccharides play a stunning variety of roles in nature. They range from food storage (starch), to structural reinforcement (cellulose), to composite interphase adhesion (hemicellulose), to soft, crosslinked structural gels (alginates)[11], to control of cellular adhesion (hyaluronic acid)[12], to control of blood clotting (heparin)[13] and neural development (chondroitin and dermatan sulfates)[14], and many, many other roles. Scientists and engineers have found that it is often useful to make semisynthetic derivatives from natural polysaccharides, in order to overcome difficulties such as poor solubility and inability to process thermally, or to enhance other specific properties (water resistance, for example). Such derivatives are important items in many fields of commerce today, and are crucial elements of products that are important to modern society, including flat screen televisions, laptop computers, foods, latex paints, pharmaceutical formulations, drugs, and dialysis systems[15]. As we know, microstructural variation can have a profound impact on properties and function of the polysaccharide. Perhaps the best example of this is the fact that a change only in anomeric stereochemistry makes the difference between cellulose and starch amylose, which have such different properties and roles in nature. Therefore, it is all the more remarkable to reflect on the following deficiencies:

- In many cases we lack adequate methodology to determine the position of substitution around the monosaccharide ring.

- Unlike in some fully synthetic polymers, in polysaccharides and derivatives it is usually quite difficult to determine the monomer sequence, or even the monomer sequence pattern. We often do not have primary evidence for whether a polysaccharide sequence is random, blocky, or occurs in some other pattern.

- We have no primary evidence for the degree of interchain variation, particularly for polysaccharide derivatives. Since so many of these polysaccharide derivatives are made by heterogeneous processes in which undissolved fibers composed of multiple polysaccharide chains are the starting material, this is a matter of concern.

- Our methods for controlling position of substitution, monomer sequence, and interchain variation in polysaccharide derivative synthesis are, to the extent that we can determine, often incomplete and imprecise.

- It will come as no surprise, then, that since we often can't adequately determine or control microstructural elements like substituent position and monomer sequence, we lack knowledge about the influence of these microstructural elements on properties.

Perhaps the easiest way to understand the importance of polysaccharide microstructure is to reflect upon the information content of polysaccharides, in comparison with other biopolymers that are more commonly considered to be information carriers (DNA, RNA, and proteins). As we know, the code of RNA and DNA is quite simple, consisting of only four "letters", corresponding to the four bases each for RNA and DNA. A better comparison for polysaccharides may be proteins. In living organisms, proteins are composed of 20 "letters", that is 20 natural amino acids. The amino acids are always linked to form the protein molecule in the same way; an amide linkage between the carboxylic acid of one amino acid and the alpha-amino group of the next. The linkage does not create an asymmetric center so does not itself have stereoisomers, and the amino acids do not have variants that contain other substituents in nature, as far as we know.

Nature uses approximately 20 common monosaccharides to construct most natural polysaccharides[16] (however, many more rare monosaccharides are found in certain natural polysaccharides, and the family continues to grow)[17]. The linkage of one monosaccharide to another does create a new asymmetric center, and often both stereochemistries appear in nature (e.g., the alpha (amylose) and beta (cellulose) linkages of 1→4 glucans). Further, while it is always the anomeric carbon of the acceptor monomer that forms the linkage, any of the free hydroxyls of the donor monomer may be involved in the linkage. In addition, substituents are possible; acetic acid esters are for example quite common in nature. An additional potential for complexity, and information content, in

6

polysaccharides is branching; this is not seen in natural polypeptides. Then there is the element of ring size; monosaccharides are cyclic hemiacetals formed from polyhydroxy aldehydes, and natural monosaccharides can (and sometimes do in nature) adopt either 5-membered ring (furanose) or 6-membered ring (pyranose) structures. Finally, there is the issue of absolute stereochemistry. In natural proteins, only the L-amino acids are observed; that is, they can be related back to L-glyceraldehyde in their absolute stereochemistry. In contrast, natural monosaccharides are found that have either the D or the L configuration as related to glyceraldehyde, and in a few cases, both configurations are observed in nature for the same monosaccharide (e.g., galactose, rhamnose, and fucose). We can make a crude quantitative comparison of complexity and information content by considering an oligomer of four monosaccharides, in comparison with a tetrapeptide (**Figure 1**).

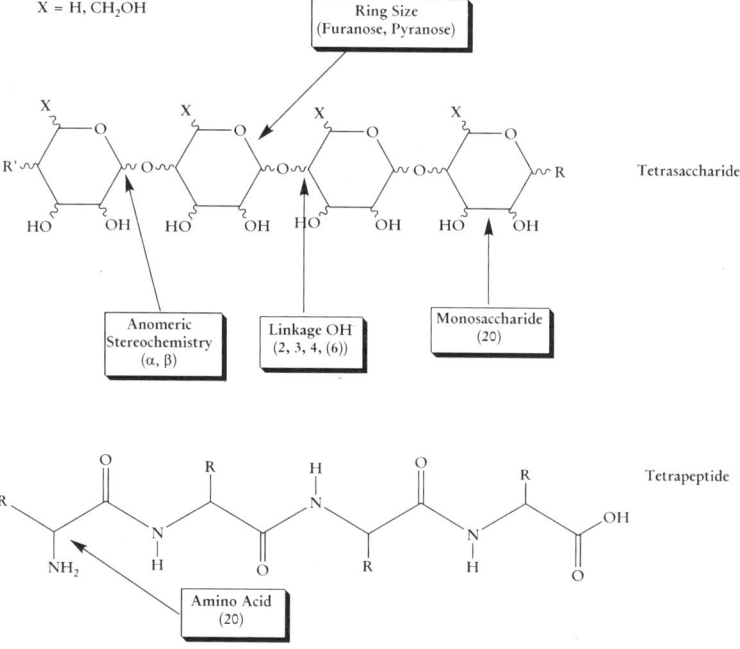

Figure 1. Potential variability of polysaccharides vs. proteins (note that potential for branching and substituents in polysaccharides is not illustrated)

If we include only the potential for amino acid variation in the oligopeptide (the only possible variation), and only the possible monosaccharide, linkage location, ring size, and absolute stereochemical variation (but not including the possibilities of branching, or of substituents) for the oligosaccharide, there are (**Table 1**) a couple of hundred thousand possible tetrapeptides, and about **84 billion possible tetrasaccharides**! Equally importantly, nature makes use of this complexity, or information content, depending on how you look at it. We know, for example, that a highly specifically sulfonated and acetylated pentasaccharide is the active portion of the heparin polysaccharide that binds the protein antithrombin III and prevents blood clotting[18]. Very recently a group of scientists determined, with remarkable speed, that batches of heparin intended for blood clotting prevention in medical situations such as surgery had been contaminated with an over-sulfonated chondroitin sulfate. These scientists thus helped to protect the public against further contamination and associated illness[19]. Specifically substituted chondroitin sulfate oligomers have been shown to guide the development of the central nervous system[20]. The highly specific structure of cell surface oligosaccharides is also exploited by pathogens, for example the pregnancy-associated malaria pathogen which recognizes certain chondroitin sulfate sequences on the surface of the placenta to guide its entry[21]. Clearly, polysaccharide microstructure matters.

Table 1. Comparative Information Density of Proteins and Polysaccharides

Possible Tetrapeptides	Possible Tetrasaccharides
160,000	83,886,080,000*

*Not including the possibilities of substituents and branching

Frontiers

Progress has been made on these problems, some of which is described elsewhere in this volume. The most fundamental issue is the lack of general analytical methods to determine position of substitution, monomer sequence, and interchain variability[22]. The methods available to analyze interchain variability, control it, or determine its effect on properties are very limited, so this aspect won't be considered further in this article.

In certain cases, we can determine the position of substitution. Of course, fully substituted polysaccharide derivatives, especially ones containing only one substituent type, are the simplest analytical problems and have in many cases been solved. For example, full spectral assignments have been carried out for the proton and carbon NMR spectra of cellulose triacetate, based on results of multidimensional NMR techniques[23]. There has also been progress on polysaccharide derivatives in which the substituent linkage is stable to the conditions required for breakdown of the polysaccharide into monosaccharides, by acidic aqueous hydrolysis of the acetal linkages. Polysaccharide ether and sulfate[24] substituents are often stable under such conditions. Hydrolysis of cellulose ethers to monosaccharides, reduction of those monosaccharides to glucitols, and analysis of the glucitol mixture (either by HPLC directly, or by GC after reaction of the free hydroxyl groups with acylating or silylating reagents) provides useful information. Each monosaccharide can be synthesized and characterized with respect to retention time or volume, response factor, and mass spectrum. This information can be used to calibrate the analysis of the polysaccharide ether, and enable quantification of the individual component monosaccharides. The result is complete and accurate knowledge of the monosaccharide composition of polysaccharide derivatives like cellulose alkyl ethers and starch sulfates. In addition, the monomer distribution can be compared with those expected from randomly substituted polysaccharide derivatives, and a statistical comparison can give some information about the

8

blocky or random nature of the monomer sequence[25]. These techniques do not provide information about the precise monomer sequence of the polymer.

Methods have also recently been developed to characterize the position of substitution of other polysaccharide derivatives whose substituent linkages are not stable to hydrolysis, such as polysaccharide esters. Partially substituted polysaccharide esters may be converted to fully substituted analogs in which the new substituent is distinct from that on the polymer to be analyzed[26]; for example, a partially substituted cellulose acetate might be reacted with propionic anhydride to create a fully substituted cellulose acetate propionate[27]. The ^{13}C NMR spectra of these fully substituted derivatives are less complex and are more easily interpreted than those of partially substituted esters, and so modern NMR techniques such as COSY, INAPT and DEPT can be used to elucidate the positions of substitution in some cases. Any technique that relies upon derivatization of a polysaccharide ester to determine its structure must confront the possible problems of ester interchange, and migration of ester groups from one hydroxyl group to another during the derivatization reaction. It is not yet clear that any of these recent derivatization/NMR techniques are proven to avoid these issues.

Recently, interesting progress has been made on the control and analysis of monomer sequence, using enzymatic catalysis. Alginates are polysaccharides that are biosynthesized in algae and in certain bacteria, originally as β-1→4 linked homopolymers of mannuronic acid (M), or mannuronan[28]. The organism then modifies the polymer according to the functional requirements of the particular tissue and the particular circumstances, by partial epimerization of the monomers at C-5[29] (**Figure 2**).

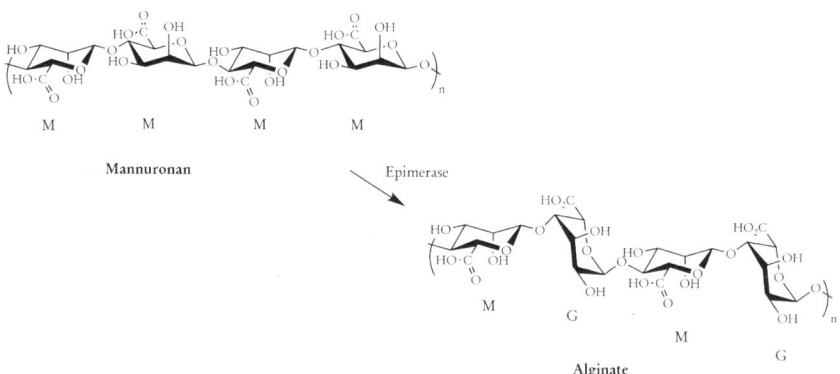

Figure 2. Epimerization of mannuronan to alginate

Upon epimerization at C-5, a new monosaccharide is formed, L-guluronic acid (G). As many as seven distinct epimerases have been isolated and identified from a particular bacterium. Each epimerase has different specificities, some converting the mannuronan into a perfect alternating M-G-M-G copolymer, others which would convert the polymer to a homoguluronic acid polymer if allowed to react to completion, and others which appear to make more random sequences. Since the G-G diads have the ability to form crosslinked chelates with calcium and a G-G diad on another molecule of alginate, such epimerization has profound consequences for the strength and flexibility of those particular tissues of the alga or bacterium[30]. Because bacteria can now be grown lacking epimerases, they can be used to produce mannuronan, and because some of the alginate epimerases have now been isolated in quantity, they can be used *in vitro* to partially epimerize the mannuronan to an alginate copolymer of M and G, in order to create specific sequence patterns, and determine the impact of sequence on properties[31]. Fortunately, the anomeric proton signals in the ^{1}H NMR spectrum are sufficiently resolved for M, G, and common diads like M-M, G-G, and M-G, that these can be quantified by proton NMR of the anomeric region. The structure-property relationships of alginic acid are of importance not only in the functioning of algae such as marine kelp, but are also of recent interest with respect to the alginate-based biofilms[32] formed by the bacterium *Pseudomonas aeruginosa*, which protect the bacterium against the immune system and antibiotics, and have been implicated in chronic infections suffered by cystic fibrosis patients[33]. Since all natural polysaccharides are made by similar enzymatic transformations, it may be that further advances in identification and isolation of the enzymes that catalyze these transformations will give scientists the tools to create and identify specific monomer sequences, and learn about the impact of sequence on properties.

Another method, *de novo* synthesis has recently become important for the creation, analysis, and determination of structure property relationships of polysaccharides with fully controlled sequence and position of substitution. Only in recent years has carbohydrate chemistry advanced sufficiently to make the synthesis of oligosaccharides of moderate DP practical[34], and also in relatively recent times, scientists have probed the question of what DP is necessary for an oligosaccharide to approach the properties of the equivalent polysaccharide. In general, oligosaccharide properties seem to approach those of the high polysaccharide when DP is in the 6-12 range[35], depending on the particular polysaccharide structure[36]. Thus, it has now become practical to use *de novo* synthesis to make polysaccharides of different monomer sequences, with fully controlled positions and extents of substitution, in order to determine the impact on properties[37]. The chapter in this book by Kamitakahara and co-workers illustrates this concept beautifully. While this methodology remains labor-intensive, and thus will be restricted for the moment to examination of the most important problems, it is the most well-controlled and reliable way we have to construct these microstructural structure-property relationships.

Finally, one must cite the recent advances in the solution chemistry of cellulose and other polysaccharides[38], and the potential impact of this solution chemistry on the ability to synthesize specific substitution patterns[39]. The ability to react cellulose in solution, rather than in a heterogeneous, topochemical process as in most commercial cellulose derivative manufacture[40], affords the possibility of using far more selective conditions. Catalysis can be milder or nonexistent[41], temperatures lower, and far more selective reagents can be used. Very interesting advances have been made in cellulose derivative synthesis in N,N-dimethylacetamide (DMAC)/LiCl, tetrabutylammonium fluoride (TBAF)/DMSO[42], and in ionic liquids[43] such as 1-N-butyl-3-methylimidazolium chloride[44]. One example, shown in **Figure 3**, is the reaction of cellulose in

DMAC/LiCl solution at both the C-6 and C-2 hydroxyls with very high selectivity, by using the sterically demanding reagent thexyldimethylsilyl chloride. This selective reaction with a sterically demanding reagent would be impossible using conventional heterogeneous polysaccharide ether synthesis methods. Alkylation of the protected cellulose at the C-3 OH, and silyl deprotection affords nearly pure 3-O-ethylcellulose. Interestingly, an aqueous solution of this pure 3-O-ethylcellulose undergoes thermal gellation at 58.5°C, as opposed to 30°C for a statistically substituted ethyl cellulose, hinting at the important impact of positional selectivity on material properties. We can expect further advances in selective polysaccharide substitution using these solution synthesis methods. As they provide us with authentic samples of selectively substituted derivatives, this will enable advancements in both analytical and structure-property studies.

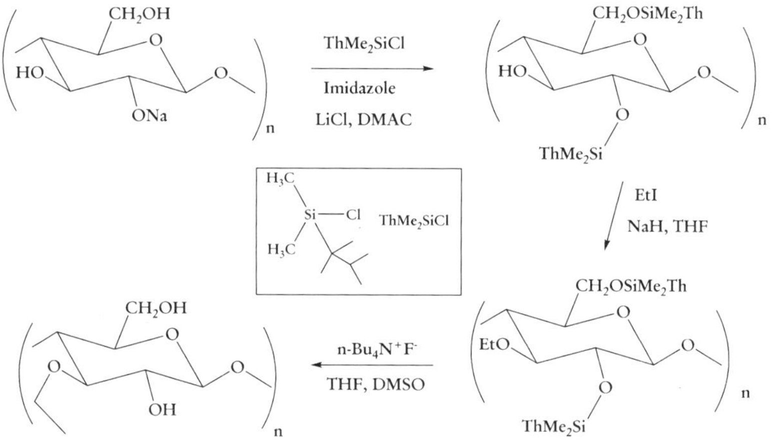

Figure 3. Selective substitution of cellulose in solution

Conclusion

Polysaccharides are perhaps the most versatile natural polymers, and certainly are the most potent carriers of information. When modified and derivatized to make semi-synthetic materials, they have enormous potential, particularly in the coming biorefinery economy, not only to displace petroleum-based materials as petroleum supply diminishes and prices rise, but to deliver high performance and value. The control of polysaccharide microstructure, the understanding of its impact on properties, and the ability to design microstructure to achieve those properties will be key elements in the emergence of that biorefinery economy. As we have shown, progress is being made towards these goals, by the combination of analytical chemistry, enzyme catalysis, solution-phase chemistry, and *de novo* synthesis. One can now imagine a future in which materials scientists possess the necessary tools to fully realize the potential of polysaccharide-based materials.

11

References

1. Linhardt, R.J. *J. Med. Chem.* **2003**, *46*, 2551-2564.
2. Dinglasan, R.R.; Alaganan, A.; Ghosh, A.K.; Saito, A.; van Kuppevelt, T.H.; Jacobs-Lorena, M. *Proc. Nat. Acad. Sci.* **2007**, *104*, 15882-15887.
3. Leong, J.M.; Robbins, D.; Rosenfeld, L.; Lahiri, B.; Parveen, N. *Infect. Immun.* **1998**, *66*, 6045-6048.
4. Holt, C.E.; Dickson, B.J. *Neuron* **2005**, *46*, 169-172.
5. Gama, C.I.; Tully, S.E.; Sotogaku, N.; Clark, P.M.; Rawat, M.; Vaidehi, N.; Goddard III, W.A., Nishi, A.; Hsieh-Wilson, L.C. *Nature Chem. Biol.* **2006**, *2*, 467-473.
6. Kerr, R.A. *Science* **2008**, *322*, 1178-1179.
7. Ragauskas, A.J.; Williams, C.K.; Davison, B.H.; Britovsek, G.; Cairney, J.; Eckert, C.A.; Frederick, W.J., Jr.; Hallett, J.P.; Leak, D.J.; Liotta, C.L.; Mielenz, J.R.; Murphy, R.; Templer, R.; Tschaplinski, T. *Science* **2006**, *311*, 484-489.
8. Himmel, M.E.; Ding, S.-Y.; Johnson, D.K.; Adney, W.S.; Nimlos, M.R.; Brady, J.W.; Foust, T.D. *Science* **2007**, *315*, 804-807.
9. Frederick Jr., W.J.; Lien, S.J.; Courchene, C.E.; DeMartini, N.A.; Ragauskas, A.J.; Iisa, K. *Biomass Bioenergy* **2008**, *32*, 1293-1302.
10. Ezeji, T.C.; Qureshi, N.; Blaschek, H.P. *Curr. Opin. Biotech.* **2007**, *18*, 220-227.
11. Mørch, Y.A.; Holtan, S.; Donati, I.; Strand, B.L.; Skjåk-Bræk, G. *Biomacromolecules* **2008**, *9*, 2360-2368.
12. Day, A.J.; Prestwich, G.D. *J. Biol. Chem.* **2002**, *277*, 4585-4588.
13. Capila, I.; Linhardt, R.J. *Angew. Chem. Int. Ed.* **2002**, *41*, 390-412.
14. Murrey, H.E.; Hsieh-Wilson, L.C. *Chem. Rev.* **2008**, *108*, 1708-1731.
15. Edgar, K.J.; Buchanan, C.M.; Debenham, J.S.; Rundquist, P.A.; Seiler, B.D.; Shelton, M.C.; Tindall, D. *Prog. Polym. Sci.* **2001**, *26*, 1605-1688.
16. Kennedy, J.F. *Carbohydrate Chemistry*, Oxford University Press, New York, **1990**, pp 3-41.
17. Mohnen, D. *Current Opinion Plant Biol.* **2008**, *11*, 266-277.
18. Lindahl, U.; Bäckström, G.; Höök, M.; Thunberg, L.; Fransson, L.-Å.; Linker, A. *Proc. Natl. Acad. Sci. USA* **1979**, *76*, 1218-1222.
19. Guerrini, M.; Beccati, D.; Shriver, Z.; Naggi, A.; Viswanathan, K.; Bisio, A.; Capila, I.; Lansing, J.C.; Guglieri, S.; Fraser, B.; Al-Hakim, A.; Gunay, N.S.; Zhang, Z.; Robinson, L.; Buhse, L.; Nasr, M.; Woodcock, J.; Langer, R.; Venkataraman, G.; Linhardt, R.J.; Casu, B.; Torri, G.; Sasisekharan, R. *Nature Biotech.* **2008**, *26*, 669-675.
20. Rawat, M.; Gama, C.I.; Matson, J.B.; Hsieh-Wilson, L.C. *J. Am. Chem. Soc.* **2008**, *130*, 2959-2961.
21. Achur, R.N.; Kakizaki, I.; Goel, S.; Kojima, K.; Madhunapantula, S.V.; Goyal, A.; Ohta, M.; Kumar, S.; Takagaki, K.; Gowda, D.C. *Biochemistry* **2008**, *47*, 12635-12643.
22. Mischnick, P. *Cellulose* **2002**, *9*, 1-13.

23. Einfeldt, L.; Günther, W.; Klemm, D.; Heublein, B. *Cellulose* **2005**, *12*, 15-24.
24. Mischnick, P.; Hennig, C. *Biomacromolecules* **2001**, *2*, 180-184.
25. Adden, R.; Müller, R.; Brinkmalm, G.; Ehrler, R.; Mischnick, P. *Macromol. Biosci.* **2006**, *6*, 435-444.
26. Liebert, T.F.; Heinze, T. *Biomacromolecules* **2005**, *6*, 333-340.
27. Tezuka, Y.; Tsuchiya, Y. *Carbohydrate Res.* **1995**, *273*, 83-91.
28. Draget, K.I.; Smidsrød, O.; Skjåk-Bræk, G. Alginates from Algae, in Polysaccharides and Polyamides in the Food Industry. Steinbuchel, A.; Rhee, S.K., eds. Wiley VCH, Weinheim, pp 1-30, **2005**.
29. Mørch, Y.A.; Donati, I.; Strand, B.L.; Skjåk-Bræk, G. *Biomacromolecules* **2007**, *8*, 2809-2814.
30. Russo, R.; Malinconico, M.; Santagata, G. *Biomacromolecules* **2007**, *8*, 3193-3197.
31. Holtan, S.; Bruheim, S.; Skjåk-Bræk, G. *Biochem. J.* **2006**, *395*, 319-329.
32. Ramsey, D.M.; Wozniak, D.J. *Molecular Microbiology* **2005**, *56*, 309-322.
33. Leid, J.G.; Willson, C.J.; Shirtliff, M.E.; Hassett, D.J.; Parsek, M.R.; Jeffers, A.K. *J. Immunology* **2005**, *175*, 7512-7518.
34. Nakatsubo, F.; Kamitakahara, H.; Hori, M. *J. Am. Chem. Soc.* **1996**, *118*, 1677-1681.
35. Benesi, A.J.; Brant, D.A. *Macromolecules* **1985**, *18*, 1109-1116.
36. Buchanan, C.M.; Hyatt, J.A.; Kelley, S.S.; Little, J.L. *Macromolecules* **1990**, *23*, 3747-3755.
37. Kamitakahara, H.; Nakatsubo, F.; Klemm, D. *Cellulose* **2006**, *13*, 375-392.
38. Liebert, T.F.; Heinze, T.J. *Biomacromolecules* **2001**, *2*, 1124-1132.
39. Koschella, A.; Fenn, D.; Illy, N.; Heinze, T. *Macromol. Symp.* **2006**, *244*, 59-73.
40. Gedon, S.; Fengl, R. *Cellulose Esters, Organic Esters* in Kirk-Othmer Encyclopedia of Chemical Technology, **1993**, *5*, Wiley & Sons, NY, NY.
41. Edgar, K.J.; Arnold, K.M.; Blount, W.W.; Lawniczak, J.E.; Lowman, D.W. *Macromolecules* **1995**, *28*, 4122-4128.
42. Kohler, S.; Heinze, T.; *Macromol. Biosci.* **2007**, *7*, 307-314.
43. El Seoud, O.A.; Koschella, A.; Fidale, L.C.; Dorn, S.; Heinze, T. *Biomacromolecules* **2007**, *8*, 2629-2647.
44. Schlufter, K.; Schmauder, H.-P.; Dorn, S.; Heinze, T. *Macromol. Rapid Comm.* **2006**, *27*, 1670-1676.

Chapter 1

Polysaccharides in Oral Drug Delivery – Recent Applications and Future Perspectives

Sandra Klein

Institute of Pharmaceutical Technology, Goethe University, Frankfurt am Main, Germany

The oral route of drug delivery is the largest, the oldest and the fastest growing segment of the overall drug delivery market. Oral drug delivery is the most preferred route for drug administration, because it represents the least invasive route. Further, oral dosage forms are easy to administer and typically come along with the highest patient compliance. However, there are only very few drugs that can be directly administered to the patient without excipients. Moreover, most of the orally administered drugs need to be adequately formulated to be delivered by means of single- or multiple-unit dosage forms. In drug formulations excipients have to fulfill very different requirements. They are used as simple filling materials, to increase the solubility and/or bioavailability of the active drug, to ease manufacturing of a dosage form, to achieve a certain release profile from the final formulation or to enhance the stability of the final drug product.

Polysaccharides represent a very important group of excipients. Particularly cellulose, starch and their derivatives have become standard excipients during the last decades. In the future the number of polysaccharide-based excipients is assumed to further increase. The present chapter gives an overview of the historical use of polysaccharide excipients and an outlook on some future applications.

Introduction

Polysaccharides represent natural molecules that are widely present in living organisms. They can be obtained from abundant renewable sources and are typically non-toxic which makes them attractive for use in pharmaceutical formulations. In addition, they offer a wide variety of valuable physicochemical properties. Thus, they have traditionally been used as pharmaceutical excipients. However, polysaccharides can also be used as starting materials for different derivatives which often possess even better physicochemical properties than their raw materials. Nowadays the pharmaceutical use of polysaccharides and their derivatives ranges from excipients through carriers and protecting agents to active substances themselves. Therefore, they are listed in many international pharmacopoeia and formularies. The United States (U.S.) National Formulary (NF) currently contains more than 30 polysaccharide-related monographs and particularly the number of polysaccharide derivatives for pharmaceutical applications is expected to further increase during the next decades.

Table 1: Polysaccharide monographs in the U.S: National Formulary 25[1]

Acacia	Guar Gum
Agar	Hydroxyethyl Cellulose (HEC)
Alginic Acid	Hydroxypropyl Cellulose (HPC)
Alfadex (α - Cyclodextrin)	Hydroxypropyl Methylcellulose (HPMC)
Betadex (β - Cyclodextrin)	Hymetellose (Methylhydroxyethylcellulose)
Carboxymethylcellulose	Hypromellose Acetate Succinate (HPMCAS)
Carageenan	Hypromellose Phthalate (HPMCP)
Carboxymethylcellulose Calcium	Methylcellulose (MC)
Carboxymethylcellulose Sodium	Pectin
Cellaburate (Cellulose Acetate Butyrate)	Sodium Starch Glycolate
Cellulose	**Starch**
· *Microcrystalline Cellulose*	· *Modified Starch*
· *Cellacefate*	· *Pregelatinized Starch*
(Cellulose Acetate Phthalate C-A-P)	· *Pregelatinized modified Starch*
· *Cellulose Acetate*	· *Corn Starch*
Croscarmellose Sodium	· *Potato Starch*
Dextrin	· *Tapioca Starch*
Ethylcellulose (EC)	· *Wheat Starch*
Galageenan	Tragacanth
Gellan Gum	Xanthan Gum

Oral drug delivery

A prerequisite for the oral bioavailability of a drug is that the drug dissolves before it reaches the site of action. Only a dissolved drug can permeate through the gastrointestinal (GI) membranes (typically those in the small intestine) to appear in the blood circulation and subsequently arrive at the site of action (see figure 1). Thus, whenever we administer a drug orally by means of a dosage form, this dosage form has to adequately release the drug, either by complete disintegration and dissolution or by diffusion of the dissolved drug through a hydrated matrix or membrane. In any case, the most important thing is to dissolve the active.

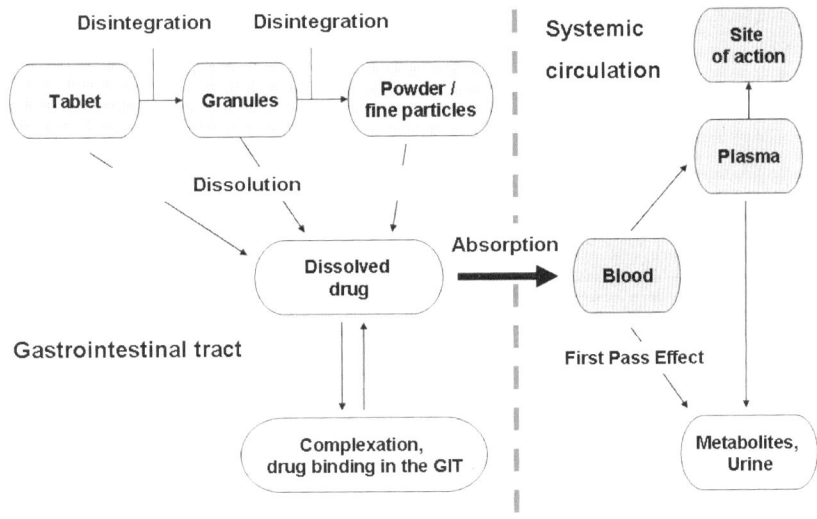

Figure 1: Schematic representation of the processes involved in the dissolution and absorption process of a drug after oral administration

After oral application, the dosage form is confronted with physiological conditions that can influence drug release. For an adequate dosage form design, it is thus essential to take these GI-parameters into account. Figure 2 and table 2 summarize some of the important processes and parameters in different sections of the human GI tract. However, in addition to the GI physiology, the choice of excipients can also be crucial for the *in vivo* behavior of the dosage forms. In the following sections therefore various polysaccharide-based excipients and their use in pharmaceutical dosage forms will be discussed.

Table 2: Segments of the GI tract and their corresponding functions [2]

Organ	Main processes	Secretions/Enzymes	pH
Mouth	mechanical digestion of food (chewing, grinding)	amylase (in saliva)	
Esophagus	passage of food from mouth through stomach	.	
Stomach	storage and mechanical digestion of food start of chemical digestion (pepsinogen secretion and activation) reduction of the number of bacteria	pepsinogen ↓ HCl pepsin	1-3 fasted 4-7 fed
Small intestine	continuation and completion of chemical digestion of carbohydrates, protein & lipids absorption of small soluble nutrients	pancreatic juice intestinal juice bile	pH-gradient along the SI ~5-7.5
Large Intestine (Colon)	water absorption from indigestible feces (proximal) (distal)		proximal 5-7 distal ~7

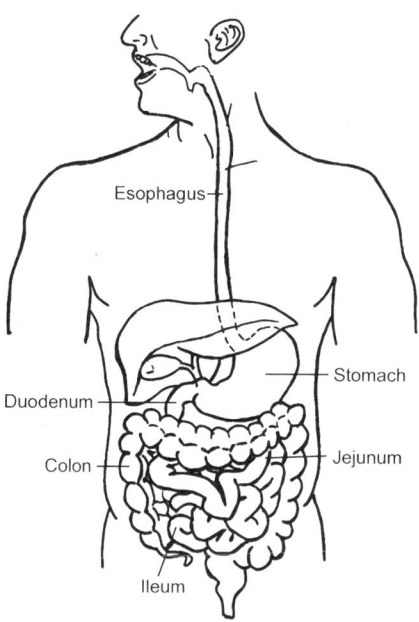

Figure 2: Anatomy of the human GI-tract relevant for oral drug absoption

Starch

Starch is a polysaccharide consisting of a large number of glucose monosaccharide units joined together by glycosidic bonds. It is predominantly present as a mixture of 15-25% amylase, 75-85% amylopectin and a minor amount of other components like proteins, a couple of inorganic compounds and water. Amylose is a linear polymer of α-1,4 linked D-glucose units. Amylopectin in contrast has an architecture where α-1,4 linked linear segments are branched together with α-1,6 linkages. This particular structure enables amylopectin to form helices. Typically, amylose and amylopectin are randomly interspersed which results in a semi-crystalline product.

Starch is by far the most consumed polysaccharide in the human diet. It is generally regarded as an essentially safe material and is therefore listed in the GRAS list of the U.S. Food and Drug Administration (FDA). The most common pharmaceutically used starches are corn-, rice- potato- and wheat starch. These are used as glidants, tablet- and capsule diluents & disintegrants and tablet binders. Particularly corn starch is one of the most common tablet disintegrants in concentrations of 3-15% (w/w). The typical composition of a corn starch-based tablet formulation is shown in table 3.

Table 3: Composition of a generic, corn starch-based tablet formulation [3]

Aminophylline Tablets, WG (100 mg)	
Formulation	
I. Aminophylline, fine powder	100 g
Corn starch	100 g
Polyvinylpyrrolidone, Kollidon® 30	6 g
II. Water	22 g
III. Magnesium stearate	1.5 g
Talc	3.0 g
Manufacturing	
Granulate mixture I with water II, dry, pass through a 0.8 mm sieve, mix with the components III and press with low compression force to tablets.	
Tablet properties	
Hardness	69 N
Disintegration	4-5 min
adapted from [3]	

The most important disadvantages that come along with the use of unmodified starch is that it has poor flow properties a high lubricant sensitivity and does not compress very well [4]. For this reason it has become common practice to substitute starch by modified starches, e.g. pregelatinized starch, hydroxyethyl starch or carboxymethyl starch which are obtained by altering the chemical or physical structure of the natural product. If modified starch is used as a binder, typical concentrations range from 5-25 % (w/w) of the overall tablet weight [5].

Cellulose

Cellulose, the skeletal substance of all vegetable tissues is the most abundant polymer on earth with an average annual production of about 50 billion tons. It consists of a linear chain of several hundred to over ten thousand D-glucose units which in contrast to starch are β-1,4 linked. These β-1,4 linkages make cellulose linear, highly crystalline, and indigestible for humans. Therefore, it is often referred to as dietary fiber. Since cellulose is not absorbed systemically following oral administration, it has little toxic potential and is thus also a GRAS listed material. Cellulose is mainly used to produce cardboard and paper, but is also one of the most important pharmaceutical excipients and food additives.

Finely ground cellulose is a frequently used tablet excipient. However, even more attractive is microcrystalline cellulose (MCC, *e.g.* Avicel®) which is very suitable for direct compression and therefore considered as the tablet excipient with the best binding properties [4]. The crystalline structures in MCC, which are obtained by acid treatment of cellulose, result in a high plasticity which does not simply come along with a high compressibility but moreover helps to achieve high tablet hardness. Because of its outstanding properties, MCC is widely used in oral dosage forms. Here, it is primarily used as a diluent in oral tablet- and capsule formulations where it is applicable in both direct compression and wet granulation processes. A typical recipe for a MCC-based generic tablet formulation is shown in table 4.

In addition to its application as a tablet and capsule diluent, tablet disintegrant and binder, cellulose can also be used as a starting material for cellulose derivatives such as cellulose ethers and cellulose esters.

Table 4: Composition of a generic, MCC-based tablet formulation [3]

Acetylsalicylic Acid Tablets (500 mg)	
Formulation	
Acetylsalicylic acid	500 g
MCC, Avicel® PH 101	200 g
Polyvinylpyrrolidone, Kollidon® 30	15 g
Polyvinylpyrrolidone (Crospovidone) Kollidon® CL	25 g
Magnesium stearate	3 g
Manufacturing	
Pass all components through a 0.8 mm sieve, mix and press with low compression force to tablets.	
Tablet properties	
Hardness	61 N
Disintegration	< 1 min
adapted from [3]	

Cellulose derivatives

Various types of cellulose derivatives are known. As mentioned before, they can be distinguished into cellulose ethers (most of which are water-soluble) and cellulose esters (many of which are nonionic substances that are insoluble in water; the exceptions being enteric cellulose esters).

Cellulose ethers

Cellulose ethers can be used for many purposes, particularly as tablet excipients. In tablet formulations their use ranges from simple fillers to more sophisticated applications, such as matrices that control drug release from the dosage form or functional coatings intended to enhance the stability of the formulation or to control drug release.

Because of their good aqueous solubility cellulose ethers like methylcellulose (MC), hydroxyethyl cellulose (HEC), hydroxypropyl cellulose (HPC) and hydroxylpropyl methylcellulose (HPMC) are mainly used either as binders in wet granulation where the granulation fluid is added to the powder starting material to produce granules, or as a coating material for protective coatings or taste masking purposes. However, ethylcellulose (EC), which does not dissolve in aqueous media, is frequently used to impart sustained action to a dosage form. With coatings of hydrated EC, drug release is *via* diffusion. This can be a slow process unless a large surface area is utilized. Aqueous dispersions tend therefore to be used to coat granules or pellets [5].

20

Sometimes combinations of the aforementioned formulation strategies are found in a single dosage form, as for example indicated in figure 3.

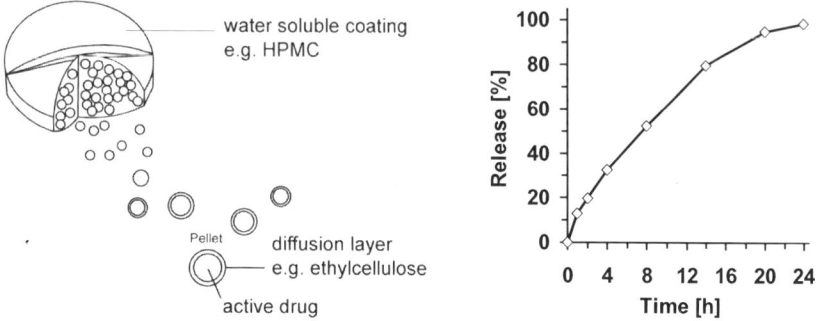

Figure 3: Design of a once-a-day formulation based on a fast-disintegrating cellulose ether and the corresponding dissolution profile of the EC-coated pellets[6]

Figure 3 shows a fast disintegrating tablet with a coating from a soluble cellulose ether (here: HPMC). This soluble polymer coating protects the tablet core from moisture, oxygen and, if pigments are suspended in the polymer, also from light. In addition, it can increase the patient compliance since it eases the swallowability of the tablet and masks undesired taste. However, it typically does not affect disintegration and dissolution of the dosage form. Thus, following ingestion, the tablet immediately disintegrates and releases a multitude of pellets. These pellets consist of the active drug, some excipients and a coating of EC. In contrast to the protective tablet coating, the EC coating is used to control drug release of the pellets and is therefore denoted a functional coating.

After tablet disintegration, the pellets move continuously along the sites of drug absorption in the human GI tract while drug release is controlled by diffusion through the hydrated EC coat. This results in a sustained action of the active drug over 12 or sometimes even 24 hours. Since drug release from the EC-coated formulation is typically independent of the pH and the composition of gastrointestinal fluids, and the controlled release of the active prevents the occurance of toxic or subtherapeutic plasma concentrations of the administered drug, such a dosage form can tremendously enhance the patient compliance. No particular dosing conditions (fasted or fed) are required, very often a once-a-day dosing is sufficient and there is only a low risk of side effects. Thus, such systems are of great benefit in the therapy of chronic diseases, for example high blood pressure or asthma.

Whereas the lower viscosity grades of the cellulose ethers are mainly used for aqueous coatings, the higher viscosity grades of the soluble cellulose ethers can also be used as matrix formers. Currently HPMC is one of the most commonly used polymers to retard the release of water soluble drugs from a matrix[5]. HPMC is available in different grades which can be distinguished by the degree of substitution (DS) and the ratio of methyl- and hydroxypropyl-substituents. One of the important mechanisms that govern the release of drugs from HPMC based matrix formulations is the diffusion of water solvent or body fluids into the dry hydrophilic polymer. During hydration of HPMC a gel layer is formed around the dry core of polymer and swelling of polymer occurs. Drug is then released *via* pure diffusion or a diffusion-erosion controlled mechanism, where the ratio and properties of the active drug, HPMC and other excipients as well as the composition of fluids and the hydrodynamics in the gastrointestinal tract will determine the detailed release profile[7-10]. Figure 4 shows the intragastric release behavior of two commercially available HPMC-based tablet formulations containing an anti-asthmatic drug. It clearly can be seen that formulation A exhibits a diffusion-erosion controlled release behavior whereas drug release from formulation B is controlled by pure diffusion through the swollen (hydrated) matrix [2].

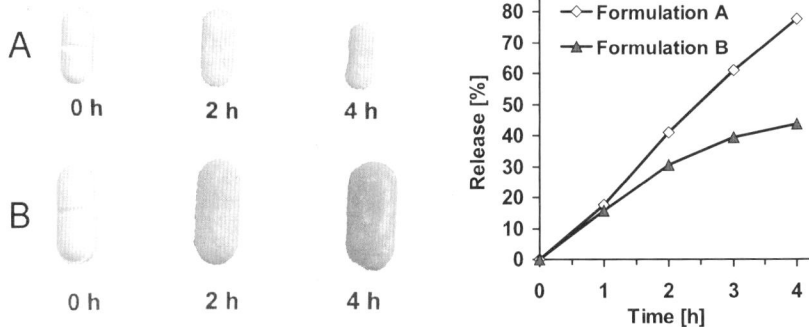

Figure 4: Pictures and corresponding drug release profiles of two HPMC-based matrix tablets before and after 2 and 4 hours gastric residence time (fed state)

Cellulose esters

Like the cellulose ethers, cellulose esters have a long history of use in the pharmaceutical industry. Based on their physicochemical properties, cellulose esters can be distinguished into two categories, non-enteric and enteric.

Non-enteric esters, like cellulose acetate, cellulose acetate butyrate and cellulose acetate propionate do not show pH-dependent solubility characteristics and (with no commercial exceptions) are insoluble in water. Because of the aforementioned properties non-enteric cellulose esters can be used to sustain drug release from oral delivery systems either by formation of a matrix or an insoluble but permeable film.

Enteric esters are those, such as cellulose acetate phthalate (cellacefate, C-A-P) or hydroxypropylmethyl cellulose phthalate (HPMCP) which are insoluble in acidic solutions but soluble in mildly acidic to slightly alkaline solutions. The pH at which the polymer dissolves can be governed by the degree of esterification. The higher the DS of carboxyl groups, the lower the dissolution pH. The different HPMCP types dissolve at pH-values in the range of pH 4.5 to

5.5 which correspond to typical pH values in the upper small intestine. C-A-P in contrast dissolves at somewhat more neutral pH conditions (pH $\geq$ 6) which indicates that drug release from C-A-P coated dosage forms occurs in the jejunum, the mid-part of the small intestine.

The pH-dependent solubility of the enteric cellulose esters has been extensively used in the manufacture of enteric coated dosage forms. Since many drugs show degradation under acidic conditions, which are typically found in the fasted stomach, they need to be protected from this environment. However, as the main site of drug absorption is the small intestine, another requirement is to quantitatively release the drug when the dosage form enters this GI segment. Such a particular release behavior can neither be achieved with pH-independent nor with water-insoluble polymers. Thus, polymers with enteric properties are the only alternative and therefore the method of choice for these requirements.

Because of the aforementioned criteria, enteric polymers are extensively used in the formulation of proton-pump inhibitors and enzymes (*e.g.* pancreatin) which would otherwise be inactivated in the stomach. However enteric polymers can also be found as a taste masking agent in food supplements like garlic or fish oil and they play an important role in the formulation of drugs that can irritate the gastric mucosa. Thus, they are also part of several formulations of nonsteroidal antiinflammatory drugs (NSAIDs) like aspirin or diclofenac sodium.

Cellulose esters for colon-specific drug delivery

With the development of a biotechnology sector that provided a large-scale availability of therapeutic proteins, the delivery of proteins has gained momentum. The important therapeutic proteins and peptides being explored for oral delivery include insulin, calcitonin, interferons, human growth hormone, glucagons, gonadotropin-releasing hormones, enkephalins, vaccines, enzymes, hormone analogs, and enzyme inhibitors. However, most of these proteins lack oral bioavailability, because of their large molecular size, low permeation through biological membranes, and susceptibility to molecular changes in both biological and physical environments. The demand for effective delivery of proteins by the oral route has brought a tremendous thrust in recent years both in the scope and complexity of drug delivery technology. Most proteins are susceptible to proteolytic degradation and deactivation in the small intestine. However, in contrast to the majority of small molecules, they are often effectively absorbed in the colon. Thus, there has recently been increasing interest in targeting peptide and protein drugs to this site. In the last decades various strategies have been described for this purpose. Amongst those, various polysaccharide-based systems can be found [11, 12]. One approach is the use of pH- and time-dependent systems which have already been established for the treatment of colon-related diseases such as ulcerative colitis, Crohn´s disease or colorectal cancer. The polymers used for colon-targeting range from enteric polymethacrylates (*e.g.* Eudragit® L, S, FS) to cellulose esters like C-A-P, HPMCP, cellulose acetate trimellitate and hydroxypropyl methylcellulose acetate succinate (HPMCAS) [13, 14].

Recently, various formulations for oral protein/peptide delivery have been described in the literature, amongst them an enteric-coated dry emulsion formulation for oral insulin delivery [15]. This edible insulin formulation is a dry microparticulate emulsion with enteric properties and consists of a surfactant, a vegetable oil, and HPMCP as the pH-responsive polymer. In an *in vitro* study, the insulin release from the formulation was shown to respond to the change in external environment in the gastrointestinal tract [15]. Overall, the study results suggest that this new enteric emulsion formulation could potentially be applied to oral delivery of several kinds of pharmaceutical peptides and proteins. Based on the results of this and many other studies, it is likely that cellulose esters, even with their long history of pharmaceutical use, will show up in a range of new formulations in the future.

As already mentioned, the use of biodegradable polymers holds great promise in terms of achieving targeted drug release to the colon. Most of these systems are based on the knowledge that anaerobic bacteria are able to recognize the various polysaccharide substrates (particularly β-glycosidic bonds) and to degrade them with enzymes. As the natural polymers fermented by the colonic microflora are categorized as GRAS materials, this makes them even more interesting for pharmaceutical applications. Therefore, a range of natural polysaccharides with different origin, like *e.g.* pectin, guar gum and inulin from plants, chitosan and chondroitin sulfate from animals and alginates from algae were studied for colon targeting during the last decades[16]. Typically those candidates being selectively degraded in the colon appeared to be very promising for this objective. However, the current challenge is to develop an optimal polysaccharide-based formulation that ensures both, a selective colon targeting and an adequate drug release in the colonic lumen.

Chitosan

Chitosan, a linear polysaccharide composed of randomly distributed β-1,4 linked D-glucosamine (deacetylated unit) and N-acetyl-D-glucosamine is a high molecular weight cationic polysaccharide derived from deacetylation of naturally occurring chitin. Chitin itself is one the most abundant natural polysaccharides next to cellulose and a structural element of crab and shrimp shells. It represents a polymer with unique physical and chemical properties that can be widely used in both the industrial and medical fields[17]. Because of its renewable sources, the compound is inexpensive and further, it is biocompatible, biodegradable and considered to be non-toxic. Chitosan is a hydrophilic, cationic and crystalline polymer that demonstrates film forming ability and gelation characteristics [18]. Another peculiar property of chitosan is the ability to adhere to cell membranes which concomitantly decreases the trans-epithelial electrical resistance of cell monolayers and increases paracellular permeability [19].

The safety of chitosan, its ability to prolong residence time in the gastrointestinal tract through mucoadhesion, and its ability to enhance absorption by increasing permeability have all been major factors contributing to its widespread evaluation as a component of oral dosage forms [20]. It can be used as a tablet excipient where in concentrations higher than 5% it displays a disintegrant power superior to that of starch and MCC[21]. In addition to its function as disintegrant and binder, it has been found useful for increasing the dissolution rate of poorly soluble acidic drugs under small intestinal conditions. This function might be a result of gel formation of the positively charged chitosan and the acidic drug. As a result of this gel formation the drug seems to be protected from the gastric environment which finally results in an enhanced dissolution rate in the small intestine[18]. The use of chitosan is not restricted to immediate release (IR) dosage forms. If used in higher concentrations, it can also control drug release. For example, it has been employed for the sustained release of drugs by a process of slow erosion of a hydrated compressed chitosan matrix[18, 22].

Chitosan is not digested by the human digestive enzymes in the upper part of the gastrointestinal tract. However, it is degraded by the microbial enzymes produced in the colon[16]. It thus can be applied for colonic-delivery systems of peptides, like insulin[23, 24], and anti-inflammatory drugs[24], like mesalazine or prednisolone which are typically used in the treatment of colonic diseases.

Increasingly, nucleic acids are also being applied as drugs, either for vaccination or for therapeutic gene expression. Many of the issues facing oral gene delivery are similar to those of oral protein delivery, including protection in the stomach and small intestine, and transport into or across intestinal epithelial cells. One proposed method to overcome these physical and degradative barriers is formulation of the drug/gene into nano- or microparticles which may partially protect the entrapped drug/gene from degradation and improve cellular uptake through endocytosis. While a variety of polymers and lipids have been employed to form drug- or gene-loaded nanoparticles, one biodegradable polymer that has received a good deal of recent attention as a component of oral drug and gene delivery systems is chitosan[20]. In numerous studies[20, 25] chitosans and their derivatives have shown to be very suitable for preparation of nano- and microparticles, some of which were able to associate large amounts of vaccines and to enhance antigen uptake by the lymphoid tissue. Thus, chitosan and its derivatives are very promising candidates for both mucosal drug delivery and vaccination.

Despite the outstanding scientific progress being made in the application of chitosan in drug delivery systems, no chitosan-based drug delivery systems have been launched to the market yet. However, since clinical trials are ongoing for a wide range of pharmaceutical formulations, chitosan-based products can be expected in the near future and it has already been proposed that chitosan may be the carrier material of the 21st century in drug delivery devices [17].

Cyclodextrins

Cyclodextrins (CDs) are a family of oligosaccharides, produced by an enzymatic process using the enzyme transglycolase and corn starch. By this process linear and cyclic dextrins are formed. CDs are non-reducing polysaccharides and form rings of different numbers of α-1,4 linked α-D-glucopyranose units. For biomedical and pharmaceutical purposes, the rings consisting of 6 (α-), 7 (β-) or 8 (γ-) glucopyranose units are the most relevant subtypes and often referred to as the "parent CDs"[26]. Due to the chair conformation of the glucopyranose units, the CDs take the shape of a truncated cone or torus rather than a perfect cylinder. The hydroxyl functions are oriented to the cone exterior which results in hydrophilic surface properties. The central cavity of the CD molecule is lined with skeletal carbons and ethereal oxygens of the glucose residue. This provides a lipophilic microenvironment into which suitably sized drug molecules may enter and form a drug:CD complex.[27] This formation of a non-covalent complex is often a 1:1 interaction but can also be of higher order[28]. In forming the complex, the physicochemical and biological properties of the drug can be altered to effect an advantage like solubilization or stabilisation of the drug. For oral drug formulations, the solubilizing effect of the CDs is of particular advantage since most of the new chemical entitities (NCEs) emerging from the pipelines of the chemists represent highly potent drugs but suffer from very poor solubility characteristics which often limits their bioavailability.

Formation of a CD inclusion complex can significantly improve the bioavailability of drugs belonging to class II of the biopharmaceutics classification scheme (BCS)[29]. These class II drugs are poorly soluble but highly permeable, so an increase in dissolution rate and apparent solubility (which can increase upon CD complexation by a factor of 10^1 to 10^3)[30] will result in an increased amount of drug permeating through the gut wall, potentially leading to higher bioavailability[31]. Figure 5 shows the possible cyclodextrin-effects on *in vitro* dissolution and *in vivo* drug absorption of a BCS class II compound[30].

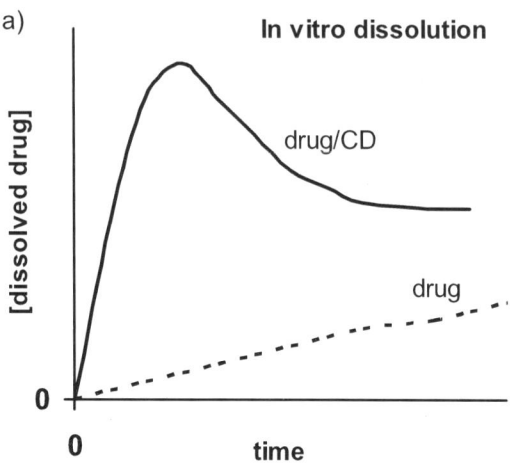

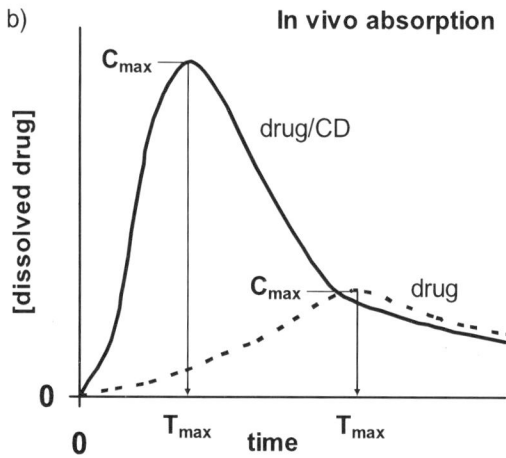

Figure 5: In vitro dissolution (a) and in vivo absorption (b) of a drug or its CD complex in function of time. The CD complexed drug attains a higher blood level peak (c_{max}) in a shorter time after administration (T_{max}) than the noncomplexed drug. The total bioavailability of the drug is represented by the area under curve (AUC) values, [g*h/cm³] (adapted from Szejtli [30])

A major drawback for the use of natural cyclodextrins, particularly for β-CD, is their limited aqueous solubility. To improve the solubility and to reduce some of the few toxicological risks (*e.g.* the nephrotoxicity after parenteral administration of β-CD) numerous CD derivatives have been prepared [27, 32-35]. Among industrially produced, standardized, and available β-CD derivatives, currently the most important ones are the heterogeneous, amorphous, highly water-soluble methylated β-CDs, 2-hydroxypropylated β-CDs (HP-β-CD) and sulfobutylether β-CD (SBE-β-CD). These hydrophilic CDs are considered non-toxic at low to moderate oral and intravenous doses and are much more water-soluble and more toxicologically benign than the natural β-CD. They are currently found in several marketed drug formulations. Figure 6 illustrates the impact of β-CD and β-CD derivatives on the dissolution rate of glyburide, a BCS class II drug in a medium simulating typical pH-conditions of the small intestine in the fasted state.

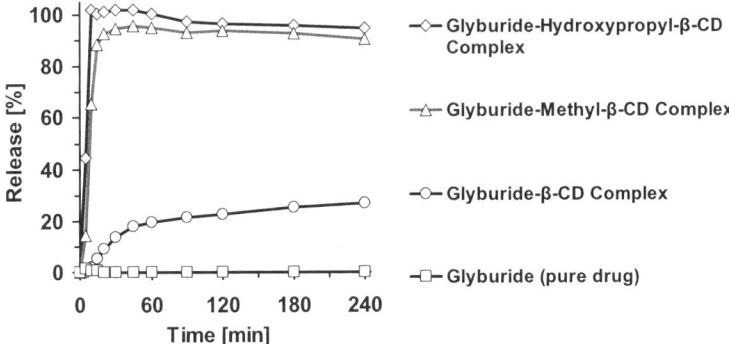

Figure 6: Dissolution profiles of glyburide (pure drug) and three glyburide-cyclodextrin complexes under conditions of the fasted state small intestine

The dissolution profiles shown in figure 6 clearly indicate the impact of complex formation on the dissolution rate of glyburide. Whereas the aqueous solubility of the pure drug is not sufficient for dissolution in a simulated intestinal fluid (SIF), the formation of glyburide-β-CD significantly improves the dissolution rate. However, due to the limited solubility of β-CD, only part of the dose (3.5 mg) can be dissolved in a physiological relevant volume of 500 ml SIF. In contrast, from hydrophilic glyburide-CD complexes obtained by complexation with methyl β-CD and HP-β-CD, the complete dose of the drug is released within a few minutes. As glyburide is a BCS class II drug, the dissolution profiles predict an increase in bioavailability when the drug would be administered as a CD complex (preferably a hydrophilic CD-complex) [27, 31].

There are a number of exciting possibilities for future applications of CDs in oral drug delivery, including new uses for existing derivatives, as well as the development of new derivatives [35, 36]. Due to the increasing number of poorly soluble NCEs, CDs will surely become important components of many oral drug formulations. Like cellulose esters and chitosan, they represent a group of polysaccharides that might be applicable for colonic delivery systems of proteins, peptides and oligonucleotides [11], since they are neither hydrolyzed nor absorbed from the stomach and small intestine, and since they are able to stabilize protein conformation through interactions with accessible hydrophobic amino-acid residues.

Overall the future of CDs in the pharmaceutical industry seems to be bright and further uses of CDs are likely to be explored as the properties of CDs are expanded and the number of commercialized and FDA-approved variants increases [36].

Conclusion

This chapter gives a brief overview of the most important polysaccharides that can be found either in currently marketed drug products or pharmaceutical research laboratories. It becomes obvious that polysaccharides and their derivatives represent a very important if not the most important group of excipients of today. Some of them, particularly cellulose and starch, have been used in pharmaceutical formulations for many years and now belong to the group of so called "standard excipients". However due to the evolving of new technologies (*e.g.* protein- and gene delivery), there is a rising need for new and specialized formulations. This concomitantly results in numerous opportunities for the use of polysaccharides, since this group of excipients offers a variety of useful physicochemical properties. As polysaccharides are available from renewable sources, represent the most abundant group of polymers on earth and are in the majority of cases non-toxic, many future patients may benefit from their use throughout the coming years.

References

1. USP, *USP 30*. USP 30 ed.; United States Pharmacopoeia Convention, Inc.: Rockville MD, 2007.
2. Klein, S., *Biorelevant Dissolution Test Methods for Modified Release Dosage Forms*. ed.; Shaker-Verlag: Frankfurt, 2005; 'Vol.' p.
3. BASF, *Pharma Ingredients - Generic drug formulations* ed.; BASF: Ludwigshafen, 2001.
4. Bolhuis, G. K.; Armstrong, N. A., Excipients for direct compaction - an update. *Pharmaceutical Development and Technology* **2006**, 11, (1), 111-124.
5. Wade, A.; Weller, P., *Handbook of Pharmaceutical Excipients*. 2nd ed.; Pharmaceutical Press: London, 1994.
6. Klein, S.; Dressman, J. B., Comparison of Drug Release From Metoprolol Modified Release Dosage Forms in Single Buffer versus a pH-Gradient Dissolution Test. *Dissolution Technologies* **2006**, 13, (1), 6-12.
7. Kavanagh, N.; Corrigan, O. I., Swelling and erosion properties of hydroxypropylmethylcellulose (Hypromellos) matrices - influence of

agitation rate and dissolution medium composition. *International Journal of Pharmaceutics* **2004,** 279, (1-2), 141-152.

8. Sako, K.; Sawada, T.; Nakashima, H.; Yokohama, S.; Sonobe, T., Influence of water soluble fillers in hydroxypropylmethylcellulose matrices on in vitro and in vivo drug release. *Journal of Controlled Release* **2002,** 81, (1-2), 165-72.

9. Mitchell, K.; Ford, J. L.; Armstrong, D. J.; Elliott, P. N. C.; Rostron, C.; Hogan, J. E., The Influence of Additives on the Cloud Point, Disintegration and Dissolution of Hydroxypropylmethylcellulose Gels and Matrix Tablets. *International Journal of Pharmaceutics* **1990,** 66, (1-3), 233-242.

10. Bettini, R.; Catellani, P. L.; Santi, P.; Massimo, G.; Peppas, N. A.; Colombo, P., Translocation of drug particles in HPMC matrix gel layer: effect of drug solubility and influence on release rate. *Journal of Controlled Release* **2001,** 70, (3), 383-91.

11. Kosaraju, S. L., Colon targeted delivery systems: review of polysaccharides for encapsulation and delivery. *Critical Reviews in Food Science and Nutrition* **2005,** 45, (4), 251-8.

12. Van den Mooter, G., Colon drug delivery. *Expert Opinion on Drug Delivery* **2006,** 3, (1), 111-25.

13. Chourasia, M. K.; Jain, S. K., Pharmaceutical approaches to colon targeted drug delivery systems. *Journal of Pharmacy and Pharmaceutical Sciences* **2003,** 6, (1), 33-66.

14. Fukui, E.; Miyamura, N.; Kobayashi, M., An in vitro investigation of the suitability of press-coated tablets with hydroxypropylmethylcellulose acetate succinate (HPMCAS) and hydrophobic additives in the outer shell for colon targeting. *Journal of Controlled Release* **2001,** 70, (1-2), 97-107.

15. Toorisaka, E.; Hashida, M.; Kamiya, N.; Ono, H.; Kokazu, Y.; Goto, M., An enteric-coated dry emulsion formulation for oral insulin delivery. *Journal of Controlled Release* **2005,** 107, (1), 91-96.

16. Jain, A.; Gupta, Y.; Jain, S. K., Perspectives of biodegradable natural polysaccharides for site-specific drug delivery to the colon. *Journal of Pharmacy and Pharmaceutical Sciences* **2007,** 10, (1), 86-128.

17. Paul, W.; Sharma, C. P., Chitosan, a drug carrier for the 21st century: a review. *Stp Pharma Sciences* **2000,** 10, (1), 5-22.

18. Illum, L., Chitosan and its use as a pharmaceutical excipient. *Pharmaceutical Research* **1998,** 15, (9), 1326-1331.

19. Dodane, V.; Khan, M. A.; Merwin, J. R., Effect of chitosan on epithelial permeability and structure. *International Journal of Pharmaceutics* **1999,** 182, (1), 21-32.

20. Bowman, K.; Leong, K. W., Chitosan nanoparticles for oral drug and gene delivery. *International Journal of Nanomedicine* **2006,** 1, (2), 117-128.

21. Fini, A.; Orienti, I., The Role of Chitosan in Drug Delivery: Current and Potential Applications. *American Journal of Drug Delivery* **2003,** 1, (1), 43-59.

22. Hejazi, R.; Amiji, M., Chitosan-based gastrointestinal delivery systems. *Journal of Controlled Release* **2003**, 89, (2), 151-165.
23. Tozaki, H.; Komoike, J.; Tada, C.; Maruyama, T.; Terabe, A.; Suzuki, T.; Yamamoto, A.; Muranishi, S., Chitosan capsules for colon-specific drug delivery: Improvement of insulin absorption from the rat colon. *Journal of Pharmaceutical Sciences* **1997**, 86, (9), 1016-1021.
24. Yamamoto, A.; Tozaki, H.; Okada, N.; Fujita, T., Colon-specific delivery of peptide drugs and anti-inflammatory drugs using chitosan capsules. *Stp Pharma Sciences* **2000**, 10, (1), 23-34.
25. van der Lubben, I. M.; Verhoef, J. C.; Borchard, G.; Junginger, H. E., Chitosan and its derivatives in mucosal drug and vaccine delivery. *European Journal of Pharmaceutical Sciences* **2001**, 14, (3), 201-207.
26. Hoepfner, E. M.; Lang, S.; Reng, A.; Schmidt, P. C., *Fiedler Encyclopedia of Excipients for Pharmaceuticals, Cosmetics and Related Areas.* 6th ed.; Editio Cantor Verlag: Aulendorf, Germany, 2007; Vol. 9
27. Brewster, M. E.; Loftsson, T., Cyclodextrins as pharmaceutical solubilizers. *Advanced Drug Delivery Reviews* **2007**, 59, (7), 645-66.
28. Stella, V. J.; He, Q., Cyclodextrins. *Toxicologic Pathology* **2008**, 36, (1), 30-42.
29. Amidon, G. L.; Lennernas, H.; Shah, V. P.; Crison, J. R., A theoretical basis for a biopharmaceutic drug classification: the correlation of in vitro drug product dissolution and in vivo bioavailability. *Pharmaceutical Research* **1995**, 12, (3), 413-20.
30. Szejtli, J., Past, present, and future of cyclodextrin research. *Pure and Applied Chemistry* **2004**, 76, (10), 1825-1845.
31. Lennernas, H.; Abrahamsson, B., The use of biopharmaceutic classification of drugs in drug discovery and development: current status and future extension. *Journal of Pharmacy and Pharmacology* **2005**, 57, 273-285.
32. Szejtli, J., Highly Soluble Beta-Cyclodextrin Derivatives. *Staerke* **1984**, 36, (12), 429-432.
33. Thompson, D. O., Cyclodextrins - Enabling excipients: Their present and future use in pharmaceuticals. *Critical Reviews in Therapeutic Drug Carrier Systems* **1997**, 14, (1), 1-104.
34. Loftsson, T.; Duchene, D., Cyclodextrins and their pharmaceutical applications. *International Journal of Pharmaceutics* **2007**, 329, (1-2), 1-11.
35. Buchanan, C. M.; Alderson, S. R.; Cleven, C. D.; Dixon, D. W.; Ivanyi, R.; Lambert, J. L.; Lowman, D. W.; Offerman, R. J.; Szejtli, J.; Szente, L., Synthesis and characterization of water-soluble hydroxybutenyl cyclomaltooligosaccharides (cyclodextrins). *Carbohydrate Research* **2002**, 337, (6), 493-507.
36. Davis, M. E.; Brewster, M. E., Cyclodextrin-based pharmaceutics: Past, present and future. *Nature Reviews Drug Discovery* **2004**, 3, (12), 1023-1035.

Chapter 2

Building new drug delivery systems: *in vitro* and *in vivo* studies of drug-hydroxybutenyl cyclodextrin complexes

Charles M. Buchanan[1], Norma L. Buchanan[1], Kevin J. Edgar[1,2], Sandra Klein[3], James L. Little[1], Michael G. Ramsey[1], Karen M. Ruble[1], Vincent J. Wacher[1,4], and Michael F. Wempe[1,5]

[1]Eastman Chemical Company, Research Laboratories, P.O. Box 1972, Kingsport, TN 37662, USA
[2]Current address: Virginia Tech, 230 Cheatham Hall, Blacksburg, VA 24061
[3]Johann Wolfgang Goethe University, Institute of Pharmaceutical Technology, 9 Max von Laue Street, Frankfurt/Main 60438, Germany
[4]Current address: 1042 N. El Camino Real, Suite B-174, Encinitas, CA 92024-1322
[5]Department of Pharmacology, East Tennessee State University, Johnson City, TN 37614, USA

In many respects, drug delivery is the perfect field for an organic chemist whose interest lies in structure-property relationships. Synthesis and characterization of a drug delivery system and subsequent introduction of a drug product incorporating this drug delivery technology into animals or humans is a complex and challenging structure-property problem. In this regard, we have developed hydroxybutenyl cyclodextrin (HBenCD) for oral and intravenous delivery of pharmaceutical actives having poor water solubility and low bioavailability. In this chapter, we summarize our recent research involving the synthesis and characterization of HBenCD, formation of drug-HBenCD complexes, *in vitro* solubility and dissolution studies of drug-HBenCD complexes, and pharmacokinetic studies in animals after oral and intravenous administration of drug-HBenCD complexes.

Many drug candidates developed today (*ca.* 40%) by the pharmaceutical industry suffer from poor water solubility (*1,2*). Poor drug solubility in a physiological environment may substantially reduce drug absorption across biological barriers, impairing drug bioavailability and introducing variability in systemic drug levels that impact the efficacy, tolerability and safety of the drug product. High bioavailability may reduce variability (*3*) and enable dose reduction, which may in turn reduce drug side effect. Therefore, it is vitally important to develop drug delivery systems that mitigate solubility, dissolution, and bioavailability issues.

In designing delivery systems for poorly water-soluble drugs, the relationship between drug solubility in a physiological environment, drug absorption across biological barriers, and bioavailability or efficacy is of paramount importance. There are many aspects to this problem; crystallinity, intrinsic aqueous solubility, drug structure, and charge state of the drug molecule are important issues. The ability of a delivery vehicle to solubilize a drug, drug loading, and interaction of the delivery vehicle with membrane transporters (*e.g.* P-glycoprotein) are also important issues. The medium in which the drug delivery system is contained is a very significant issue.

Cyclodextrins are cyclic oligomers of glucose, which typically contain 6, 7, or 8 glucose monomers joined by α-1,4 linkages. These oligomers are commonly called α-CD, β-CD, and γ-CD, respectively, and they differ in cavity size and water solubility. Relative to β-CD, α-CD has a smaller cavity (5.3 versus 6.6 Å) and higher water solubility (14.5 versus 1.9 g/mL at 25 °C). Similarly, γ-CD has a larger (8.4 Å) and more flexible cavity and a higher water solubility (23.2 g/mL at 25 °C) (*4*). The cavity size of β-CD is suitable for complex formation with many pharmaceutical actives. Topologically, cyclodextrins form a torus that has a hydrophobic interior and a hydrophilic exterior. This allows the CD to be dissolved in water, where it acts as a host molecule and forms inclusion complexes with hydrophobic guest molecules. This feature has led to the use of CDs in pharmaceutical formulations in order to improve aqueous solubility of poorly soluble compounds (*5,6*). Unmodified cyclodextrins, particularly β-CD, are relatively crystalline and have limited aqueous solubility. In parenteral formulations this limited solubility is a very serious issue as renal concentration of the unmodified CD can lead to crystallization of the CD and necrotic damage (*7*). Fortunately, the solubility of unmodified cyclodextrins in water can be significantly increased by the addition of a small number of substituents to the hydroxyl groups of the anhydroglucose monomers (*8*). It is important to note that cyclodextrin derivatives that are similar in structure can provide significantly different solubilization rates, dissolution rates, and stabilization of drugs under otherwise similar conditions (*8*).

In this context, we have developed a new cyclodextrin derivative, hydroxybutenyl cyclodextrin (HBenCD) that is highly soluble in both water and organic media such as polyethylene glycol (*9*). We have shown that HBenCD is very effective in solubilizing a broad spectrum of drugs (*9-12*) and that formulations incorporating drug-HBenCD complexes or mixtures can significantly increase oral bioavailability in animal models (*13-17*).

Synthesis and Characterization of HBenCD

In designing a hydroxybutenyl cyclodextrin as a drug delivery vehicle, it was important that the degree of substitution (DS, number of CD hydroxyls bearing a butenyl substituent) and the molar substitution (MS, total number of substituents. MS can be higher than DS due to chain extension occurring by subsequent reaction of a butenyl substituent with the epoxide) be high enough so as to eliminate the possibility of unsubstituted cyclodextrin in the reaction product. This approach eliminates the need for subsequent purification to remove unsubstituted cyclodextrin that might raise toxicity concerns. Furthermore, it was important to obtain a random distribution of substituents as this leads to a multi-component mixture which reduces the possibility of crystallization and increases water solubility.

Reaction (9), of the parent CDs with 3,4-epoxy-1-butene in the presence of a base (KOH) at 60-100 °C smoothly provided HBenCD with the desired attributes (Figure 1). Temperature, reaction time, equivalents of 3,4-epoxy-1-butene, and concentration of base influenced the product DS and MS. Following the reaction, we found that the HBenCD could be easily purified from the crude reaction mixture by nanofiltration using a 500 molecular weight cut-off cellulose acetate membrane. This process allows the low molecular weight by-products to pass through the membrane with the water while the higher molecular weight HBenCD is held as the retentate allowing concentration of the product in the aqueous medium. Following nanofiltration, the product was isolated as an amorphous white solid either by freeze-drying or by co-evaporation of the water with ethanol.

As illustrated by the MALDI-TOF-MS spectrum of HBenBCD (Figure 1), at the proper DS and MS (MS = 5.84) unsubstituted β-CD was not observed in the product. The bell shaped molecular-weight distribution is characteristic of a random substitution pattern. It should be noted that MALDI-TOF-MS cannot distinguish a CD substituted with two 3-hydroxy-1-butenyl monomers from a CD substituted with one 3-hydroxy-1-butenyl dimer. Fortunately, both DS and MS can be obtained easily by ^{1}H NMR.

W determined that the desired ranges of MS and DS for HBenBCD were 4.4-5.4 and 2.3-3.2, respectively. Typically, we targeted a MS for HBenBCD of 4.9 (Mw = 1478). This product is a white, amorphous solid with a glass transition temperature of 210 °C and is thermally stable to 275 °C by thermogravimetric analysis. The pH of a 10 wt% solution in water is typically 6-8. HBenBCD is highly soluble in both water (>500 g/L) and water-soluble organic media (>400 g/L) such as polyethylene glycol.

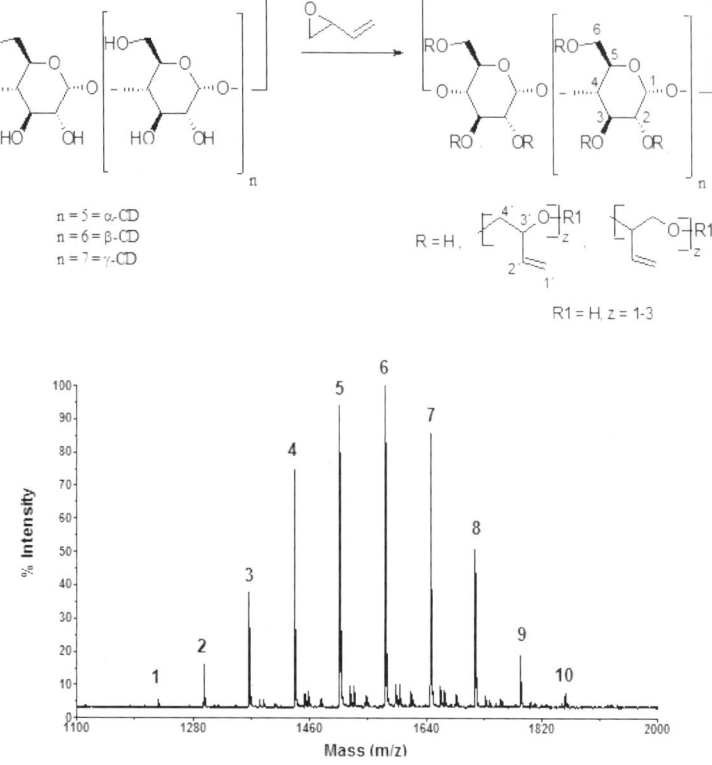

Figure 1. Scheme for the preparation of hydroxybutenyl cyclodextrins and a typical MALDI-TOF spectrum of HBenBCD.

Solubility and Dissolution Studies

During the course of our research, we have determined the solubility and dissolution profiles for numerous drug-HBenCD complexes. In this account we restrict ourselves to a summary of 3 classes of drugs: antifungals (itraconazole, ketoconazole, clotrimazole, voriconazole); estrogen modulators (tamoxifen, letrozole, raloxifene); and protease inhibitors (saquinavir, ritonavir). Full details of the work can be found in the references cited (*9-18*). Typically, solubility measurements were made using a modified version of the traditional shaking flask method. Standard dissolution studies were obtained according to USP 28-NF 23 711. Dissolution studies in biorelevant media were performed with a mini-paddle apparatus (modified DT 606 HH, Erweka, Heusenstamm, Germany) (12,14).

Antifungals

Fungal infections can be a serious medical problem, in particular as a complicating factor in underlying diseases and conditions, such as haematological malignancy, HIV infection, and neutropenia. Although there are different forms of treatments, a preferred treatment regime for cutaneous and systemic fungal infections involves administration of antifungal azole compounds. Antifungal azole compounds are structurally diverse and are characterized by having imidazole or triazole functionalities. This class includes such drugs as itraconazole, ketoconazole, clotrimazole, voriconazole and several others (Figure 2). As a class these compounds generally have low water solubility, which in most cases translates into low bioavailability.

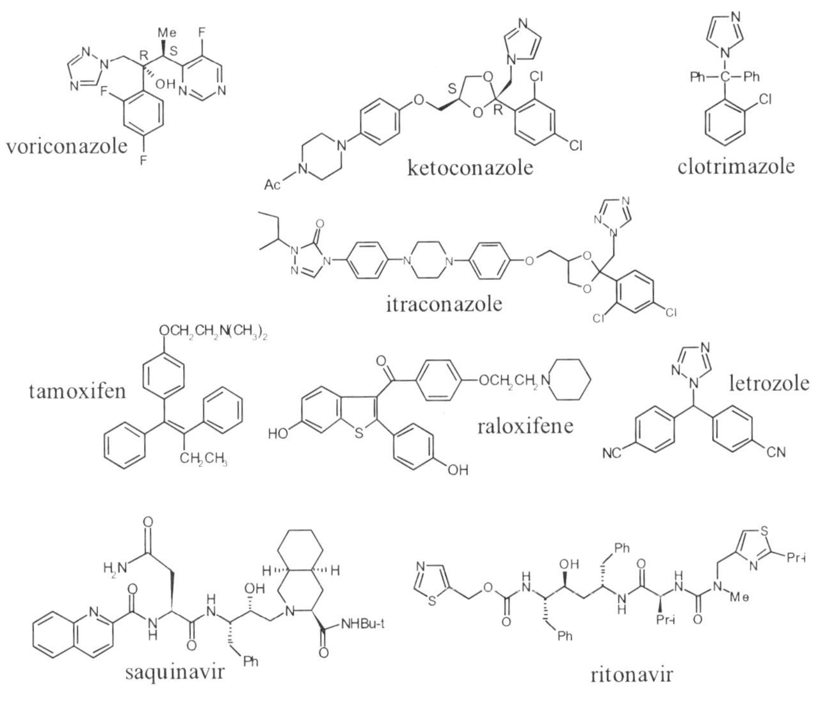

Figure 2. Drug structures discussed in this account.

As a specific example, itraconazole is a broad-spectrum triazole agent available for the treatment of histoplasmosis, blastomycosis, onychomycosis, and amphotericin B-refractory aspergillosis. Itraconazole is highly effective *in vitro* against *Candida albicans* and other *Candida* species. The effectiveness of itraconazole is in part due to the fact that hydroxyitraconazole, the main metabolite of itraconazole, also has considerable antifungal activity.

Itraconazole is a weakly basic drug (pKa *ca.* 3.7) with very poor water-solubility (intrinsic solubility, $(S_o) = 30$ µg/mL). The drug is highly crystalline having a T_m of 170 °C and a log P value of 5.66 at pH 8.1. Itraconazole can only be ionized and solubilized in water at very low pH. Itraconazole has limited solubility in organic solvents with CH_2Cl_2 being the preferred solvent. Itraconazole also has limited stability at elevated temperatures and at low pH.

Table I presents the equilibrium solubility of four antifungal drugs in the absence (S_0) and presence of 10 wt% HBenBCD (S_t) as well as the increase in drug solubility (S_t/S_0). As this data illustrates, the solubilities of these antifungal drugs in the absence of HBenBCD in water are very low, ranging from 0.7 mg/mL for voriconazole to 3 µg/mL for clotrimazole. However, in the presence of 10 wt% HBenBCD, the increase in the solubilities of these drugs in water is very significant. For example, the solubility of ketoconazole in water increases from 10 µg/mL to 7.21 mg/mL (S_t/S_0 = 600).

Table I. Equilibrium solubility (mg/mL) of antifungal drugs in the presence of 10 wt% HBenBCD (S_t) and the increase in drug solubility (S_t/S_0) relative to when HBenBCD is absent (S_0).

	voriconazole		ketoconazole		clotrimazole		itraconazole	
	S_t	S_t/S_0	S_t	S_t/S_0	S_t	S_t/S_0	S_t	S_t/S_0
Water	7.13	9.8	7.21	600.4	0.17	57.3	0.51	17.1
pH 3 phosphate	8.11	13.0	8.95	4.3	1.31	4.5	0.49	16.7
pH 3 citrate	6.52	8.4	17.88	3.3	3.77	9.6	0.37	14.2
pH 3 D-tartrate	7.37	9.8	22.64	3.2	4.57	11.0		
pH 3 L-tartrate	9.31	8.9	23.76	2.8	4.47	9.9		
pH 3 D,L-tartrate	3.88	7.8	21.36	2.7	4.80	10.8		
pH 4.8 acetate	10.45	9.6	10.98	95.5	0.32	17.1	0.41	8.2
pH 7.4 phosphate	3.73	9.6	7.69	548.9	0.11	35.7		

In the pH 3 buffers, S_0 and S_t generally increased for all of the antifungal drugs except for itraconazole; in the pH 4.8 and 7.4 buffers the changes in S_0 and S_t were small or nonexistent. These observations are related to the specific pKa of the individual drugs. Ionization of a drug can both increase S_0 and modify the driving force (inclusion of a hydrophobic guest with exclusion of water from the cyclodextrin cavity) for drug complexation. For example, in water (little or no ionization) S_t/S_0 for ketoconazole (pKa ca. 6.5) was 600 and in the pH 3 buffers (ionized form of the drug) S_t/S_0 for ketoconazole was 2.7-4.3. In the case of itraconazole (pKa ca. 3.7), there is virtually no change in S_0 and S_t/S_0 is larger (14-17) regardless of the buffer media.

Of particular interest is the impact of the pH 3 tartrate buffers on the solubilities of voriconazole (single diastereoisomer), ketoconazole (racemic mixture of diastereoisomers), and clotrimazole (non stereogenic) relative to other pH 3 buffers. In the case of voriconazole and ketoconazole the magnitude of S_0 and S_t increased in the order of D,L-tartrate<D-tartrate<L-tartrate. In the case of clotrimazole, there was not a significant difference between the tartrates. In general, the pH 3 tartrate buffers were superior to the other pH 3 buffers examined. Briefly, we concluded that an electrostatic interaction is first established between the tartrate and the drug and that the stereochemistry of the drug and the tartrate contributes to both the strength of the electrostatic interaction and complexation of the drug-tartrate aggregate by HBenBCD (chiral recognition).

Figure 3 provides representative equilibrium solubility data for 3 antifungal drugs versus concentration of HBenBCD. Voriconazole solubility in 50 mM pH 3 succinate buffer increased linearly to *ca*. 10 wt% HBenBCD concentration before reaching a plateau. This type of behavior is indicative of A_n phase solubility (19). Solubilization of voriconazole by HBenBCD was essentially identical to that obtained with hydroxypropyl-β-cyclodextrin (HPBCD, data not shown, *cf*. ref 11). Similar behavior is observed with clotrimazole in pH 3 L-tartrate buffer. The initial part of the solubility curve is linear and a negative curvature is observed at higher HBenBCD concentrations. The amounts of clotrimazole solubilized by HBenBCD and HPBCD were similar. In the case of itraconazole, a much different behavior is observed. The initial part of the solubility curve is linear and the amount of drug solubilized is very small. At higher HBenBCD concentrations a positive curvature in the solubility curve is observed. This type of curvature is classified as type A_p suggesting the formation of higher-order complexes (19). In contrast to voriconazole and clotrimazole, the amounts of itraconazole solubilized by HBenBCD and HPBCD were significantly different. For example, at *ca*. 25 wt% CD, the S_t/S_0 ratio for HBenBCD was 132, which was 4x greater than that observed for HPBCD.

These results illustrate that the amount of drug solubilized can vary significantly even though the drugs are structurally similar. Comparing the solubilization of the three drugs using HBenBCD to that obtained using HPBCD, we find that in two cases they are the same while in one case they are much different. These observations illustrate that cyclodextrin derivatives with seemingly small structural differences can provide significantly different solubilization, dissolution rates, and stabilization of drugs under otherwise similar conditions. Hence screening experiments are invaluable to determining the best cyclodextrin and other formulation components for a particular drug, making high through-put measurement methods an important tool in such work (18).

Figure 4 provides representative dissolution data (37 °C, standard USP 28-NF 23 711 method) for HBenBCD complexes of voriconazole and itraconazole, as well as for voriconazole without HBenBCD. In the absence of HBenBCD, dissolution of voriconazole was slow and incomplete with only 80% being dissolved after 6 h. In contrast, when complexed with HBenBCD, voriconazole dissolved rapidly and completely at all pH levels. It should be noted that the resulting voriconazole- HBenBCD solutions were stable with respect to voriconazole crystallization for the lifetime of the experiment. Dissolution of itraconazole from the capsules filled with itraconazole:HBenBCD complex was rapid and complete at pH 1.2 (*ca*. 100% in 15 min), while at pH 4.5 dissolution was rapid but incomplete (*ca.* 40%). In the absence of HBenBCD, essentially no itraconazole dissolves.

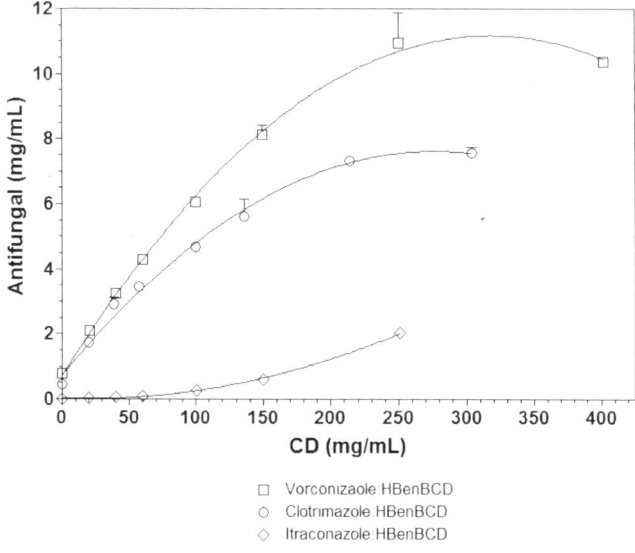

Figure 3. The equilibrium solubility of voriconazole (pH 3 succinate buffer), clotrimazole (pH 3 L-tartrate buffer), and itraconazole (pH 3 phosphate buffer) at variable HBenBCD concentrations.

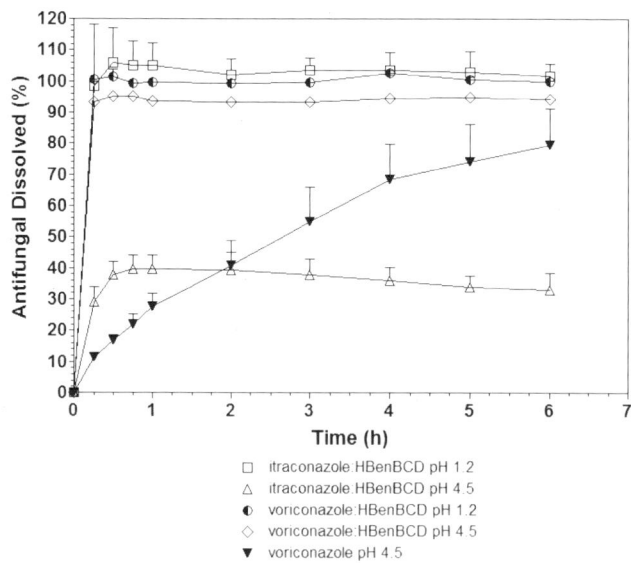

Figure 4. Dissolution of voriconazole and itraconazole at 37 °C from gelatin capsules filled with solid formulations (USP 28-NF 23 711).

We have also investigated the dissolution of itraconazole formulated with HBenBCD in biorelevant media (data not shown, *cf.* ref 14). In simulated gastric fluid at pH 1.2 rapid release of the entire dose was observed. The resulting drug solution was stable over the test duration of 4 h indicating that the drug should not precipitate in the stomach. Increasing the gastric pH to a value of 1.8 also resulted in rapid drug release but, as expected, the maximum dissolution was *ca.* 45%. At pH 6.8, the least favorable pH for dissolution of the poorly soluble weak base, minimal dissolution was observed with only 1 wt% itraconazole being dissolved after 4 h. In biorelevant dissolution media simulating the physiological conditions in the upper small intestine in both the fed and fasted states, a maximum release of *ca.* 7% itraconazole was observed. This indicates that drug release from itraconazole:HBenBCD complexes would be slower in the small intestine than in the stomach, but would be unaffected by food consumption.

Estrogen Modulators

Breast carcinoma is a major health problem affecting as many as one in eight women. Although treatment of breast cancer has advanced tremendously in recent years, approximately 25% of women with breast cancer will eventually die from the disease. It has long been understood that the growth of some cancers of the breast is stimulated or maintained by estrogens. Thus, treatment of breast cancers that are hormonally responsive has included a variety of approaches to decrease estrogen levels including surgery (*e.g.* ovariectomy) and chemotherapy (*e.g.* antiestrogens). Therapeutic agents that can reduce the effects of estrogen include selective estrogen receptor modulators, such as tamoxifen, which compete with estrogen at the receptor level. Aromatase inhibitors have more recently been introduced as a new class of therapeutic agents. Aromatase inhibitors act through inhibition of the cytochrome P450 enzyme, aromatase, which catalyses the conversion of androgens to estrogens. Letrozole, a Type II aromatase inhibitor, binds reversibly to the heme of the cytochrome P450 subunit of the enzyme, excluding both estrogen and oxygen from the enzyme. Raloxifene hydrochloride was approved by the Food and Drug Administration in 1997 as a treatment for osteoporosis. Raloxifene is a bone and liver estrogen agonist, which increases bone mineral density and decreases low density lipoprotein (LDL)-cholesterol. Recently, it has been found that raloxifene is a breast and uterus estrogen receptor antagonist and thereby may decrease the risk of invasive breast cancer.

Table II presents the equilibrium solubility of tamoxifen base and tamoxifen citrate in water and various buffers (0.05 M) in the absence (S_0) of HBenBCD and in the presence of 10 wt% HBenBCD (S_t) as well as the CD-mediated increase in drug solubility (S_t/S_0). The intrinsic solubility (S_0) of tamoxifen base in water was quite low (24 $\mu g/mL$) and the increase in solubility of tamoxifen base due to complexation with HBenBCD (S_t) was only marginal ($S_t/S_0 = 5$). At pH 1.2, HBenBCD significantly increased tamoxifen base solubility ($S_0 = 37$ $\mu g/mL$, $S_t/S_0 = 241$). At pH 3, the solubility of tamoxifen was not only dependent upon the pH of the medium but, was also apparently upon the type of buffer. At pH 3, the lowest solubility was observed with the phosphate buffer ($S_t/S_o = 49$) while the highest solubility was observed with the tartrate buffer ($S_t/S_o = 175$). As would be expected, as the pH of the medium (4.8 and 7.4) approached the pKa of tamoxifen the solubility of tamoxifen base in the presence of HBenBCD was greatly diminished. The intrinsic solubility ($S_o = 329$ $\mu g/mL$) of the citrate salt of tamoxifen in water was much higher than that of tamoxifen base but was somewhat diminished in the buffer solutions due to increased ionic strength of the media. In the presence of HBenBCD, the solubility of tamoxifen citrate was significantly increased in all media; S_t/S_o ranged from 54 to 293.

Table II. Equilibrium solubility (mg/mL) of tamoxifen base and tamoxifen citrate in the absence of HBenBCD (S_0), in the presence of 10 wt% HBenBCD (S_t) and the increase in drug solubility (S_t/S_0).

	tamoxifen base			*tamoxifen citrate*		
	S_0	S_t	S_t/S_0	S_0	S_t	S_t/S_0
Water	0.024	0.120	5	0.329	17.775	54
pH 1.2 HCl/KCl	0.037	8.915	241	0.050	14.749	295
pH 3 phosphate	0.089	4.394	49	0.180	12.874	72
pH 3 citrate	0.047	9.497	202	0.079	16.996	215
pH 3 D,L-tartrate	0.069	12.10	175	0.133	17.886	135
pH 4.8 acetate	0.141	1.163	8	0.150	17.483	117
pH 7.4 phosphate	0.028	0.668	24	0.056	10.896	195

Figure 5 provides representative equilibrium solubility data for tamoxifen base, tamoxifen citrate, letrozole, and raloxifene hydrochloride versus varying concentrations of HBenBCD. With increasing HBenBCD concentration, the solubility curve for tamoxifen base increases rapidly to about 10 wt% HBenBCD after which the slope of the curve is diminished. This behavior is indicative of formation of tamoxifen base:HBenBCD inclusion complexes with finite solubility (B_s solubility) (19). In contrast, the equilibrium solubility curve for tamoxifen citrate was very nearly linear over the entire range of HBenBCD concentration. At 400 mg/mL HBenBCD, 41.84 mg ± 0.29 of tamoxifen citrate was solubilized versus 6.51 mg ± 0.07 for tamoxifen base. The solubility curve for letrozole was linear over the HBenBCD concentration range examined. However, the solubility of letrozole was significantly lower than the solubility of tamoxifen base and citrate. At 300 mg/mL HBenBCD, 2.50 mg ± 0.03 of letrozole was solubilized. Although the absolute solubility is less relative to tamoxifen base and citrate, this is a significant increase over the intrinsic solubility of letrozole ($S_0 = 62$ $\mu g/mL$). The solubility curve for raloxifene was similar to that of tamoxifen citrate. The solubility of raloxifene hydrochloride ($S_0 = 448$ $\mu g/mL$) in pH 2 tartrate buffer was increased to 27.0 mg/mL when the concentration of HBenBCD was 298 mg/mL.

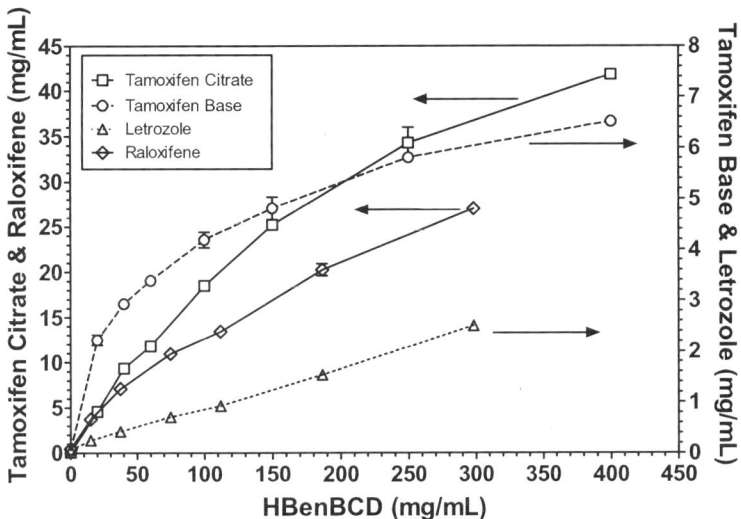

Figure 5. The equilibrium solubility of tamoxifen base and citrate (pH 3 phosphate buffer), letrozole (pH 2 L-tartrate buffer), and raloxifene (pH 2 L-tartrate buffer) at variable HBenBCD concentrations.

Figure 6 provides the dissolution profiles for native tamoxifen base and tamoxifen citrate (no HBenBCD) at pH 1.2 and 4.5 as well as for tamoxifen base and citrate formulated with HBenBCD at pH 1.2, 6.0, and 7.45. The tamoxifen formulations were prepared so that each contained equivalent amounts of tamoxifen on the basis of tamoxifen base (80 mg/mL, 1:3 molar ratio of tamoxifen:HBenBCD). The initial dissolution of tamoxifen base at pH 1.2 was relatively slow (16% at 15 min) and a maximum dissolution of *ca.* 45% was reached in *ca.* 3 h. At pH 4.5, the dissolution of tamoxifen base was minimal with only 5 wt% being dissolved after 6 h. The dissolution profiles for tamoxifen citrate were relatively insensitive to pH. At pH 1.2 and 4.5, the dissolution profiles overlapped and the maximum dissolution reached was *ca.* 5%.

A much different result was obtained with the tamoxifen:HBenBCD formulations. At pH 1.2, 100% dissolution of tamoxifen base was achieved in *ca.* 30 min versus 85% for tamoxifen citrate. At pH 6.0, the dissolution profiles of tamoxifen base and citrate were equivalent reaching *ca.* 92% dissolution within 30 min. At pH 1.2 and 6.0, precipitation of tamoxifen was not observed during the course of the experiment. At pH 7.5, the dissolution profiles for tamoxifen base and citrate were similar. The dissolution rate of tamoxifen citrate was only slightly faster than that of tamoxifen base; the total amount of tamoxifen dissolved was essentially equivalent. In both cases, the percentage of drug dissolved reached a maximum of 26% then decreased as the drug precipitated. Precipitation of a basic drug can often be expected as the pH of the medium approached its pKa. It is noteworthy that, in the presence of HBenBCD, the dissolution profile of tamoxifen base was equivalent to the salt.

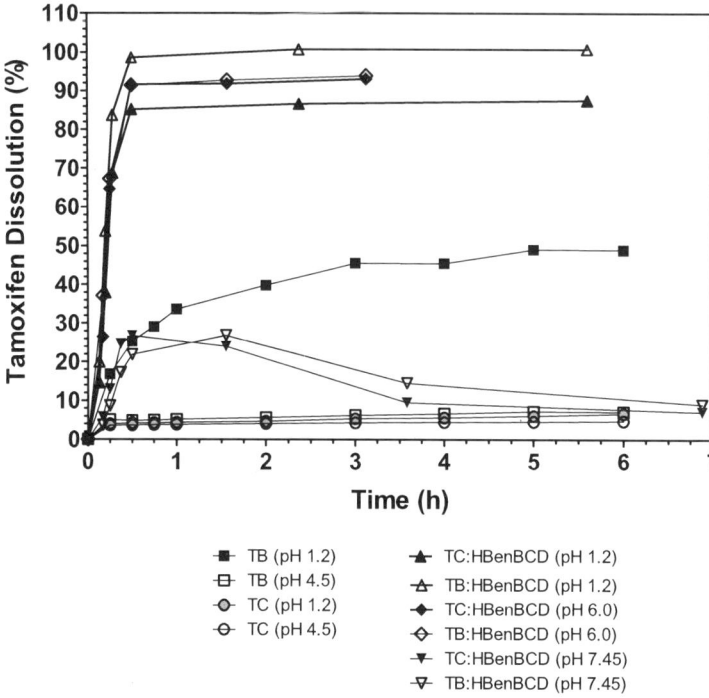

Figure 6. Dissolution of tamoxifen base (TB), tamoxifen citrate (TC), a TB:HBenBCD formulation, and a TC:HBenBCD formulation at 37 °C from solid filled gelatin capsules (USP 28-NF 23 711).

The dissolution profiles for letrozole and letrozole:HBenBCD formulations were much different than those of tamoxifen (data not shown, *cf.* ref 15). In the case of the letrozole:HBenBCD formulation, all of the drug was dissolved within 15 min at all pHs examined (1.2, 4.5, and 6.8). In the absence of HBenBCD, 17.5-22.0% of letrozole was dissolved at 15 min at pH 1.2, 4.5, and 6.8. After 6 h, the amount of letrozole dissolved was 77%, 67%, and 71% at pH 1.2, 4.5, and 6.8, respectively.

The dissolution profiles for raloxifene hydrochloride and raloxifene hydrochloride:HBenBCD formulations (Figure 7) illustrate the ability of HBenBCD to provide for rapid release formulation with basic drugs. In the absence of HBenBCD, the dissolution of raloxifene hydrochloride was slow and the extent of dissolution over 6 h was very sensitive to pH. At pH 6.8, 1.2, and 4.5, 12%, 41%, and 73% of raloxifene hydrochloride, respectively, dissolved over 6 h. Formulating with HBenBCD essentially eliminated the pH effect. After 45 min, 100% of the drug was dissolved at pH 1.2 and 4.5 and 93% of the drug was dissolved at pH 6.8. No crystallization of the drug was observed over the remaining time of the experiment.

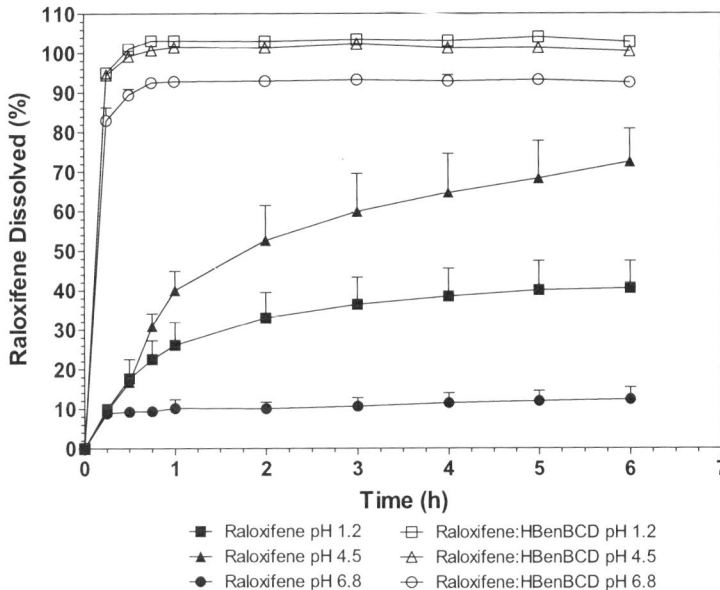

Figure 7. Dissolution of raloxifene hydrochloride and a raloxifene hydrochloride:HBenBCD formulation at 37 °C from powder filled gelatin capsules (USP 28-NF 23 711).

Protease Inhibitors

Human immunodeficiency virus (HIV) infection, which leads to acquired immunodeficiency syndrome (AIDS), remains a serious worldwide health problem. The discovery of HIV protease inhibitors introduced new and effective first-line therapies for HIV/AIDS. Helping to combat HIV-related diseases and prolong survival, protease inhibitors are commonly administered with reverse transcriptase inhibitors. However, poor patient compliance, noxious side-effects, and viral resistance have led to a recommendation to treat with two different kinds of protease inhibitors.

The most important HIV protease inhibitors in clinical use are saquinavir, nelfinavir, indinavir, lopinavir, ritonavir (as a CYP3A4 inhibitor to "boost" plasma concentrations of other protease inhibitors), atazanavir, and amprenavir. Many protease inhibitors have poor aqueous solubility which produces very low and variable bioavailability. Poor solubility combined with metabolism and transport issues result in HIV/AIDS patients requiring frequent dosing of large capsules and pills. Commonly, the patients are unable to adhere to these treatment regimes.

The solubilities of saquinavir base, saquinavir mesylate, and ritonavir in different media in the presence and absence of HBenBCD are summarized in Table III. Saquinavir base has poor water solubility ($S_o = 207 \pm 5$ μg/mL) which was significantly increased by 10 wt% HBenBCD ($S_t/S_o = 26$). Saquinavir base solubility at pH 1.4 ($S_o = 87 \pm 6$ μg/mL) was also poor but dramatically improved by complexation with HBenBCD ($S_t/S_o = 37$). Saquinavir base was more soluble in pH 3.0 buffer, particularly the tartrate buffers, without ($S_o = 6.5$-7.0 mg/mL) or with HBenBCD ($S_t = 29.7$-31.3 mg/mL). Saquinavir base solubility at pH 4.8 or higher was diminished relative to pH 3.0 buffer. However, the HBenBCD did help to maintain significant drug solubility in these higher pH media ($S_t = 6$-12 mg/mL) relative to water or pH 1.4 buffer.

As expected, the intrinsic solubility of saquinavir mesylate in water (2.1 ± 0.3 mg/mL) was higher than saquinavir base. However, unlike saquinavir base, 10 wt% HBenBCD increased saquinavir mesylate solubility in all media over a broad pH range. In the presence of HBenBCD, saquinavir mesylate solubility was essentially the same (11-13 mg/mL) in all media except for pH 1.4 buffer where the drug solubility dropped to 3.8 ± 0.3 mg/mL ($S_t/S_o = 14$) and with pH 7.3 phosphate buffer where drug solubility dropped to 9.4 ± 0.4 mg/mL ($S_t/S_o = 43$).

The solubility of ritonavir in the different media was generally less than that of saquinavir. In water, 10 wt% HBenBCD increased the solubility of ritonavir from 106 μg/mL (S_o) to 1.2 mg/mL ($S_t/S_o = 11.3$). In the pH range of 3.0 to 7.3, the increase in solubility of ritonavir in buffer solutions containing 10 wt% HBenBCD was similar, but slightly lower to that observed for 10 wt% HBenBCD in water. This observation is not unexpected as ritonavir has two pKa values, 3.5 and 11.5, and adjustment of pH within a pharmaceutically relevant pH range is not expected to have a significant effect on ritonavir solubility. Indeed, this experiment indicates that ritonavir is sensitive to the ionic strength of the buffer media.

The equilibrium solubility curves for saquinavir base, saquinavir mesylate, and ritonavir in water at variable concentrations of HBenBCD are shown in Figure 8. The shapes of the curves for the two forms of saquinavir are the same; both have a slight negative curvature at higher drug concentrations. As would be expected, more saquinavir mesylate is solubilized than saquinavir base at any given concentration of HBenBCD. In the case of ritonavir, the initial part of the solubility curve is linear and the amount of drug solubilized is very small. At higher HBenBCD concentrations a positive curvature in the solubility curve is observed suggesting the formation of higher-order complexes.

The dissolution profiles for saquinavir base:HBenBCD and saquinavir mesylate:HBenBCD powder filled capsules were similar. In the case of saquinavir mesylate:HBenBCD powder filled capsules, dissolution was rapid at each pH examined with *ca.* 100% of the drug being released into the medium within 30 min at pH 1.2 and pH 4.5; *ca.* 90% was released within 30 min at pH 6.8 (Figure 9). Once dissolved, the drug remained dissolved in the presence of HBenBCD over the time course of the experiment. In contrast, saquinavir mesylate (no HBenBCD) dissolution was slow and incomplete with *ca.* 4-11% being dissolved after 30 min. After 6 h, the maximum percentages of saquinavir mesylate (no HBenBCD) dissolved were 9.4, 19.2, and 4.4% at pH 1.2, 4.5, and 6.8, respectively.

Table III. Equilibrium solubility (mg/mL) of protease inhibitors in the absence of HBenBCD (S_0), in the presence of 10 wt% HBenBCD (S_t) and the increase in drug solubility (S_t/S_0).

	saquinavir base			saquinavir mesylate			ritonavir		
	S_0	S_t	S_t/S_0	S_0	S_t	S_t/S_0	S_0	S_t	S_t/S_0
Water	0.2	5.5	26	2.1	11.6	6	0.1	1.2	11
pH 1.4 HCl/KCl	0.1	3.2	36	0.3	3.8	14	ND	ND	ND
pH 3 phosphate	2.8	9.3	3	2.9	12.7	4	0.1	0.9	9
pH 3 D,L-tartrate	7.0	31.3	5	3.7	12.4	3	ND	ND	ND
pH 3 D-tartrate	6.5	29.7	5	3.6	12.0	3	ND	ND	ND
pH 3 L-tartrate	7.0	30.3	4	3.7	11.7	3	ND	ND	ND
pH 3 citrate	5.0	24.5	5	3.1	11.9	4	0.2	0.8	4
pH 4.8 citrate	1.1	11.9	11	1.8	12.4	7	0.2	1.0	6
pH 4.8 propionate	1.0	9.8	10	2.2	12.8	6	0.1	1.0	9
pH 4.8 acetate	1.3	11.4	9	2.1	13.5	6	0.1	1.1	10
pH 6.5 phosphate	0.2	6.1	28	0.3	12.2	44	0.1	0.9	8
pH 7.3 phosphate	0.2	6.2	30	0.2	9.4	43	0.1	0.8	7

ND: Not Determined.

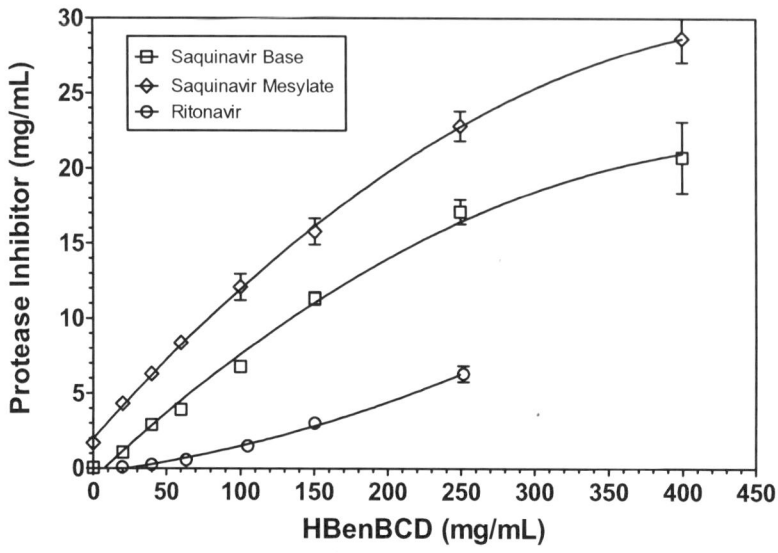

Figure 8. The equilibrium solubility (water) of saquinavir base, saquinavir mesylate, and ritonavir at variable HBenBCD concentrations.

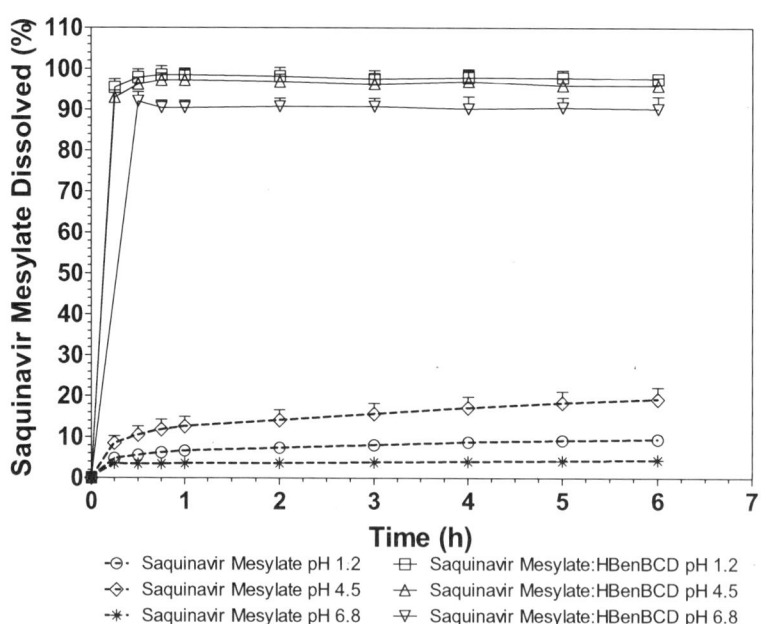

Figure 9. Dissolution of saquinavir mesylate and a saquinavir mesylate:HBenBCD formulation at 37 °C from powder filled gelatin capsules (USP 28-NF 23 711).

In the case of saquinavir base:HBenBCD powder filled capsules, drug dissolution was rapid at each pH examined with approximately 100% of the drug being released into the medium within 30 min at pH 1.2 and pH 4.5; 95% was released within 30 min at pH 6.8 (data not shown, cf. ref 17). Once dissolved, the drug concentration in the presence of HBenBCD remained stable over the time course of the experiment; the HBenBCD effectively prevented crystallization of the drug even at concentrations that would have been supersaturated in the absence of HBenBCD. In contrast, saquinavir base (no HBenBCD) dissolution was slow and incomplete with 15-20% being dissolved after 30 min. After 6 h, the maximum percentages of saquinavir base (no HBenBCD) dissolved were 16.2 and 24.9% at pH 1.2 and 6.8, respectively.

Pharmacokinetic Studies

During the course of our work, we have conducted a number of preclinical trials involving intravenous and oral administration of drug:HBenBCD formulations and drug (no HBenBCD) to Sprague-Dawley or Wistar-Hannover rats. In general, we were interested in directly comparing the *in vivo* results obtained with these formulations and drugs, and to determine the correlation between *in vitro* and *in vivo* results. We were particularly interested in probing the effects of factors such as CD structure, formulation, food, gender, and chronopharmacokinetics on absorption, metabolism, elimination, and oral bioavailability of drugs in these animal models. The results are interesting and complex; since this account is intended as a summary, the reader is urged to see the references cited herein for a full account.

Antifungals: Itraconazole

Itraconazole is currently available in multiple formulations under the brand-name Sporanox® (Janssen Pharmaceutica N.V.). One such formulation is an oral solution (10 mg of itraconazole/mL) containing 400 mg/mL of hydroxypropyl-β-cyclodextrin (HPBCD). Sporanox® Oral Solution is clear and yellowish in color with a target pH of 2.

Itraconazole is also available as an intravenous (IV) dosage form which is similar to that of the oral liquid form with the notable exception that the target pH is *ca.* 4.5. Itraconazole is not given as an IV bolus but rather infused over a one hour period. There is also a hard gelatin capsule form of Sporanox® (100 mg of itraconazole coated on sugar spheres). The solid capsule form does not contain any cyclodextrin. The recommended dose of both the solid formulation and the oral liquid is 200 mg (2 capsules or 20 mL) given once daily.

In humans, the bioavailability of itraconazole from the oral solution has been reported to be 55%, which is 37% higher than that observed from the capsule form. Hence, the solution and capsule dosage forms cannot be used interchangeably. The capsule formulation is supposed to be taken after a full meal, which limits its utility as many patients are unable to ingest solid food and gastrointestinal events may limit food intake. Higher oral bioavailability with the oral liquid dosage form is observed in humans when the patient has fasted. Pharmacokinetic (PK) studies of itraconazole in Sprague-Dawley rats and in mice indicate that the oral bioavailability of itraconazole was influenced by the dosage form. The oral bioavailabilities of itraconazole when administered as an oral solution or as a capsule were 35 and 10%, respectively. One should also note that itraconazole is both a CYP3A substrate and inhibitor, and a P-gp substrate. Fortunately, hydroxyitraconazole, the main metabolite of itraconazole, also has considerable antifungal activity. Also, the AUC and C_{max} for hydroxyitraconazole and itraconazole are essentially equal.

Pharmacokinetic parameters for IV, oral solution, and solid filled capsule dosing of male Sprague-Dawley rats are summarized in Table IV. Figure 10a provides typical plasma concentration-time curves for itraconazole after intravenous doses of Sporanox® injection solution (Group 1) and itraconazole:HBenBCD aqueous solution (Group 2). The most notable feature of the data presented in Figure 10a is the apparent secondary absorption of itraconazole at *ca.* 12 h. This observation suggests hepatic recirculation of itraconazole mediated by HBenBCD. The net result is that the area under the curves (AUC) for itraconazole and hydroxyitraconazole obtained after administration of the itraconazole:HBenBCD aqueous solution is slightly greater (*ca.* 18%) than those obtained with the Sporanox® injection solution (Table IV).

Table IV. Pharmacokinetic parameters for itraconazole (ITZ) and hydroxyitraconazole (HITZ).

Group	C_{max} (ng/mL)		T_{max} (h)		AUC$_{0-48 h}$ (ng•h/mL)		%F
	ITZ	HITZ	ITZ	HITZ	ITZ	HITZ	ITZ
Sporanox® IV (G1) fasted	2951.0	580.0	0.5	4.0	27751	17439	---
HBenBCD IV (G2) fasted	2755.0	609.5	0.3	12.0	32687	20344	---
Sporanox® Oral Solution (G3) fasted	443.7	419.8	5.0	3.0	10330	10425	31.6
HBenBCD Oral Solution (G4) fasted	620.9	510.3	6.0	5.0	14525	13823	44.4
Sporanox® Capsules (G5) fed	228.3	270.0	5.0	5.0	3533	5652	10.8
HBenBCD Capsules (G6) fed	326.3	406.8	9.0	9.0	6209	9619	19.0
HBenBCD Capsules (G7) fasted	329.3	345.0	4.0	9.0	6607	8510	20.2

Abbreviations: C_{max}, maximum plasma concentration; T_{max}, time required to reach C_{max}; AUC$_{0-48}$, total area under the plasma concentration-time curve from 0 to 48 h; F, oral bioavailability. Values are means ± standard deviations.

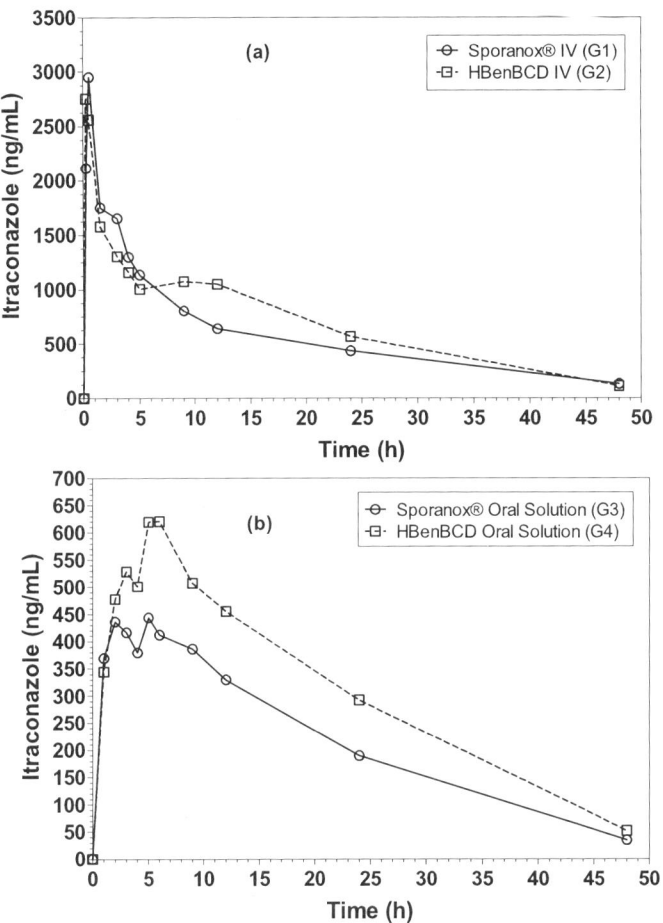

Figure 10. Plasma concentration-time curves for itraconazole. (a) After intravenous doses of Sporanox® injection solution and itraconazole:HBenBCD aqueous solution. (b) After oral gavages of Sporanox® solution (contains HPBCD) and aqueous itraconazole:HBenBCD solution.

Figure 10b displays plasma concentration versus time profiles for itraconazole after dosing of Sporanox® (contains HPBCD) and itraconazole:HBenBCD oral solutions. The itraconazole plasma concentration versus time profiles for these formulations are similar. Following a rapid increase in itraconazole plasma concentration, itraconazole plasma concentrations begin to level off or decline between *ca.* 2-4 h followed by a secondary increase in plasma concentrations between *ca.* 5-6 h (T_{max}, Table IV). In the case of the itraconazole:HBenBCD oral solution, the second rise in itraconazole plasma concentrations is larger than that observed for the itraconazole:HPBCD oral solution. The net effect is that the itraconazole AUC for the itraconazole:HBenBCD oral solution (14526 ng•h/mL) is larger than the

Sporanox® (10330 ng•h/mL) oral solution. The plasma concentration versus time profiles for hydroxyitraconazole are similar to those observed for itraconazole.

The plasma concentration versus time profiles for itraconazole and hydroxyitraconazole after oral dosing with capsules of Sporanox® beads (group 5, no CD, fed animals) and itraconazole:HBenBCD powder (groups 6 and 7, fed and fasted animals) were similar to those shown in Figure 10b (cf. ref 14). In the case of Sporanox® beads, the itraconazole AUC was 3533 ng•h/mL and the observed T_{max} was 4 h. Fed animals dosed orally with itraconazole:HBenBCD powder had an itraconazole AUC of 6209 ng•h/mL, with T_{max} apparently shifted to 9 h. However, the apparent shift in T_{max} may be due to a secondary absorption phase; there is an earlier maximum at 4 h. Fasted animals dosed orally with the same itraconazole:HBenBCD powder had an AUC of 6607 ng•h/mL, with a T_{max} of 4 h. That is, in agreement with our in vitro dissolution studies, there are apparently no significant food effects on the PK parameters after dosing rats with itraconazole:HBenBCD powder. It is interesting to note that, while both the C_{max} and the AUC for the HBenBCD powder formulations were ca. 2-times greater than the Sporanox® beads formulation, T_{max} for the 2 formulations were the same. In general, CD-containing drug formulations tend to have reduced T_{max} relative to drug formulations that do not contain CDs.

The oral bioavailability of itraconazole obtained with each dosage form is also provided in Table IV. The highest oral bioavailability was obtained with the oral solutions. Dosing of the Sporanox® liquid solution (group 3, contains HPBCD) gave an oral bioavailability of 32%. Dosing with the itraconazole:HBenBCD oral solution (group 4) resulted in an oral bioavailability of 44%, or 1.4-times that observed with the HPBCD. Again, this result correlates with our in vitro solubility studies which indicated that HBenBCD was more effective than HPBCD in increasing the solubility of itraconazole. In the case of the itraconazole:HBenBCD powder filled capsules, the observed oral bioavailabilities were similar (19-20%) regardless of the dietary state of the animals, and were 2-times that obtained by dosing with Sporanox® beads (no CD). The results obtained in this study are comparable to values obtained in other laboratories, which have found that the oral bioavailabilities of itraconazole obtained with Sporanox® oral and capsule dose forms were 31-35 and 10%, respectively.

Estrogen Modulators

Tamoxifen is used clinically as a nonsteroidal antiestrogen for first-line endocrine treatment as well as adjuvant therapy in early and metastic breast cancers in postmenopausal women. Tamoxifen is also approved as a prophylactic agent in women at high risk of developing breast cancer. The commercial form of tamoxifen is tamoxifen citrate (Nolvadex®) and is offered as 10 and 20 mg tablets (15.2 and 30.4 mg of the citrate salt). Tamoxifen is a low dose therapeutic agent typically given to patients for long periods of time.

The Z-isomer of tamoxifen is biologically active but the E-isomer has limited biological activity. After oral dosing of tamoxifen, over 50% of tamoxifen is converted to two major metabolites, 4-hydroxytamoxifen and N-desmethyltamoxifen. 4-Hydroxytamoxifen has biological activity similar to that of tamoxifen and is cleared rapidly, having a half-life slightly longer than that of tamoxifen. N-Desmethyltamoxifen has diminished bioactivity relative to that of tamoxifen but has a half-life of days.

In initial PK studies involving IV and oral administration of tamoxifen:HBenBCD formulations to male Sprague-Dawley rats, we found clear indications of apparent food and chronopharmacokinetics effects on tamoxifen oral bioavailability. In view of these surprising observations, we elected to conduct a PK study in which food and chronological effects could be independently probed. Also embedded within this study were comparisons of the form of tamoxifen (tamoxifen base versus tamoxifen citrate:HBenBCD complexes) and type of formulation (pre-formed tamoxifen:HBenBCD complex in water versus tamoxifen and HBenBCD dissolved in common organic solvents).

The pharmacokinetic parameters for 16 treatment groups dosed with tamoxifen formulated with HBenBCD are summarized in Table V (tamoxifen only. See ref 13 for metabolite data). Both food and time of dosing had a pronounced effect on tamoxifen pharmacokinetics after oral administration. Figure 11 provides plasma concentration-time curves for tamoxifen for the 4 treatment groups which received an oral aqueous dose of tamoxifen base:HBenBCD complex. Administration of oral aqueous doses of tamoxifen base:HBenBCD complexes to fasted rats in the morning gave the highest C_{max} (217 ng/mL) and $AUC_{0 \to \infty}$ (2668 ng·h/mL). Relative to the AM/fasted group, C_{max} for the AM/fed group was lower (135 ng/mL) and the $AUC_{0 \to \infty}$ (1937 ng·h/mL) was smaller. Comparing the two groups dosed in the afternoon, the observation is that the fasted rats had a higher C_{max} and AUC than the fed rats. This is indicative of a food effect on tamoxifen oral bioavailability. Similarly, comparison of the AM/fed group to the PM/fed group and the AM/fasted group to the PM/fasted group reveals that the rats dosed in the morning always had the higher C_{max} and AUC regardless of their dietary status. Detailed inspection of the data contained in Table V for the remaining groups reveals similar food and chronological effects on the oral bioavailability of tamoxifen. Neither the form of tamoxifen (base versus citrate) nor the type of solution (aqueous versus PG:PEG400:H_2O) had an impact on the observed food or chronological effects. It is apparent that food and chronological effects must be considered in pharmacokinetic studies involving tamoxifen.

Table V. Pharmacokinetic parameters of tamoxifen after a single oral administration of tamoxifen base/HBenBCD complexes to Sprague-Dawley rats (n = 3).

Formulation	Time/Food	T_{max} h	C_{max} ng/mL	$t_{1/2}$ h	$AUC_{0\to24}$ ng·h/mL	$\%F_{0\to24}$
tamoxifen base: HBenBCD aqueous oral solution	AM/Fed	2	135.0 ± 47.8	8.5 ± 2.2	1626 ± 326	35.3 ± 7.3
	PM/Fed	3	104.4 ± 25.8	9.2 ± 2.4	1052 ± 353	22.7 ± 7.6
	AM/Fasted	3	216.7 ± 71.8	7.3 ± 0.9	2378 ± 652	51.4 ± 14.2
	PM/Fasted	4 (3-6)	164.7 ± 11.5	8.4 ± 1.3	1714 ± 283	37.0 ± 6.1
tamoxifen citrate: HBenBCD aqueous oral solution	AM/Fed	4 (3-4)	127.3 ± 39.1	8.1 ± 0.5	1716 ± 302	37.5 ± 7.4
	PM/Fed	3 (2-4)	142.9 ± 27.6	8.6 ± 3.5	1399 ± 171	30.2 ± 3.8
	AM/Fasted	4 (2-6)	174.8 ± 17.5	7.3 ± 1.1	2260 ± 65	49.0 ± 1.4
	PM/Fasted	3 (2-4)	154.7 ± 98.4	12.6 ± 7.7	1550 ± 508	33.5 ± 11.0
tamoxifen base- HBenBCD PG:PEG400 solution	AM/Fed	2.3 ± 1.5	209.9 ± 52.7	9.6 ± 2.9	2118 ± 120	47.3 ± 3.1
	PM/Fed	2	129.4 ± 24.4	13.8 ± 5.1	1515 ± 155	34.6 ± 3.5
	AM/Fasted	3 (2-4)	321.0 ± 184.7	6.1 ± 0.8	2929 ± 1485	71.9 ± 193
	PM/Fasted	4 (a)	194.6 (a)	7.0 (a)	2064 (a)	49.3 (a)
tamoxifen citrate- HBenBCD PG:PEG400 solution	AM/Fed	2 (1-3)	151.2 ± 19.2	9.5 ± 2.7	1836 ± 264	39.3 ± 4.3
	PM/Fed	3 (2-4)	113.1 ± 44.9	9.6 ± 3.5	1263 ± 605	28.0 ± 13.5
	AM/Fasted	4 (2-6)	176.9 ± 56.8	8.2 ± 1.6	1830 ± 226	39.9 ± 4.4
	PM/Fasted	3 (2-4)	186.1 ± 83.3	10.3 ± 4.8	2143 ± 443	46.2 ± 8.2

Abbreviations: C_{max}, maximum plasma concentration; T_{max}, time required to reach C_{max}; AUC_{0-24}, total area under the plasma concentration-time curve from 0 to 24 h; F, oral bioavailability. Values are means ± standard deviations except for T_{max} which are median ranges.
(a) n = 1.

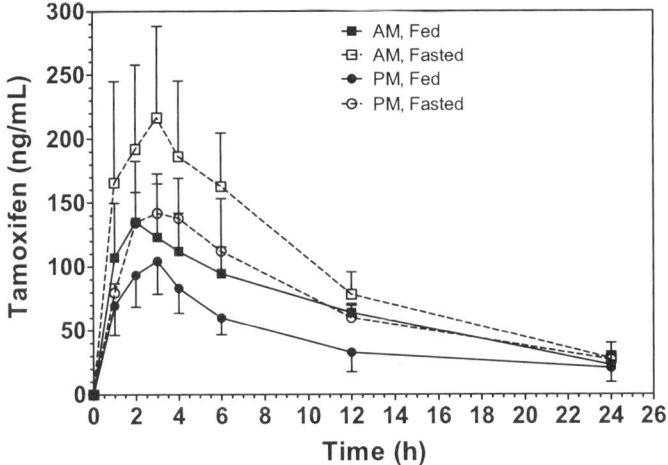

Figure 11. Plasma concentration-time curves after oral doses of aqueous solutions of tamoxifen base:HBenBCD complexes. Error bars indicate one standard deviation.

Figure 12 illustrates that oral administration of tamoxifen:HBenBCD complexes provided a very significant increase in tamoxifen oral bioavailability relative to oral administration of tamoxifen base or tamoxifen citrate (no HBenBCD). The oral bioavailability of tamoxifen base (AM/fed) was *ca.* 5% while the oral bioavailabilities obtained with the tamoxifen base:HBenBCD complex aqueous solution and the tamoxifen base-HBenBCD PG:PEG400:H$_2$O solution (AM/fed) were *ca.* 35% and 47%, respectively. That is, the tamoxifen base-HBenBCD formulations provided a 7- to 9-fold increase in oral bioavailability relative to that of tamoxifen base alone. Recognizing there is a difference in dietary state, if the same comparison is made between tamoxifen base (AM/fed) and the HBenBCD formulations (AM/fasted), the increase in oral bioavailability was 10- to 14-fold.

Comparing the tamoxifen citrate groups, the oral bioavailability of tamoxifen citrate (AM/fed) was *ca.* 13% while the oral bioavailabilities obtained with the tamoxifen citrate:HBenBCD complex aqueous solution and the tamoxifen citrate-HBenBCD PG:PEG400:H$_2$O solution (AM/fed) were *ca.* 37% and 39%, respectively. In this case, the tamoxifen citrate-HBenBCD formulations provided a 3-fold increase in oral bioavailability relative to that of tamoxifen citrate alone. If the same comparison is made between tamoxifen citrate (AM/fed) and the tamoxifen citrate-HBenBCD formulations (AM/fasted), the increase in oral bioavailability was 3-to 4-fold. With tamoxifen citrate, the differences in oral bioavailabilities between the two HBenBCD formulations were only marginal.

Comparing tamoxifen base-HBenBCD and tamoxifen citrate-HBenBCD formulations, it is clear that the tamoxifen base-HBenBCD formulations provided the larger increase in oral bioavailability relative to the parent drug (10- to 14-times) versus 3- to 4-fold for the tamoxifen citrate formulations. However, direct comparison of oral bioavailabilities of the tamoxifen base and

54

citate-HBenBCD formulations reveals that they were essentially the same. That is, when formulated with HBenBCD, it is not necessary to convert the base form of the drug to the salt form to achieve higher solubility and oral bioavailability.

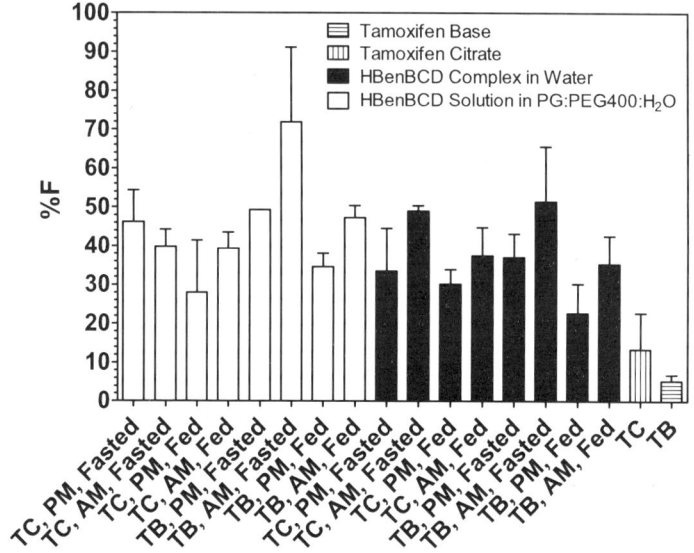

Figure 12. Oral bioavailability (%F) of tamoxifen obtained with Sprague-Dawley rats after oral administration of tamoxifen base, tamoxifen citrate, and tamoxifen:HBenBCD formulations. Abbreviations: TB, tamoxifen base; TC, tamoxifen citrate. Error bars indicate one standard deviation.

For purposes of statistical analysis, a stepwise least squares regression factorial model was constructed in which the form of tamoxifen (base versus citrate) in the complex, time of dosing (AM versus PM), dietary status (fed versus fasted), and gavage solution (aqueous versus PG:PEG400:H_2O) were allowed to vary independently. The model converged when the calculated tamoxifen oral bioavailabilities gave the best fit to the experimental oral bioavailability. The form of tamoxifen (base versus citrate) and interaction terms were found not to be significant. The model showed that oral bioavailability of tamoxifen administered as HBenBCD formulations increased when the gavage solution was PG:PEG400:H_2O (P = 0.0014), when the animals were dosed in the morning (P < 0.0001), and when the animals were fasted prior to dosing (P < 0.0001).

Letrozole

Letrozole, manufactured by Novartis under the trade mark Femara[®], was approved by the FDA in 1998 for the treatment of advanced breast cancer in postmenopausal women, with hormone receptor positive or unknown breast cancer, who had failed one prior antiestrogen treatment. Subsequently, the FDA approved letrozole tablets (2.5 mg/day) for first-line treatment of postmenopausal women with hormone receptor-positive or hormone receptor-unknown locally advanced or metastatic breast cancer. In postmenopausal women, letrozole decreases plasma concentrations of estradiol, estrone, and estrone sulfate by 75-95% from baseline with maximal suppression achieved within 2-3 days of treatment initiation. Recent reports indicate that there may be significant gender differences in letrozole pharmacokinetics in rats.

Pharmacokinetic parameters for IV, oral solution, and solid filled capsule dosing of male and female Sprague-Dawley rats are summarized in Table VI. Figure 13 provides typical plasma concentration-time curves for letrozole after oral (capsules) dosing. Inspection of Table VI and Figure 13 shows clearly that there are gender differences. The two male IV dose groups (with and without HBenBCD) were significantly different (AUC = 2441 and 3669 ng·h/mL, respectively) while the two female IV dose groups (0-24 h) were not statistically different from one another. In male rats, HBenBCD increased the peak blood levels (C_{max}) from 332 ng/mL to 489 ng/mL; in female rats, HBenBCD also increased C_{max} values (from 493 to 605 ng/mL) and delayed T_{max}.

Table VI. Pharmacokinetic parameters of letrozole after administration of letrozole and letrozole/HBenBCD complexes to male and female Sprague-Dawley rats (n = 3).

Gender	Route/ Formulation	AUC (ng·h/mL)	T_{max} (h)	C_{max} (ng/mL)	%F
Male	iv/solution	2541 ± 58	1.5 ±0.3	332 ± 20	100%
Male	iv/HBenBCD solution	3669 ± 333	1.8 ±0.3	489 ± 42	100%
Female	iv/solution	24439 ± 504	2.5 ±0.5	493 ± 25	100%
Female	iv/HBenBCD solution	27366 ± 789	4.0 ±0.0	605 ± 24	100%
Male	oral/suspension	1061 ± 166	6.0 ±1.4	71 ± 13	42 ± 5
Male	oral/HBenBCD solution	2323 ± 110	2.3 ±0.3	157 ± 7	63 ± 3
Male	oral/CMC capsule	970 ± 95	8.4 ±0.0	87 ± 2	38 ± 3
Male	oral/HBenBCD capsule	1687 ± 83	6.3 ±0.0	140 ± 5	66 ± 2
Female	oral/CMC capsule	23217 ± 821	16.4 ±3.8	326 ± 13	95 ± 2
Female	oral/HBenBCD capsule	27640 ± 1464	5.4 ±1.7	464 ± 39	101 ± 3

Abbreviations: C_{max}, maximum plasma concentration; T_{max}, time required to reach C_{max}; AUC, total area under the plasma concentration-time curve; F, oral bioavailability. Values are means ± standard deviations.

When administered orally as a capsule preparation to male rats, HBenBCD increased the apparent absorption rate (k_a = 0.50 and 0.66 for letrozole and letrozole:HBenβCD, respectively). The difference in the rates of absorption in male rats with and without HBenBCD was most profound during the first 4 h following oral administration. The use of HBenBCD in male rats resulted in a 74% increase in AUC_{0-36} (1687 ± 83 versus 970 ± 95 ng·h/mL), a 60% increase in C_{max} from 87 to 140 ng/mL, and a decrease in T_{max} from 8.4 h to 6.3 h. The effects of orally administering letrozole with and without HBenBCD were not as pronounced in female rats. In female rats, the use of HBenBCD resulted in a 19% increase in AUC_{0-126} (27640 ± 1464 versus 23217 ± 821 ng·h/mL), a 43% increase in C_{max} from 326 to 464 ng/mL, and a decrease in T_{max} from 16.4 ± 3.8 h to 5.4 ± 1.7 h.

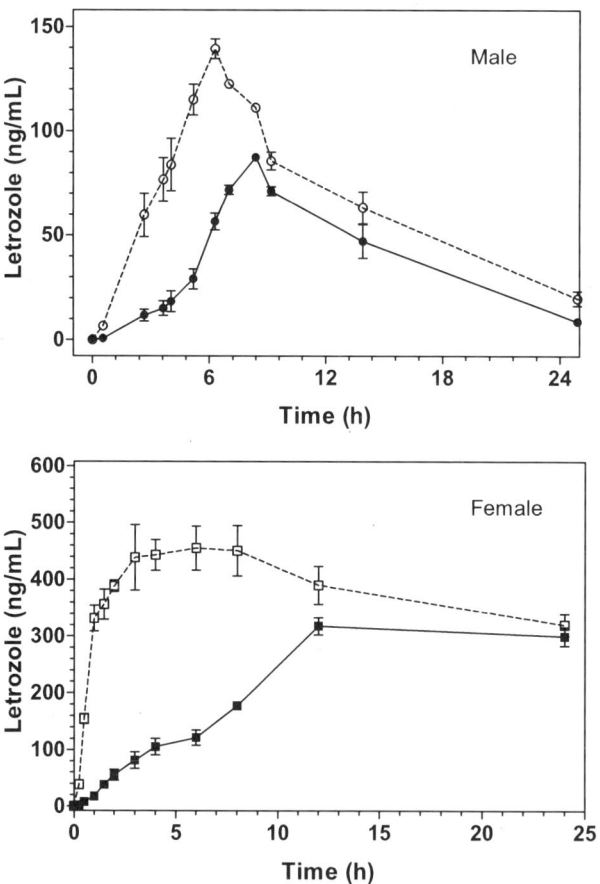

Figure 13. Pharmacokinetics of letrozole administered as a capsule formulation in the presence of either carboxymethylcellulose (filled symbols) or HBenBCD (open symbols).

In both male and female rats, the increase in letrozole blood levels produced by HBenBCD was present for 12 h following administration. To illustrate the profound differences in absorption by the presence of HBenBCD, an 80% increase in the AUC_{0-12} for male rats and a 180% increase in the AUC_{0-12} for female rats were observed. Differences in letrozole blood levels between animals receiving letrozole/CMC and letrozole/HBenBCD capsules were resolved by 24 h.

The presence of HBcnBCD improved the oral bioavailability of letrozole without altering apparent terminal half-life. In male rats, the absolute oral bioavailability of letrozolc in male rats obtained with the letrozole:HBenBCD capsule formulation was $46 \pm 2\%$ compared to $38 \pm 3\%$ with a letrozole/CMC capsule formulation. If AUC for the IV solution without HBenBCD is used, the absolute bioavailability in male rats becomes $66 \pm 2\%$. Although the presence of HBenBCD did increase absorption in female rats, the absolute bioavailability of letrozole was >90% regardless of the formulation.

Collectively, the study indicates that complexation of letrozole with HBenBCD doubled the absorption in male rats and enhanced the absorption in female rats. In male rats, letrozole oral absorption was limited by both solubility and uptake; when solubility is increased by complexation with HBenBCD, oral absorption is more rapid and bioavailability improves. However, in female rats, letrozole solubility limits the rate but not the extent of absorption. Complexation with HBenBCD, while it does produce more rapid initial oral absorption, does not improve letrozole bioavailability. The data on oral absorption of letrozole and the effect of HBenBCD support the hypothesis that uptake of letrozole by female rats has higher affinity and/or greater capacity than in male rats. Following absorption, the effects of HBenBCD were similar in both sexes, with slightly higher blood concentrations achieved for 8-12 h following administration.

Raloxifene

Clinical use of raloxifene has been somewhat hampered by the poor water solubility of the drug (340 µg/mL) and the low oral bioavailability of raloxifene. In humans, it has been reported that *ca.* 60% of a normal raloxifene dose is rapidly absorbed and that the oral bioavailability of raloxifene is only *ca.* 2% (20). However, presystemic glucuronide conjugation is extensive. Oral bioavailability and time to maximum plasma concentration are dependent upon systemic interconversion and enterohepatic cycling of raloxifene and its glucuronide metabolites. Raloxifene is metabolized by UDP-glucuronosyl-transferases to afford raloxifene 6-β-glucuronide in the liver or raloxifene 4'-β-glucuronide in the liver and intestine. Raloxifene glucuronides are excreted into the intestine in the bile, converted back to raloxifene via intestinal β-glucuronidase and reabsorbed or excreted in feces. It should be noted that raloxifene glucuronides exhibit biological activity although somewhat lower than that of the parent raloxifene.

Table VII summarizes pharmacokinetic parameters for raloxifene and raloxifene glucuronides after oral administration of raloxifene, raloxifene:HBenBCD aqueous solution, and powder and liquid fill capsules to male Wistar-Hannover rats. Figure 14 provides typical plasma concentration-time curves for raloxifene and raloxifene glucuronides after oral administration of powder filled capsules. For simplicity, the AUCs for the different raloxifene glucuronides have been combined in both the figure and table.

Table VII. Pharmacokinetic parameters of raloxifene after oral administration of raloxifene and raloxifene-HBenBCD formulations to male Wistar-Hannover rats.

Formulation	Analyte	$AUC_{0\text{-}72\,h}$ (ng•h/ml)	T_{max} (h)	C_{max} (ng/ml)	Total Contact	F (%)
Raloxifene (capsules)	Ral.	300 ± 41	4.0 ± 0.5	42.9 ± 4.2	55 ± 10	2.6 ± 0.4
	Gluc.	153 ± 55	4.0 ± 0.5	24.3 ± 15.6		
Raloxifene-HBenBCD (capsules)	Ral.	901 ± 270	2.5 ± 0.5	107.6 ± 42.6	231 ± 40	7.7 ± 2.2
	Gluc.	1013 ± 130	0.75 ± 0.25	297.8 ± 80.0		
Raloxifene-HBenBCD (aq. gavage)	Ral.	749 ± 27	4.0 ± 0.5	81.4 ± 25.7	193 ± 18	6.4 ± 0.8
	Gluc.	996 ± 153	0.75 ± 0.25	324.4 ± 73.2		
Raloxifene-HBenBCD-PEG400-PG (capsules)	Ral.	668 ± 149	4.0 ± 0.5	136.3 ± 124.4	192 ± 37	5.7 ± 1.3
	Gluc.	887 ± 216	0.50 ± 0.25	243.5 ± 172.6		
Raloxifene-HBenBCD-PG (capsules)	Ral.	476 ± 154	5.0 ± 0.5	55.3 ± 40.0	227 ± 80	4.1 ± 1.2
	Gluc.	990 ± 648	4.0 ± 0.5	112.6 ± 40.7		
Raloxifene-PG (capsules)	Ral.	321 ± 76	5.0 ± 0.5	46.5 ± 41.8	94 ± 23	2.7 ± 0.6
	Gluc.	391 ± 152	5.0 ± 0.5	54.3 ± 40.9		

Abbreviations: C_{max}, maximum plasma concentration; T_{max}, time required to reach C_{max}; AUC, total area under the plasma concentration-time curve; F, oral bioavailability; Ral., raloxifene; Gluc., combine glucuronides; aq. aqueous. Values are means ± standard deviations. Total Contact = AUC raloxifene + AUC glucuronides/raloxifene dose.

After intravenous administration of a raloxifene:HBenBCD solution, the $AUC_{0\text{-}72\,h}$ for raloxifene dosed at 2.5 mg/kg (2965 ± 386 ng•h/mL) and for the combined raloxifene glucuronides (182 ± 27 ng•h/mL) indicate that raloxifene hepatic metabolism is not as significant (raloxifene glucuronides AUC/raloxifene AUC = 0.062 ± 0.012) as intestinal metabolism after oral dosing. After the distribution phase, the initial rates of elimination for both raloxifene and raloxifene glucuronides were rapid ($t_{1/2}$ 2.5 ± 0.3 h). After *ca.* 8 h, extensive enterohepatic recycling causes the $t_{1/2}$ to become significantly longer (data not shown, *cf.* ref 16).

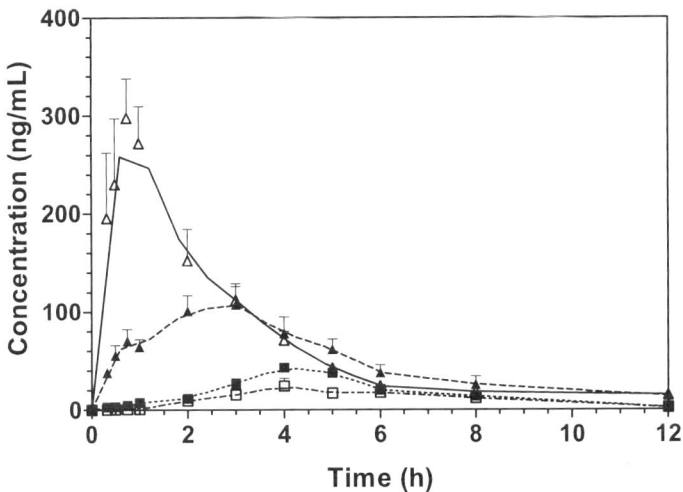

Figure 14. Mean concentration (ng/mL; n=4 ± SEM) of raloxifene and raloxifene glucuronides from oral administration of solid raloxifene [(■) raloxifene, (□) raloxifene glucuronides]; raloxifene:HBenBCD capsules [(▲) raloxifene, (△) raloxifene glucuronides].

In the case of oral raloxifene capsule dosing, T_{max} values for both drug and metabolite were 4 h. Relative to IV dosing, drug metabolism was far more significant via oral administration (raloxifene glucuronides AUC/raloxifene AUC = 0.52 ± 0.20). When the animals were dosed with raloxifene:HBenBCD powder filled capsules, the $AUC_{0-72 h}$ values for both raloxifene and raloxifene glucuronides were significantly larger than that observed when dosing oral raloxifene capsules. In addition, the T_{max} values for raloxifene and raloxifene glucuronides were smaller which, indicate a more rapid rate of absorption and metabolism in the presence of HBenBCD. After oral (capsule and aqueous gavage) dosing of raloxifene:HBenBCD, raloxifene metabolism was more extensive (raloxifene glucuronides AUC/raloxifene AUC = 1.21 ± 0.41 and 1.33 ± 0.21) than control (no HBenBCD).

Dosing with a solution of raloxifene dissolved in propylene glycol containing no HBenBCD (Table VII) afforded lower AUC values for raloxifene and raloxifene glucuronides than those of animals dosed with HBenBCD containing liquid formulations in either PEG400/PG or PG. It is interesting to note that relative to dosing with raloxifene/PG liquid filled capsules, the AUC for raloxifene-HBenBCD-PG was essentially unchanged while the AUC for the raloxifene glucuronide was significantly increased. The pharmacokinetic parameters for the animals dosed with raloxifene and HBenBCD dissolved in PEG400/PG were very similar to those obtained by dosing the animals with an aqueous gavage of a raloxifene:HBenBCD complex. However, when the animals were dosed with raloxifene and HBenBCD dissolved in propylene glycol (group 6), the pharmacokinetic parameters were different. In particular, T_{max} for raloxifene glucuronides increased to 4 h and the ratio of raloxifene glucuronides AUC/raloxifene AUC (2.27 ± 1.77) was higher.

Raloxifene oral bioavailability after dosing male Wistar-Hannover rats with raloxifene powder filled capsules was $2.6 \pm 0.4\%$. When the animals were dosed with raloxifene:HBenBCD powder filled capsules, the oral bioavailability of raloxifene was $7.7 \pm 2.2\%$ (a 3-fold increase). Likewise, the oral bioavailability of the other 3 HBenBCD formulations were $6.4 \pm 0.8\%$ (a 2.5-fold increase), $5.7 \pm 1.3\%$ (a 2.2-fold increase), and $4.1 \pm 1.2\%$ (a 1.6-fold increase), respectively.

Since raloxifene undergoes extensive presystemic metabolism, measurement of raloxifene levels alone may not provide the best indication of the extent of raloxifene dissolution and uptake from the intestine into the portal blood. A better measure of the effect of the HBenBCD on total raloxifene absorption may be 'total raloxifene exposure', measured as (AUC raloxifene + AUC metabolites)/raloxifene dose. Using this combined measure, it can be seen that administration of raloxifene as raloxifene:HBenBCD powder filled capsules caused a 4.2-fold increase in total raloxifene systemic delivery.

If one compares the oral bioavailability of raloxifene formulated as a liquid in propylene glycol (no HBenBCD) to that obtained with the other liquid fill formulations (HBenBCD-PEG400-PG, HBenBCD-PG), the importance of HBenBCD in maintaining raloxifene solubility was clearly evident; the oral bioavailability of raloxifene was 1.5-2.1 times greater and the total raloxifene exposure was twice that observed in the absence of HBenBCD.

Protease Inhibitors: Saquinavir

It has been reported that saquinavir is metabolized by cytochrome P450 3A (CYP3A) enzymes and is an efflux transporter substrate (*i.e.* P-glycoprotein, P-gp). The combination of these metabolism and transport issues with poor water solubility causes low and variable oral saquinavir bioavailability.

Pharmacokinetic parameters for saquinavir after administration of saquinavir formulations to male Wistar-Hannover rats are summarized in Table VIII. Figure 15 provides typical plasma concentration versus time profiles for saquinavir after oral dosing with saquinavir mesylate-MCC and saquinavir base:HBenBCD powder filled capsules.

Following oral dosing with saquinavir base:HBenBCD powder filled capsules, initial drug absorption was rapid (T_{max} *ca.* 0.9 h, $C_{max} = 268.2 \pm 125.4$ ng/mL) and was followed by rapid drug elimination. At *ca.* 4 h, a second saquinavir absorption peak was observed followed again by rapid elimination. In the case of saquinavir mesylate oral dosing (no HBenBCD), the observed C_{max} (14.4 ± 10.7 ng/mL) and T_{max} (*ca.* 5.3 h) were not truly distinct as the drug plasma concentration appeared to be nearly constant from about 4 to 8 h. Given that the absorption and elimination of saquinavir was apparently rapid, this observation was consistent with slower drug dissolution and/or overlapping absorption-elimination-reabsorption. The net effect was that the $AUC_{0-24\,h}$ for the group dosed with saquinavir base:HBenBCD (675.8 ± 264.4 ng•h/mL) was more than nine-fold larger than for the group dosed with saquinavir mesylate (73.4 ± 25.1 ng•h/mL).

Table VIII. Pharmacokinetic parameters of saquinavir after administration of saquinavir and saquinavir-HBenBCD formulations to male Wistar-Hannover rats.

Group	Formulation	C_{max} (ng/mL)	T_{max} (h)	$AUC_{0\text{-}24\,h}$ (ng•h/mL)	%F
1 (iv)	SB-HBenBCD	464.2 ± 90.4	0.3	910.3 ± 211.9	---
2 (oral)	SM capsules	14.4 ± 10.7	5.3 (4.0-8.0)	73.4 ± 25.1	2.0 ± 0.7
3 (oral)	SB-HBenBCD capsules	268.2 ± 125.4	0.9 (0.7-1.0)	675.8 ± 264.4	18.6 ± 7.3[2]
4 (oral)	SM-HBenBCD gavage	103.0 ± 59.5	0.3	156.5 ± 102.6	4.3 ± 2.8
5 (oral)	SB-HBenBCD-PEG400 capsules	127.5 ± 144.1	1.2 (0.7-2.0)	276.1 ± 124.6	7.6 ± 3.4
6 (oral)	SB-PEG400 capsules	128.2 ± 65.6	1.9 (0.7-3.0)	599.2 ± 509.9	16.5 ± 14.0
7 (oral)	SB-HBenBCD gavage	70.3 ± 34.2	0.6 (0.3-1.0)	141.7 ± 31.8	3.9 ± 0.9

Abbreviations: saquinavir base, SB ; saquinavir mesylate, SM C_{max}, maximum plasma concentration; T_{max}, time required to reach C_{max}; $AUC_{0\rightarrow24}$, total area under the plasma concentration-time curve from 0 to 24 h; F, bioavailability. Values are means ± standard deviations. The oral bioavailability for group 3 was significantly different from groups 2, 4, 7 ($p<0.01$) and from group 5 ($p<0.05$).

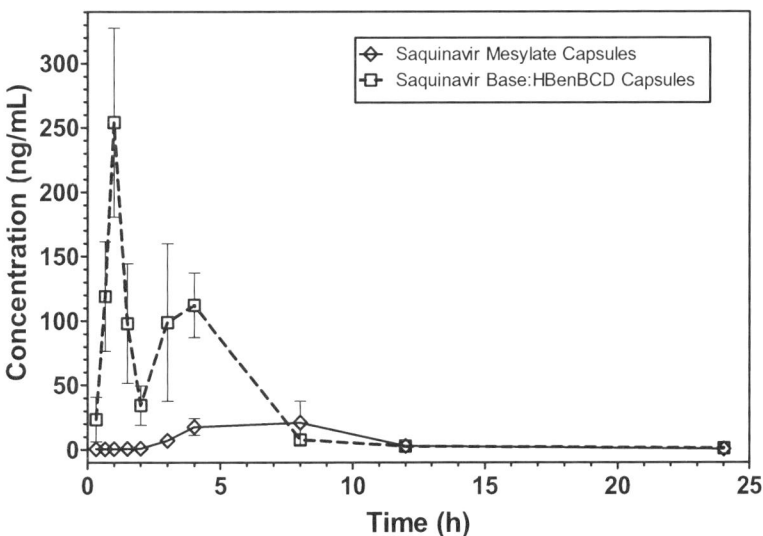

Figure 15. Plasma concentration-time curves after oral administration of saquinavir and saquinavir:HBenBCD powder filled capsules. Error bars are standard error of the mean.

When the animals were dosed with an oral aqueous gavage of saquinavir mesylate:HBenBCD and saquinavir base:HBenBCD (Table VIII), T_{max} was extremely short (*ca.* 0.3-0.6 h) and the initial elimination was rapid ($t_{1/2}$ *ca.* 0.5 h). As was observed with dosing of saquinavir base:HBenBCD oral capsules, a second absorption peak appeared at *ca.* 3 h (data not shown. *cf.* ref 17). The values of $AUC_{0-24\,h}$ for the two aqueous gavage groups (156.5 ± 102.6 ng•h/mL) were not statistically different.

After oral dosing of the animals with capsules filled with PEG400 solutions of saquinavir base-HBenBCD (group 5) and saquinavir base (group 6), T_{max} (1.2-1.9 h) and C_{max} (group 5 = 127.5 ± 144.1 ng/mL, group 6 = 128.2 ± 65.6 ng/mL) were similar for both groups. However, a 2nd absorption peak at *ca.* 3 h for group 6 was far more pronounced relative to group 5 (*cf.* ref 17). Also, a 3rd absorption peak was observed with group 6 but absent in group 5. The net effect was that the $AUC_{0-24\,h}$ for group 6 (599.2 ± 509.9 ng•h/mL) was larger than group 5 (276.1 ± 124.6 ng•h/mL). Due to the variability in the $AUC_{0-24\,h}$ for group 6, this difference was not statistically significant.

Saquinavir oral bioavailability (F) after dosing with saquinavir mesylate-MCC powder filled capsules was 2.0 ± 0.7% (group 2). The three liquid dosage forms (groups 4, 5, and 7) afforded similar oral bioavailability, with no or only marginal improvements relative to group 2. Also relative to group 2, the T_{max} values for groups 4, 5, and 7 were significantly reduced and C_{max} was increased and saquinavir elimination rates were much higher. The result was that the $AUC_{0-24\,h}$ and oral bioavailabilities were quite similar and not significantly different. Oral bioavailabilities for group 3 (18.6%, saquinavir base:HBenBCD powder filled capsules) and group 6 (16.5%, capsules filled with saquinavir base PEG400 solution) were 8-9 fold higher than control group 2. Groups 3 and 6 were different from the other groups in the study, the C_{max} was much larger and both groups had a significant 2nd absorption peak (120-200 ng/mL) sustained over a longer period of time. Consequently, the $AUC_{0-24\,h}$ and F for these 2 groups (3 and 6) were greater. However, due to the $AUC_{0-24\,h}$ and F variability for group 6, the $AUC_{0-24\,h}$ and F for group 6 were not significantly different from group 2. In the case of group 3, F was significantly different relative to groups 2, 4, 7 ($p < 0.01$), and 5 ($p < 0.05$).

Conclusions

Hydroxybutenyl cyclodextrin has proven to have significant value in drug delivery systems. We have found that we can use HBenBCD to improve the solubility of a structurally diverse array of poorly water-soluble drugs. Due to the solubility of HBenBCD in water and in certain organic solvents, HBenBCD provides significant flexibility in formulating drugs for intravenous and oral administration. In general, inclusion of HBenBCD in the pharmaceutical formulation leads to rapid and extensive dissolution of drug. In most cases, drug solubility is maintained even when the dissolution medium is supersaturated with drug. Most gratifyingly, we have found that the improved drug solubility and dissolution rates provided by HBenBCD in *in vitro* studies were reflected in increased oral bioavailability in *in vivo* studies involving animal models. For example, our *in vitro* dissolution studies indicated that food consumption should not influence oral bioavailability of itraconazole when dosing with itraconazole:HBenBCD oral formulations. Our *in vivo* studies showed that there are no significant food effects on the PK parameters after dosing rats with itraconazole:HBenBCD powder. Similarly, our *in vitro* solubility studies indicated that HBenBCD was more effective than HPBCD in increasing the solubility of itraconazole. Our *in vivo* studies showed that dosing rats with itraconazole:HBenBCD oral solution resulted in an oral bioavailability of 44%, or 1.4-times that observed with the HPBCD. Some aspects of our *in vivo* studies could not be predicted from the *in vitro* studies we conducted. The initial finding that there was both a food and a chronopharmacokinetic effect on tamoxifen oral bioavailability was a surprise. Although we do not know if the results translates across species, the results are very significant in that they suggest that tamoxifen oral bioavailability could be enhanced if the female patient took her dose in the morning before breakfast. Similarly, we could not predict from our *in vitro* studies that there would be a difference in letrozole absorption rates between male and female rats. It is this type of experimentally observed details that makes drug delivery technology such a fascinating structure-property problem.

References

1. Wong, S.M.; Kellaway, I.W.; Murdan, S. Enhancement of the dissolution rat and oral absorption of a poorly water soluble drug by formation of surfactant-containing microparticles. *Int. J. Pharm.* **2006**, *317*, 61-68.
2. Naseem, A.; Olliff, C.J.; Martini, L.G.; Lloyd, A.W. Effects of plasma irradiation on the wettability and dissolution of compacts of griseofulvin. *Int. J. Pharm.* **2004**, *269*, 443-450.
3. Rowland, M.; Tozer, T.N. Clinical Pharmacokinetics Concepts and Applications, third edition, 1994, Lippincott Williams and Wilkins.
4. Szente, L.; Szejtli, J. *Advanced Drug Delivery Reviews*, **1999**, *36*, 17-28.
5. Connors, K.A. The stability of cyclodextrin complexes in solution. *Chem. Rev.* **1997**, *97*, 1325-1357.
6. Szejtli J.. Selectivity/structure correlation in cyclodextrin chemistry. *Supramolecular Chemistry* **1995**, *6*, 217-223.
7. Uekama K, Hirayama F, Irie T. Cyclodextrin drug carrier systems. *Chem. Rev.* **1998**, *98*, 2045-2076.
8. Szejtli J. Cyclodextrins in drug formulations: Part II. *Pharmaceutical Technology* **1991**, *15*, 24-38.
9. Buchanan, C.M.; Alderson, S.R.; Cleven, C.D.; Dixon, D.W.; Ivanyi, R.; Lambert, J.L.; Lowman, D.W.; Offerman, R.J.; Szejtli, J.; Szente, L. Synthesis and characterization of water-soluble hydroxybutenyl cyclodextrins. *Carbohydrate Research* **2002**, *327*, 493-507.
10. Buchanan, C.; Buchanan, N.; Edgar, K.; Lambert, J.; Posey-Dowty, J.; Ramsey, M.; Wempe, M. Solubilization and Dissolution of Tamoxifen-Hydroxybutenyl Cyclodextrin Complexes *J. Pharmaceutical Sciences* **2006**, *95*, 2246-2255.

11. Buchanan, C.; Buchanan, N.; Edgar, K.; Ramsey, M. Solubility and Dissolution Studies of Antifungal Drug:Hydroxybutenyl-β-Cyclodextrin Complexes *Cellulose* **2007**, *14*, 35-47.

12. Klein, S.; Wempe, M.F.; Zoeller, T.; Buchanan, N.L.; Lambert, J.L.; Ramsey, M.G.; Edgar. K.J.; Buchanan, C.M. Improving Glyburide Solubility and Dissolution by Complexation with Hydroxybutenyl-β-Cyclodextrin. *Journal of Pharmacy and Pharmacology* Submitted.

13. Buchanan, C.; Buchanan, N.; Edgar, K.; Little, J.; Malcolm, M.; Ruble, K.; Wacher, V.; Wempe, M. Pharmacokinetics of tamoxifen after intravenous and oral dosing of tamoxifen – hydroxybutenyl-β-cyclodextrin formulations *J. Pharmaceutical Sciences* **2007**, *96*, 644-660.

14. Buchanan, C.M.; Buchanan, N.L.; Edgar, K.J.; Klein, S.; Little, J.L.; Ruble, K.M.; Wacher, V.J.; Wempe, M.F. Pharmacokinetics of itraconazole after intravenous and oral dosing of itraconazole - cyclodextrin formulations. *J. Pharmaceutical Sciences* **2007**, *96*, 3100-3116.

15. Wempe, M.F.; Rice, P.J.; Buchanan, C.M.; Buchanan, N.L.; Edgar, K.J.; Hanley, G.A.; Ramsey, M.G.; Skotty, J.S. Letrozole:Hydroxybutenyl-β-cyclodextrin (HBenBCD) complexes: A Pharmacokinetic study in Sprague-Dawley Rats *Journal of Pharmacy and Pharmacology* **2007**, *59*, 795-802.

16. Wempe, M.F.; Wacher, V.J.; Ruble, K.M.; Ramsey, M.G.; Edgar, K.J.; Buchanan, N.L.; Buchanan, C.M. Pharmacokinetics of Raloxifene in Male Wistar-Hannover Rats: Influence of Complexation with Hydroxybutenyl-beta-cyclodextrin. *International J. Pharmaceutics* **2008**, *346*, 25-37.

17. Buchanan, C.M.; Buchanan, N.L.; Edgar, K.J.; Little, J.L.; Ramsey, M.G.; Ruble, K.M.; Wacher, V.J.; Wempe, M.F.. Pharmacokinetics of saquinavir after intravenous and oral dosing of saquinavir-hydroxybutenyl-beta-cyclodextrin formulations *Biomacromolecules* **2008**, *9*, 305-313.

18. Buchanan, N.L.; Buchanan, C.M. High Throughput screening method for determination of equilibrium drug solubility. This book.

19. Higuchi ,T.; Conners, K.A. Phase-solubility techniques. *Adv. Anal. Chem. Instr.* **1965**, *4*, 117-212.

20. Hochner-Celnikier, D. Pharmacokinetics of raloxifene and its clinical application Eur. *J. Obstet. Gynecol. Reprod. Bio.* **1999**, *85*, 23-29.

Chapter 3

High Throughput Screening Method for Determination of Equilibrium Drug solubility

Norma L. Buchanan, Charles M. Buchanan

Eastman Chemical Company, Research Laboratories, P.O. Box 1972, Kingsport, TN 37662, USA

Using a UV platereader, multichannel pipettes and disposable 96-well plates, we have developed a method which allows high throughput screening of the equilibrium solubility of drugs in aqueous media. The method was developed as part of an effort to evaluate the potential of hydroxybutenyl-β-cyclodextrin (HBenCD) for the delivery of poorly soluble drugs. This method differs from other reported high throughput screening methods in that thermodynamic or equilibrium solubility is measured at greatly increased throughput and without the use of an organic solvent. Conservatively a 100 fold increase in test throughput can be achieved. Additionally, a high throughput method for measuring solution stability was developed.

In the pharmaceutical or drug delivery industry, the solubility of a drug in biologically relevant media is often the key factor determining the success or failure of a new drug candidate. The drug must dissolve in order to be absorbed or be bioavailable when oral or intravenous delivery is desired. Often the drug may have such limited solubility that a therapeutic dose is either difficult or impossible to deliver effectively. Therefore, efficient methods for screening the solubility of a given drug or the dissolution of the drug from a given formulation in relevant media are desired. This chapter describes methods developed to allow equilibrium solubility and solution stability (*i.e.*, how long after dissolution does the drug stay in solution) to be determined with reasonably high throughput using commonly available and inexpensive laboratory equipment.

Generally, drug solubility has been measured as either kinetic or thermodynamic (equilibrium) solubility (1). Kinetic solubility is determined by first dissolving the drug in a water miscible organic solvent such as DMSO; then the organic solution is added to an aqueous buffer. The dilution will result in precipitation of excess drug from the supersaturated and now mostly aqueous solvent system. After equilibrium is established, the drug remaining in solution is measured, typically using HPLC techniques. Equilibrium solubility is determined by adding an excess of drug to an aqueous buffer or relevant medium, mixing until the drug's equilibrium concentration is achieved (usually 24-72 hours are required), and then filtering excess solids (drug) from the solution prior to measurement; again typically HPLC is the preferred measurement technique. This approach is known as the 'shake flask' method and is described by *ASTM International test method E 1148-02*.

In screening new drug candidates for suitability in oral drug formulations, industry tends to use kinetic solubility measurements rather than thermodynamic because the kinetic solubility approaches provide for a more rapid result. However, kinetic solubility inherently provides a result that is altered by the presence of the organic solvent such that solubility is often overestimated.

Once a formulation has been developed, the formulated drug is tested for its dissolution behavior in relevant media. That is, how much of the drug contained in a dose will dissolve and over what timeline, and then how long will the drug stay in solution once dissolved. The industry standard method for measuring dissolution requires very specific instrumentation and is detailed by the *United States Pharmacopeia 2004 guideline: USP 28-NF 23 711*. Dissolution measurements are labor intensive, time consuming, and require a significant quantity of drug.

Equilibrium Solubility Method

When we began to explore methods for obtaining solubility data, we were evaluating the use of hydroxybutenyl-β-cyclodextrin (HBenCD) as a vehicle for drug delivery. We chose not to pursue kinetic solubility as an approach largely because we did not want to overestimate the impact of HBenCD nor confuse results with the use of the organic co-solvent which would also likely interact with cyclodextrin (CD). For those interested, more details of the utility of HBenCD as a drug delivery vehicle can be found in an earlier chapter of this book and associated references.

Phase 1: Miniaturized shake flask method

We began with the shake flask method. Our first effort at streamlining the shake flask method involved simply changing the scale of the experiment in order to spend less money purchasing the drugs we wanted to test with HBenCD. The drug of interest was weighed (2-10 mg depending on the reported inherent solubility of the drug) into a 2 mL vial containing a miniature Teflon coated stir bar. Then 500 μL of water, aqueous buffer or known solution of cyclodextrin was transferred into the vial using a pipette and sealed with a Teflon lined cap. We used a 25 position stir plate for mixing. Each test vial was stirred at ambient temperature, usually for 60 hours. Laboratory temperature was kept at 22 °C ± 2 °C. After a minimum of 24 hours, the drug solutions (containing excess drug in all cases) were taken up into 2 mL gas tight syringes. The solids were filtered from the drug solutions using PALL Life Sciences IC Acrodisc (PES, *www.pall.com*) 13 mm diameter syringe filters with 0.2 micron pore size. Prior to measurement by UV, the filtered solutions were diluted at least 10X with 1/1 ethanol/water in order to increase the solution volume such that measurement in a 1.0 cm pathlength cell was possible. Dilution with ethanol/water also provided a means to adjust the measured absorbance to the linear range of the spectrometer and provided a mechanism to break the 'drug-CD' complexes so that an accurate drug concentration could be measured. The ethanol allowed drug solubilized by the CD to move out of the CD cavity and remain in solution during the measurement while the water kept the CD in solution.

For this initial work, a Perkin-Elmer Lambda 25 UV-VIS spectrometer was used to measure drug concentration by measuring absorbance in 1 cm pathlength quartz cells. There was no need to incur the expense of HPLC method development since there were no unknowns in our experiment and we could easily account for the contribution of HBenCD to the UV spectra. For any given drug, its absorptivity and linearity of response were determined. Measurements were made at wavelengths chosen based on the characteristic absorbance of the

68

given drug. Figure 1 shows a typical absorptivity measurement and Figure 2 below shows solubility data typical of our miniaturized shake flask method.

For this one illustrative case (Figure 2), there were 42 independent equilibrium solubility results. That translates to weighing ibuprofen into 42 vials, adding a stir bar to each vial, preparing CD solutions (42 including blanks), transferring test medium into each of the 42 vials, sealing and labeling each vial, briefly sonicating each vial to disperse the solid drug in the aqueous media, stirring at 22 °C ± 2 °C for at least 60 hours, transferring the saturated solution into a gas tight syringe and filtering each of the 42 test solutions into clean labeled vials, then diluting each sample as needed (two or more dilutions were required for a least a fraction of the samples to stay within the linear response range of the spectrometer), then transferring the 42 or more diluted samples into UV cells, measuring absorbance, cleaning the gas tight syringes and cells between samples, then calculating, recording and reporting results.

We needed to explore the impact of HBenCD upon the solubility of different drugs, the impact of aqueous buffer composition, pH and ionic strength, along with cyclodextrin concentration and cyclodextrin composition upon the amount of drug that could be solubilized. The miniaturized shake flask method quickly became impractical. One person could on average generate 50-60 independent results per week. While this rate of data generation compared reasonably to the miniaturized shake flask method reported by A. Glomme, *et al.* (2), it was simply not fast enough to meet our needs.

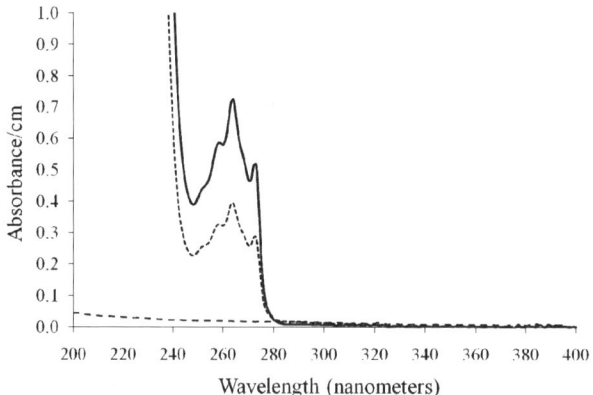

Figure 1. Absorptivity of 0.025 g/L and 0.050 g/L ibuprofen solutions in 1/1 V/V ethanol/water solution was 1.0716 abs/cm/(g/L) at 264 nanometers.

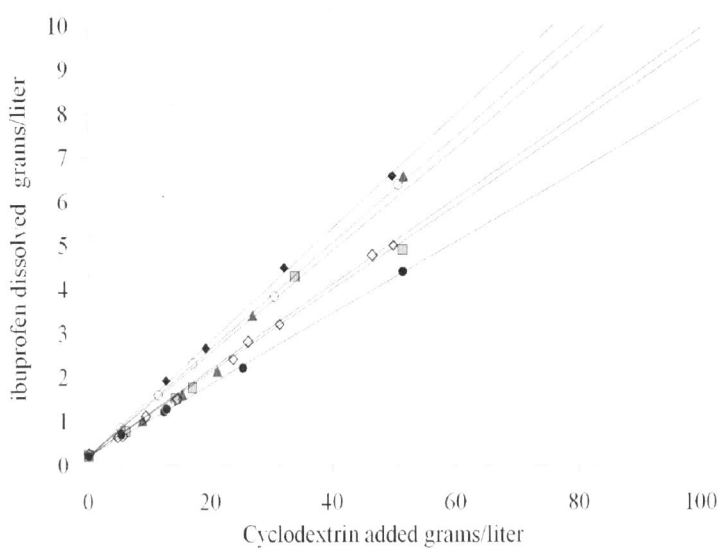

Figure 2. Equilibrium solubility of ibuprofen as a function of CD composition and concentration. Measurements were made for CD solutions prepared in millipore water from 0 to 50 g/L for a series of HBenCDs and HPCDs.

Phase 2: High throughput equilibrium solubility screening method

In reviewing the work of others who had undertaken the problem of developing useful and reliable high throughput screening methods, (2-10), we determined that our best option was to employ a UV platereader and move as much as possible to 96-well microplates. We chose a Molecular Devices SpectraMax 384 Plus UV platereader which has a spectral range of 190 to 1000 nanometers. Figure 3 shows a comparison of results for a set of samples measured with both the standard double-beam UV spectrometer and the UV platereader. The data presented in Figure 3 are for a set of tamoxifen:HBenCD complexes of varying tamoxifen content that were dissolved in Millipore water. Measurements were made by dilution with 1/1 v/v ethanol/water in 1.0 cm pathlength FUV quartz cells in the case of the UV spectrometer or with 200 microliters per well (Greiner UV-STAR 96-well 300μL volume disposable microplate *www.greinerbioone.com*) using the UV platereader. The correlation coefficient from this comparison was 0.9923 and the slope was 1.0. We therefore judged the platereader capable of providing reliable UV measurements for our screening work.

While using the platereader significantly reduced the labor and time required carrying out the absorbance measurements, this change alone was not enough. We needed to streamline the overall task of determining equilibrium solubility as much as possible. Our method needed to include the basic steps of mixing, filtering, diluting, and measuring. We looked for more efficient ways to carry out each step.

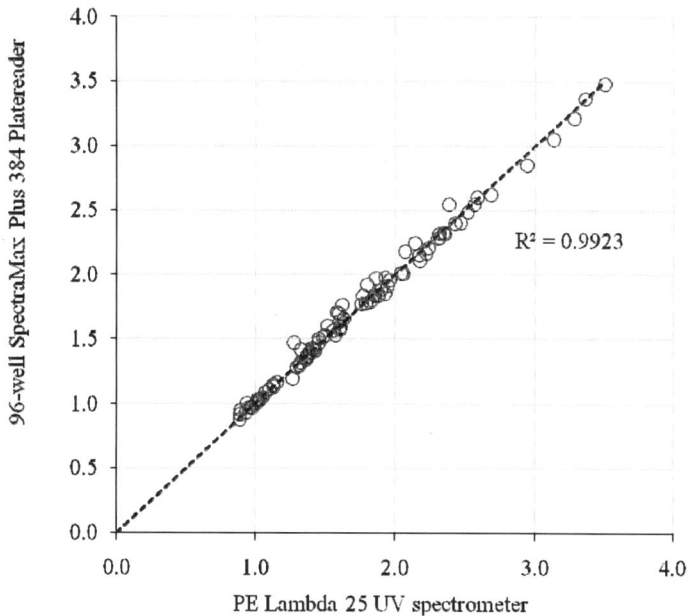

Figure 3. Comparison of results using the 96 well plate reader (Greiner UV 96-well disposable plate versus results using the Perkin-Elmer Lambda 25 spectrometer (1.0 cm quartz cells)). Data are plotted as wt% tamoxifen content in tamoxifen-HBenCD complexes that were dissolved in Millipore water.

Mixing Issues

While we could not change the elapsed time required to reach equilibrium, we wondered if we could cut the labor involved by switching from small vials for the 'shake flask' portion of the task to 96-well microplates. We believed that the geometry of 2 ml round bottom polypropylene 96-well plates might work using a rotary shaker to carry out the mixing portion of the experiment. We used a simple Excel worksheet to create a map of each plate to simplify both the ease of setup and labeling for each 96-well plate. Figure 4 provides an example of a 'plate map' where CD concentration was varied by row, cyclodextrin composition was varied by column, all wells contained tamoxifen and all solutions were prepared in 0.05M pH 3.0 citrate buffer. In the example of Figure 4, CD solutions were prepared and labeled for ease in filling the 96-well plates. For example, HB02 is HBenCD at 2 wt% concentration. Figure 5 shows solubility results for solutions comparing mixing in the 96-well plate to stirring in individual vials. Note that although the results are not identical for the two approaches, agreement is reasonable considering that the cyclodextrin solutions used were not identical and extensive sample handling was required for each solubility data point.

Industry tends to carry out solubility experiments at 37 °C, average human body temperature. Our goal was to screen for the impact of the drug delivery agent, and working at ambient temperature is operationally simpler than at 37 °C. Therefore we examined whether carrying out the mixing at 37 °C provided significant advantage. Figure 6 shows solubility data collected as a function of mixing temperature. As would be expected, there is an increase in drug solubility at elevated temperatures. The data show that, for screening purposes, there is no significant advantage in running at elevated temperature. In fact, doing so only complicates the required activities, *i.e.*, if one carries out the mixing phase of the experiment at 37 °C, then it would also be necessary to carry out the filtration, dilution and measurements at 37 °C. Otherwise, precipitation of drug could occur during a lower temperature phase of sample handling, adding to error and uncertainty. What is important is that all measurements be made at the same temperature, permitting direct comparison.

In addition to mixing temperature, the ionic strength of the buffers used to prepare the cyclodextrin solutions is important. If the ionic strength of the buffer is too high, there can be a salting out effect, effectively limiting the amount of drug that can dissolve. However, if the ionic strength is too low then solution pH cannot be maintained. Figure 7 shows data that illustrate this point. We found that buffers of 0.05M concentration generally provided enough buffer capacity without significantly impairing drug solubility.

Drug: tamoxifen base
Buffer: 0.05M citrate at pH 3.0

wt% CD		1	2	3	4	5	6	7	8	9	10	11	12
		HBenBCD			HBenCD2			HPBCD			SulfoHBenBCD		
0	A	HB0	HB0	HB0	HB20	HB20	HB20	HP0	HP0	HP0	SHB0	SHB0	SHB0
2	B	HB02	HB02	HB02	HB202	HB202	HB202	HP02	HP02	HP02	SHB02	SHB02	SHB02
4	C	HB04	HB04	HB04	HB204	HB204	HB204	HP04	HP04	HP04	SHB04	SHB04	SHB04
6	D	HB06	HB06	HB06	HB206	HB206	HB206	HP06	HP06	HP06	SHB06	SHB06	SHB06
10	E	HB10	HB10	HB10	HB210	HB210	HB210	HP10	HP10	HP10	SHB10	SHB10	SHB10
15	F	HB15	HB15	HB15	HB215	HB215	HB215	HP15	HP15	HP15	SHB15	SHB15	SHB15
25	G	HB25	HB25	HB25	HB225	HB225	HB225	HP25	HP25	HP25	SHB25	SHB25	SHB25
40	H	HB40	HB40	HB40	HB240	HB240	HB240	HP40	HP40	HP40	SHB40	SHB40	SHB40

Figure 4. Shows a 'plate map' for a typical experiment.

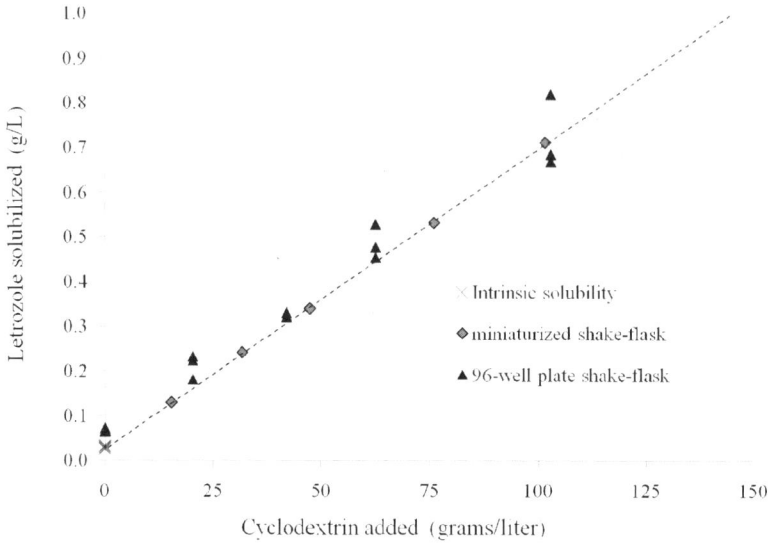

Figure 5. Solubility results comparing mixing carried out in 96-well 2 mL round bottom plate versus mixing in individual vials.

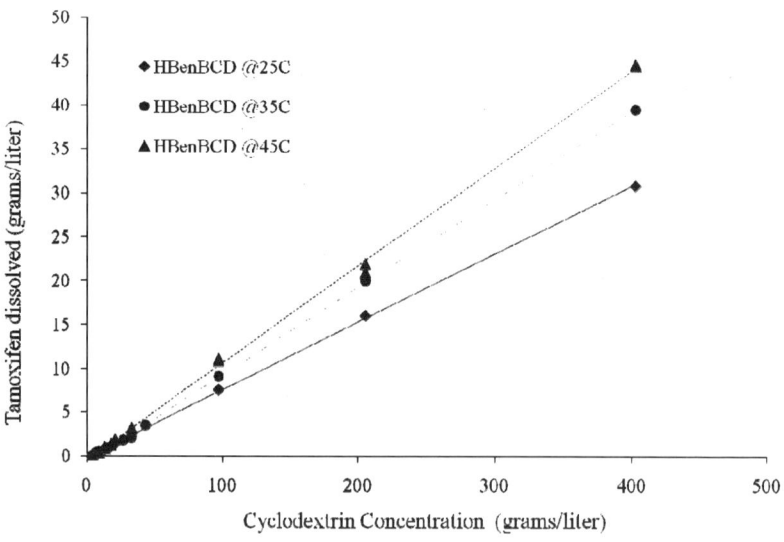

Figure 6. Equilibrium solubility of tamoxifen in HBenCD solutions is shown as a function of experiment temperature and CD concentration.

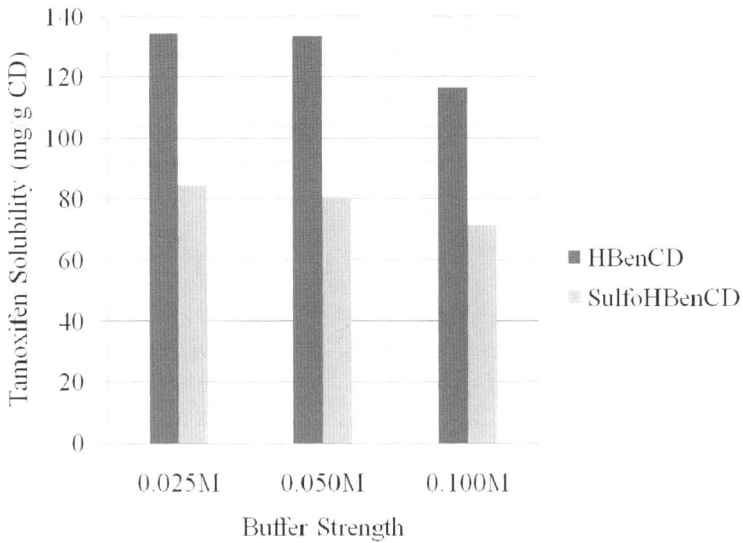

Figure 7. Effect of buffer molarity on the amount of drug dissolved in the presence of 5 wt% cyclodextrin at pH 3.0 in phosphate buffers.

Filtering

After the mixing period, the contents from the mixing plate have to be filtered. One at a time filtering with syringe filters is time and labor intensive. Since we were now carrying out the mixing in 96-well mixing plates, we tested the use of 96-well multiscreen filter plates to replace one at a time filtration. We used multiscreen filter plates with hydrophilic membrane filters (*www.millipore.com*) along with a vacuum manifold designed for use with the 96-well microplate platform. The amount of drug absorbed onto the filter is known to be influenced by the choice of membrane material and should be tested for any given drug. Figure 8 illustrates the potential impact of filter type. 96-Well multiscreen plates with polyvinylidene fluoride (PVDF), hydrophilic polytetrafluoroethylene (PTFE), and hydrophilic polypropylene (GHP) were tested. Aliquots (200 µL) of drug:HBenCD complex solutions containing no solids were filtered through the 96-well multiscreen plates one time. The resulting measured drug content is shown for solutions of tamoxifen base (47 mg/L), saquinavir base (250 mg/L), itraconazole (7 mg/L), and raloxifene HCl (32 mg/L). Clearly, choice of membrane material is a critical element.

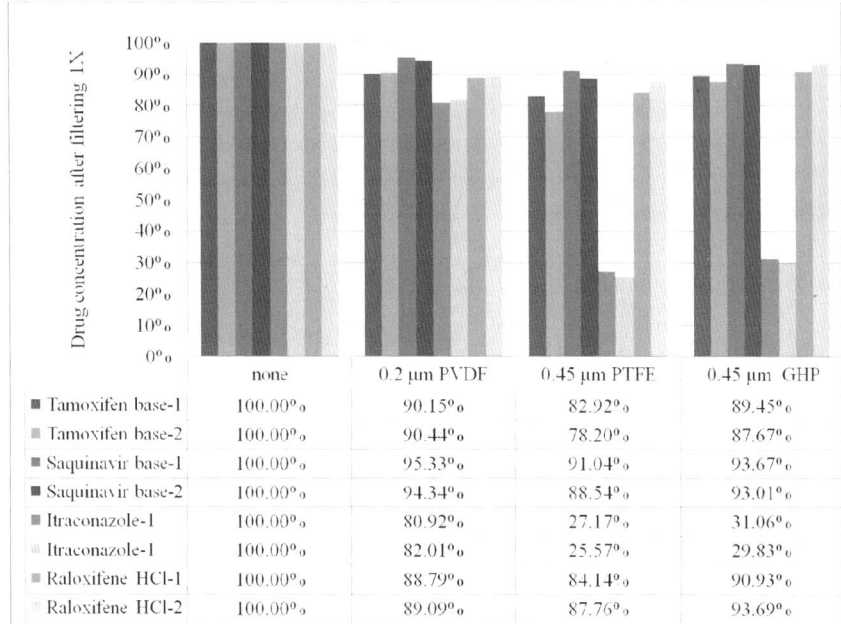

	none	0.2 µm PVDF	0.45 µm PTFE	0.45 µm GHP
■ Tamoxifen base-1	100.00%	90.15%	82.92%	89.45%
■ Tamoxifen base-2	100.00%	90.44%	78.20%	87.67%
■ Saquinavir base-1	100.00%	95.33%	91.04%	93.67%
■ Saquinavir base-2	100.00%	94.34%	88.54%	93.01%
■ Itraconazole-1	100.00%	80.92%	27.17%	31.06%
Itraconazole-1	100.00%	82.01%	25.57%	29.83%
■ Raloxifene HCl-1	100.00%	88.79%	84.14%	90.93%
Raloxifene HCl-2	100.00%	89.09%	87.76%	93.69%

Figure 8. Effect of filter membrane choice on the amount of drug recovered after filtration.

Dilution

Multichannel pipettes were valuable for transfer of test media into the mixing plates, and of equilibrated drug solutions into the 96-well filter plates. The pipettes were also valuable in carrying out the dilutions needed to prepare the equilibrated solutions for measurement. We did not explore other diluents but rather continued our use of 1/1 v/v ethanol/water because this solvent mixture had proved to be successful in our miniaturized shake flask method. Use of the 96-well microplates with 8 or 12 channel pipettes reduced the time required for this step from hours of tedious repetitive manual solution transfer to less than 5 minutes elapsed time to transfer both the equilibrated test solutions and the diluent into the wells of a 96-well UV measurement plate.

Measurement

Not only does a 96-well UV platereader provide (Figure 3) the measurement accuracy needed for solubility screening, but it also can be configured to scan either a chosen spectral range or a specific set of wavelengths for the entire 96-well plate in a matter of seconds. A single sample measurement takes approximately 3 minutes scan time using a standard UV spectrometer.

While it is possible to scan each empty UV measurement plate to account for variations in UV absorbance due to the plate itself, we found that plate-to-plate and well-to-well variation for Greiner UV-Star plates were extremely small in comparison to the absorbance measured when any amount of CD was present. Therefore it was not necessary to measure the blank plate absorbance of each plate for our screening work although we did use an average absorbance correction for a selected plate blank containing 200 µL of diluent per well. Additionally, we used an average correction factor for pathlength based on the plate blank containing 200 µL of diluent per well. Because we could now run many more experiments in the same timeline, we generally ran each test point in triplicate.

Results

We report results as grams drug dissolved per liter of solution, but often one sees the impact of cyclodextrin in drug formulations reported in terms of binding constant (11). Binding constant (K) is defined as $K = slope/S_o(1-slope)$ when the drug concentration is plotted versus the cyclodextrin concentration in units of moles/L. Intrinsic solubility (S_o) is defined as the solubility of the drug alone in pure water. If one reports solubility in terms of binding constant then the measured absorbance per well needs to be as accurate as possible because the binding constant equation uses the intrinsic solubility, S_o, in the denominator of the equation. One can readily appreciate that error in determining the intrinsic solubility strongly impacts the calculated value for the binding constant. This is especially significant because of the low values of S_o (sub µg/L in many cases) for poorly soluble drugs. In contrast, comparison of drug delivery systems using g/L solubility measurements in the presence of the drug delivery systems provides an independent result for each test condition, eliminating dependence on the experimentally difficult S_o values. We believe that this method of reporting results is more appropriate in most cases, because the solubility data can be directly converted to amount soluble in the normal human plasma volume and compared to the desired dose.

If, rather than screening for the impact of potential drug delivery agent, the goal is to measure the intrinsic solubility of poorly soluble drugs; it should be noted that because the 96-well platform provides efficiency and cost effectiveness, enough replicates can be run to meet this goal. Since the test volume for mixing and the volume of the 96-well UV measurement plate are well matched, we recommend no dilution in measuring the absorbance of S_o test wells. Use of the equilibrated drug-water solution without dilution will help to move the measured absorbance out of the baseline and provide a more reproducible and accurate value.

General experimental: 6-well equilibrium solubility screening method

Step 1 – Mixing: The drug of interest is loaded into a 96-well 2 mL polypropylene mixing plate according to the experimental plate map. The drug is loaded into the plate using a column loader. For any row or column where it is desired that no drug be included, tape off that row or column when filling the column loader. Once each well of the column loader is full, place an empty 96-well mixing plate on top of the column loader, then invert both parts such that each well of the mixing plate is aligned with the corresponding well of the column loader. Transfer drug to the mixing plate by tapping the column loader causing the drug to drop into the wells of the mixing plate. (We used a small rubber mallet to facilitate that step).

Transfer test solutions (300-500 µL volumes per well) into the wells of the mixing plates using either 8 or 12 channel pipettes (*www.rainin.com*). Seal the plate with aluminum sealing tape, briefly mix the plate contents using a vortex mixer, then place on a rotary shaker (Helidolph Titramax 1000, *www.helidolph.com*) and shake at 800-1000 rpm at 22 °C ± 2 °C for 48-72 hours. Inspect the mixing plate at intervals to insure that each well continues to contain excess (undissolved) drug. Add more drug to any well as needed.

Step 2 – Filtration: After the equilibration period, transfer the mixing plate contents either row by row or column by column into a 96-well filter plate (we generally used a plate with PVDF filter membranes of 0.2 µ pore size). Place the filter plate on a vacuum manifold designed for 96-well plates (*www.millipore.com*) with a corresponding storage plate positioned to collect the filtrates. Apply 10-25 mm Hg vacuum for 10-30 seconds to filter the well contents. Should any well not filter within 30 sec; do not continue to pull vacuum, because such continued application of vacuum can concentrate the filtrates, rather move the contents of that well to a syringe filter for manual filtration.

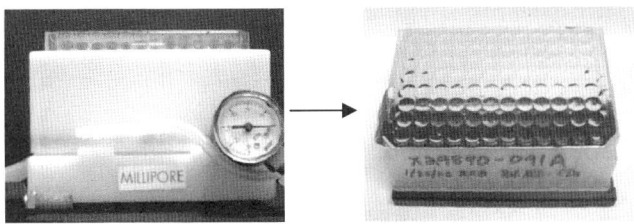

Step 3 – Dilution: Transfer aliquots (10-100 µL) from each well of the storage plate *via* multichannel pipette into the corresponding wells of a UV measurement plate either row by row or column by column. Dilute with an appropriate quantity of 1/1 V/V ethanol/water such that each well contains 200 µL total volume.

Step 4 – Measurement: Select assay wavelengths based on the drug's absorbance along with one or more baseline point(s). Baseline absorbance is used to monitor for scattering (that will occur if there are solids in the test solution, for example, due to filter breakthrough) and to correct for plate or CD contribution to the measured absorbance. Use the shake feature of the UV platereader to mix the filtrate and diluent until a homogeneous solution is attained (can be determined by repetitively measuring the plate). The amount of shake time needed will vary depending on the viscosity of the filtrate. However, shake for a minimum of 60 seconds prior to making the absorbance measurement. Keep the UV measurement chamber temperature at 22 °C. If measured absorbance is outside the linear range of the platereader, then prepare a second measurement plate using a higher dilution factor. Use the calculation function of the platereader to convert the measured result from absorbance/well to grams drug dissolved per liter test medium. Equilibrium solubility (S) is calculated according to the equation $S = (f(A_w)/a)$, where (a) is the absorptivity of the drug in abs/cm/(g/l) and (f) is the correction factor for pathlength to convert abs/well (A_w) to abs/cm.

Solution Stability Method

During the course of developing the solubility test method, it occurred to us that since the UV platereader had a shaking function and could easily be configured to track concentration versus time, we might be able to obtain solution stability (precipitation from a super saturated solution as a function of pH) information which is in part information similar to that obtained from USP2 dissolution testing. We had results for several drug:HBenCD complexes using the USP2 Dissolution apparatus and test method and knew that the drug:CD complexes consistently reached maximum drug solution concentration within a matter of minutes. Our concept was to dissolve drug:CD complexes in a suitable aqueous buffer, then rapidly add a diluent to shift solution pH to the desired value, for example, pH 1.2 to simulate gastric pH or pH 6.8 to simulate mid small intestinal pH. We refer to the test condition buffer as the shift pH buffer.

78

We expected to see 'noise' in the UV absorbance spectrum as a result of precipitation that would likely occur when the pH was shifted. However, because we could collect so many readings over the course of a typical dissolution timeline, accurate concentration information was still obtained. Measurement of solution pH both before initiating a test and at the completion of a test ensured that pH had been stable over the course of the experiment. In order to maintain adequate buffer capacity, we limited the volume of the drug solution to be diluted rather than change the buffer strength (we used buffers as defined by USP2 test method). The concentration of drug after dilution was usually kept within the range of therapeutic dosage for that drug.

For example, 10 µL of tamoxifen base:HBenCD complex solution was preloaded into rows B-H of a 96-well plate. Row A was loaded with the shift buffer only. Each column of the plate was diluted with 190 µL of a selected shift pH buffer at timed intervals of 15 seconds such that at the conclusion of the experiment one could correct each column of data in time to its zero time point. After dilution with the shift pH buffers, the test wells contained 35 mg/L tamoxifen base as the initial or 100% theoretical drug concentration. Absorbance readings were taken at 60 second intervals. Figure 9 shows the experimental data of this example for one column of the 96-well plate where the shift pH buffer was a 0.05 M phosphate buffer at pH 7.4 and illustrates that the average drug concentration can be observed even in the presence of the noise spikes that occur whenever a solid particle is in the UV beam. These spikes can easily be removed by software if desired, but clearly they do not prevent observation of the drug concentration versus time profile (solution stability).

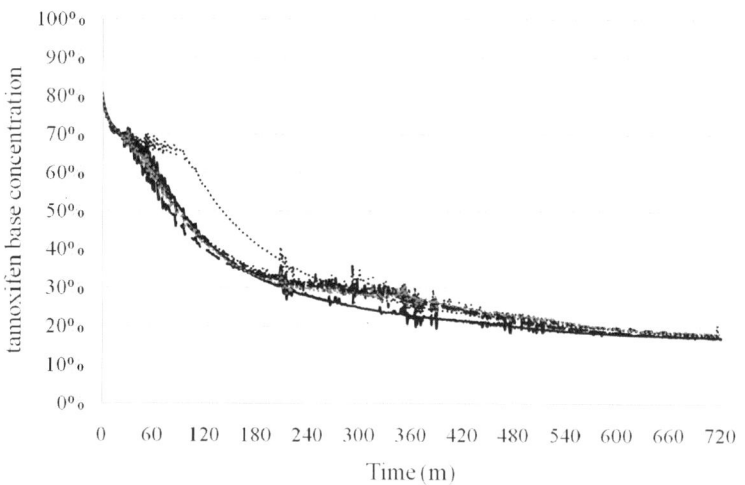

Figure 9. Solution stability profile showing occasional spikes for 35 mg/L tamoxifen base:HBenCD complex in 0.05 M phosphate buffer at pH 7.4.

Data from another example is summarized in Figure 10. Here solution stability was monitored for 120 minutes at 60 second intervals with 8 replicates per test pH value. Data are shown as the average of 8 replicates per pH value without any data filtering (spike removal). The pH test range shown in Figure 10 was from gastric pH of 1.2 to mid small intestine pH of 6.8. Drug concentration at 100% was 50 mg itraconazole/liter. This example 'dissolution' experiment was completed in 125 minutes using 48 wells or ½ of one 96-well plate. The same data would have required *ca.* 12 days to collect with only three replicates per test pH using a standard USP2 apparatus and test method.

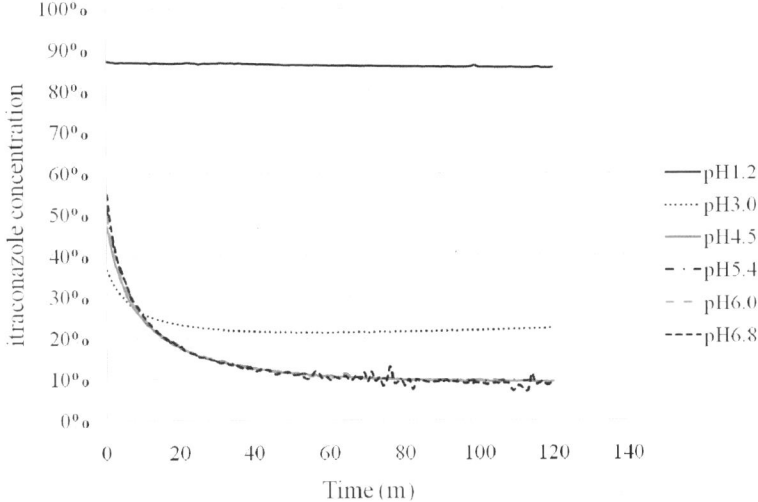

Figure 10. Concentration of itraconazole in solution vs. time for 10 μL of solution at pH 1.2 diluted to 200 μL total solution with shift pH buffer.

Conclusions

Methods have been developed that allow either equilibrium solubility or solution stability profiling to be carried out at high throughput using commonly available laboratory equipment. In our laboratory, the limitation in test points that could be processed per week was determined by the rotary shaker used. It could shake 6 microplates simultaneously. Since we allowed each plate 48-60 hours to equilibrate, we could run about 2000 test points per week, a vast improvement over shake flask methods. Cost per test point was also reduced because the amount of drug required for a 200 μL test volume and the volume of buffers and supplies were all minimized. Overall significant savings in time, labor, and cost of materials are realized. The methods described are perhaps most useful in screening the impact of drug delivery agents on the solubility of specific drugs; but can be broadly applied where solubility or solution stability are to be measured.

References

1. Lipinski, C.A.; Lombardo, L.; Dominy, B.W.; and Feeney, P.J.; *Experimental and Computational Approaches to Estimate Solubility and Permeability in Drug Discovery and Development Settings,* Advanced Drug Delivery Reviews, Vol. 23, 1997; pp 3-25.
2. Glomme, A.; Marz, J.; Dressman, J.B.; *Comparison of a Miniaturized Shake-Flask Solubility Method with Automated Potentiometric Acid/Base Titrations and Calculated Solubilities,* J. Pharmaceutical Sciences, Vol. 94, No. 1, January 2005; pp 1-16.
3. Weiss, Alan; Lynch, John; *Screening Compounds for Aqueous Solubility: A New Automated, High-Throughput Method for Solubility Determination,* Preclinica, Vol. 2, No. 2, March/April, 2004; pp 119-121.
4. Wahlich, John C.; *The Automation of Dissolution Testing,* Pharmaceutical Technology, September, 1980; pp 92-101.
5. Box, Karl; Bevan, Christopher; Comer, John; Hill, Alan; Allen, Ruth; Reynolds, Derek; *High-Throughput Measurement of pKa Values in a Mixed-Buffer Linear pH Gradient System,* Anal. Chem., Vol. 75, No. 4, February 15, 2003; pp 883-892,.
6. Quarterman, Charmaine P.; Bonham, Nicholas M.; Irwin, Alan K.; *Improving the Odds – High Throughput Techniques in New Drug Selection,* European Pharmaceutical Review, pp 27-32.
7. Kerns, Edward H.; Di, Li; *Automation in Pharmaceutical Profiling,* JALA, April, 2005; pp 114-122.
8. Pan, Lin; Ho, Quynh; Tsutsui, Ken; Takahashi, Lori; *Comparison of Chromatographic and Spectroscopic Methods Used to Rank Compounds for Aqueous Solubility,* J. of Pharmaceutical Sciences, Vol. 90, No. 4, April, 2001; pp 521-529.
9. Ruell, Jeffery; Avdeef, Alex; *A Measured Solution,* Modern Drug Discovery, June, 2003; pp 47-49.
10. Chen, Teng-Man; Shen, Hong; Zhu, Chegnyue; *Evaluation of a Method for High Throughput Solubility Determination using a Multi-wavelength UV Plate Reader,* Combinatorial Chemistry and High Throughput Screening, Vol. 5, No. 7, 2002; pp 575-581.
11. Conners, K. A.; *The Stability of Cyclodextrin Complexes in Solution,* Chem. Rev. 97, 1997; pp 1325-1357.

Chapter 4

Cellulose Nanocrystals for Drug Delivery

Maren Roman[1], Shuping Dong[1], Anjali Hirani[2], and Yong Woo Lee[2]

[1]Macromolecules and Interfaces Institute and Department of Wood Science and Forest Products, Virginia Polytechnic Institute and State University, Blacksburg, VA 24061
[2]Department of Biomedical Sciences and Pathobiology, School of Biomedical Engineering and Sciences, Virginia Polytechnic Institute and State University, Blacksburg, VA 24061

Since the first investigation of liposomes as drug carrier systems in chemotherapy, nanoscale carrier systems have attracted increasing attention in therapeutic and diagnostic medicine. Cellulose nanocrystals have attractive properties as nanoscale carriers for bioactive molecules in biomedical applications. To test whether cellulose nanocrystals could be used as carriers in the targeted delivery of therapeutics, their toxicity to human brain microvascular endothelial cells was measured. Cellulose nanocrystals were found to be non-toxic to the cells. For cellular uptake studies, cellulose nanocrystals were labeled with fluorescein-5'-isothiocyanate. The uptake studies showed minimal uptake of untargeted cellulose nanocrystals. The lack of toxicity and untargeted uptake support the potential of cellulose nanocrystals as carriers in targeted drug delivery applications.

Introduction

Since the first investigation of liposomes as drug carrier systems in chemotherapy in 1974 (*1*), nanoscale carrier systems have attracted increasing attention in therapeutic and diagnostic medicine. In search of the "magic bullet", an expression coined by Nobel laureate Paul Ehrlich in the early 1900s, many different types of nanoscale systems are under evaluation, including metal, inorganic, and polymer nanoparticles, quantum dots, carbon nanotubes, polymer micelles, dendrimers, and liposomes (*2*). To achieve the desired benefits, nanoscale carrier systems have to be non-toxic, biodegradable, able to overcome the physiological barriers in the body, and able to withstand the immune system long enough to carry out their mission. Neither of the currently studied systems is optimal. Frequently encountered problems include toxicity, toxic degradation products, low stability in the bloodstream, and accumulation over time in certain organs such as the kidneys, liver, or spleen. According to Couvreur and Vauthier, "we are still far from the magic bullet"(*2*).

This study investigates a novel nanoscale carrier system that is based on an abundant, benign biopolymer, namely cellulose, and that has great potential to be the "magic bullet" for nanobiomedicine. Cellulose nanocrystals are ideally suited as nanoscale carriers for bioactive molecules in biomedical applications due to the following attributes:

(i) Cellulose is biocompatible and does not trigger an immune response when embedded in bodily tissue (*3*).

(ii) Cellulose nanocrystals are rodlike and have a size range between 50 and 200 nm with the majority of the particles between 100 and 150 nm long. Thus, they are too large for removal from the bloodstream by the renal system (*i.e.* the kidneys) but still small enough that the rate of clearance from the bloodstream by the mononuclear phagocytic system is sufficiently delayed (*4*).

(iii) Being entirely composed of polysaccharide molecules, cellulose nanocrystals are highly hydrophilic in nature. A hydrophilic surface has been shown to impede adsorption of opsonin proteins, a critical step before phagocytosis during removal of nanoparticles from the bloodstream (*5*). Thus, cellulose nanocrystals are expected to have an inherently prolonged blood circulation half-life as compared to hydrophobic nanoparticles.

(iv) The surface chemistry of cellulose nanocrystals is governed by hydroxyl groups, which can be easily converted into other functional groups for covalent and non-covalent binding of bioactive molecules at high densities to the surface of the nanoparticles.

Also noteworthy in this context is the fact that cellulose can be broken down by certain fungal and bacterial enzymes to glucose (*6, 7*), a readily metabolized biochemical. Thus, though not an endogenous compound *per se*, it might be possible to achieve endogenous removal of cellulose from the body through *e.g.* systemic administration of cellulolytic enzymes and *in situ* degradation to glucose.

Many human diseases, including Alzheimer's disease, hypertension, and type 2 diabetes, are associated with an inflammation of the blood vessels (*8, 9*). Vascular inflammation in the brain plays a role in the pathogenesis of multiple sclerosis and traumatic brain injury, to name a few (*10, 11*). Inflammation of the blood vessels is characterized by a dysfunction of vascular endothelial cells, the cells that line the inside of the blood vessels. The presented work is part of a larger effort to develop a drug delivery system that selectively targets inflammation-activated human brain microvascular endothelial cells (HBMECs) in the therapy of cerebrovascular inflammatory diseases. Here we report the results of our preliminary studies on the *in vitro* effect of cellulose nanocrystals on resting, non-activated HBMECs. To test whether cellulose nanocrystals exhibit any toxicity towards HBMECs, the cytotoxicity of cellulose nanocrystals was measured by MTT assay. Cellular uptake studies, using nanocrystals labeled with fluorescein-5'-isothiocyanate (FITC), were carried out to assess the uptake of untargeted cellulose nanocrystals.

Materials and Methods

Preparation of Cellulose Nanocrystals

Cellulose nanocrystals were prepared by sulfuric acid hydrolysis of dissolving-grade softwood sulfite pulp. Lapsheets of the pulp (Temalfa 93A-A), kindly provided by Tembec, Inc., were cut into small pieces of approximately 1 cm by 1 cm and milled in a Wiley mill (Thomas Wiley Mini-Mill) to pass a 60 mesh screen. The milled pulp was hydrolyzed under stirring with 64 wt % sulfuric acid (10 mL/g cellulose) at 45 °C for 60 min. The hydrolysis was stopped by diluting the reaction mixture 10-fold with cold (~4 °C) deionized water (Millipore Direct-Q 5, 18.2 MΩ·cm). The nanocrystals were collected and washed once with deionized water by centrifugation for 10 min at 4 °C and 4,550 × g (Thermo IEC Centra-GP8R) and then dialyzed (Spectra/Por 4 dialysis tubing) against deionized water until the pH of fresh dialysis medium stayed constant over time. The nanocrystal suspension was sonicated (Sonics & Materials Model VC-505) for 10 min at 200 W under ice-bath cooling and filtered through a 0.45 μm polyvinylidene fluoride (PVDF) syringe filter (Whatman) to remove any aggregates present. Immediately prior to use, the cellulose nanocrystal suspension was sterilized by filtration through a 0.2 μm PVDF syringe filter (Whatman).

Characterization of Cellulose Nanocrystals by Atomic Force Microscopy

Atomic force microscopy (AFM) was performed with an Asylum Research MFP-3D mounted onto an Olympus IX 71 inverted fluorescence microscope. One drop of a 0.001 wt % suspension of cellulose nanocrystals in water was deposited onto a microscopy slide and allowed to dry in air under ambient conditions. Samples were scanned in intermittent contact mode in air with Olympus OMCL-AC160TS tips (nominal tip radius: <10 nm).

Fluorescent Labeling of Cellulose Nanocrystals

Cellulose nanocrystals were labeled with fluorescein-5'-isothiocyanate (FITC) via a three-step reaction (Figure 1) (*12*).

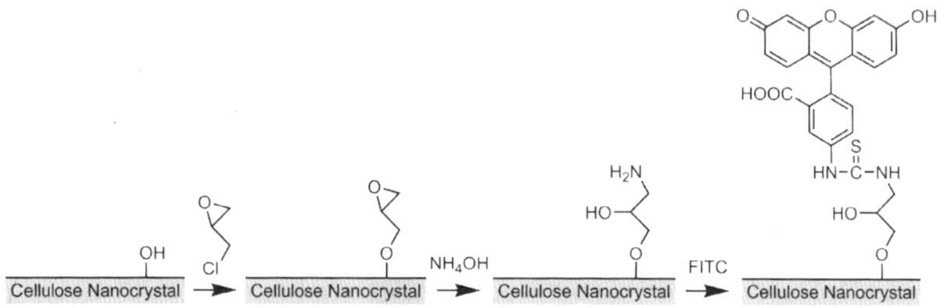

Figure 1. Reaction pathway for the fluorescent labeling of cellulose nanocrystals. (Reprinted with permission from ref 12. Copyright 2007 American Chemical Society)

First, the surface of the nanocrystals was decorated with epoxy functional groups via reaction with epichlorohydrin (6 mmol/g cellulose) in 1 M sodium hydroxide (132 mL/g cellulose) according to the method of Porath and Fornstedt (*13*). After 2 h at 60 °C, the reaction mixture was dialyzed (Spectra/Por 4 dialysis tubing) against deionized water (Millipore Direct-Q 5, 18.2 MΩ·cm) until the pH was below 12. Next, the epoxy ring was opened with ammonium hydroxide to introduce primary amino groups. After adjusting the pH to 12 with 50% (w/v) sodium hydroxide, ammonium hydroxide (29.4%, 5 mL/g cellulose) was added and the reaction mixture heated to 60 °C for 2 h. The reaction mixture was dialyzed until the pH was 7. Finally, the primary amino group was reacted with the isothiocyanate group of FITC to form a thiourea. Following the method of Swoboda and Hasselbach (*14*), FITC (0.32 mmol/g cellulose) was added to the aminated nanocrystals in 50 mM sodium borate buffer solution (50 mL/g cellulose), containing ethylene glycol tetraacetic acid (5 mM), sodium chloride (0.15 M), and sucrose (0.3 M). The reaction mixture was stirred overnight in the dark and then dialyzed against deionized water until FITC was no longer detected in the dialysate by UV/vis spectroscopy. The suspension was sonicated (10 min, 200 W, ice bath cooling), centrifuged (10 min, 4,550 × g, 25 °C), and filtered through a PVDF syringe filter (0.45 μm) to remove any aggregates. The final suspension (0.5 wt %) had a pH of 6. Immediately prior to use, the fluorescently-labeled cellulose nanocrystal suspension was sterilized by filtration through a 0.2 μm PVDF syringe filter (Whatman).

Fluorescence of Labeled Cellulose Nanocrystals by Microplate Reader

Different amounts of a suspension of FITC-labeled cellulose nanocrystals in water at a concentration of 0.5 mg/mL were added to phosphate-buffered saline (PBS) in wells of a 96-well plate to produce the following concentrations: 0.5, 1, 5, 10, 50, 100, and 500 μg/mL and a total volume of 100 μL in each well. The fluorescence intensity from each well was determined with a fluorescence microplate reader (Molecular Devices SpectraMax M2e) at excitation and emission wavelengths of 485 and 530 nm, respectively.

Cell Culture

The cytotoxicity and cellular uptake of cellulose nanocrystals were studied using human brain microvascular endothelial cells (HBMECs). HBMECs were isolated, cultivated, and purified as previously described (15). The cells were positive for factor VIII-Rag, carbonic anhydrase IV, Ulex Europeus Agglutinin I, took up fluorescently labeled low-density lipoprotein, and expressed γ-glutamyl transpeptidase, demonstrating brain endothelial cell characteristics (15). Contamination with non-endothelial cells, such as pericytes and glial cells, was below 1%. HBMECs were cultured in RPMI 1640 medium with 10% fetal bovine serum, 10% NuSerum, 30 $\mu g/mL$ of endothelial cell growth supplement, 15 U/mL of heparin, 2 mM L-glutamine, 2 mM sodium pyruvate, nonessential amino acids, vitamins, 100 U/mL of penicillin, and 100 $\mu g/mL$ of streptomycin at 37 °C in a humid atmosphere of 5% CO_2 and 95% air. For plating of the cells, the culture medium was removed from the flask and the flask rinsed with Hanks' Balanced Salt Solution (Mediatech, Inc.). Cells were detached by adding 1.5 mL of trypsin/EDTA (Mediatech, Inc.). When the cells had detached, medium was added to stop trypsinization, the cells were resuspended by repeated pipetting, and then transferred to a centrifuge tube for counting and plating.

Cytotoxicity Assay

The cytotoxicity of cellulose nanocrystals was measured by MTT (3-[4,5-dimethyl-2-thiazol]-2,5-diphenyl-2H-tetrazolium bromide) assay with HBMECs. Briefly, HBMECs were plated at 50,000 cells/well in a 48-well plate. After 24 h of incubation at 37 °C in a humid, 5% CO_2 and 95% air atmosphere, cells in different wells were treated with increasing concentrations of cellulose nanocrystals (0, 10, 25, and 50 $\mu g/mL$). After 24, 48, or 72 h of additional incubation, the culture medium was removed and replaced with RPMI 1640 containing 0.5 mg/mL of 3-[4,5-dimethyl-2-thiazol]-2,5-diphenyl-2H-tetrazolium bromide (MTT). The cells were incubated for another 4 h, after which the medium was aspirated, and the formazan product was solubilized with dimethyl sulfoxide. The absorbance at 570 nm was determined for each well with a microplate reader (Molecular Devices SpectraMax M2e). Results are presented in percent of control culture results.

Cellular Uptake Assay

For cellular uptake studies, HBMECs, plated in a 48-well plate and incubated for 72 h at 37 °C in a humid, 5% CO_2 and 95% air atmosphere, were treated with 50 $\mu g/mL$ of FITC-labeled cellulose nanocrystals during further incubation for 1, 2, 4, 8, 16 or 24 h. After washing the cells three times with PBS, to remove any cellulose nanocrystals not taken up, and covering the cells with fresh PBS, the fluorescence intensity from each well was determined with a fluorescence microplate reader (Molecular Devices SpectraMax M2e) at excitation and emission wavelengths of 485 and 530 nm, respectively.

Statistical Analysis

Statistical analysis of data was completed using SigmaStat 3.5 (Systat Software, Inc., Point Richmond, CA). One-way ANOVA was used to compare mean responses among the treatments. For each endpoint, the treatment means were compared using a Bonferroni least significant difference procedure. Differences among the means were considered significant at $P < 0.05$.

Results and Discussion

Cellulose Nanocrystals

Figure 2 shows an AFM amplitude image of the nanoparticles obtained by sulfuric acid hydrolysis of bleached wood pulp. The image, among other, similar ones, confirmed that the obtained particles were rodlike and a few tens to a few hundreds of nanometers in length.

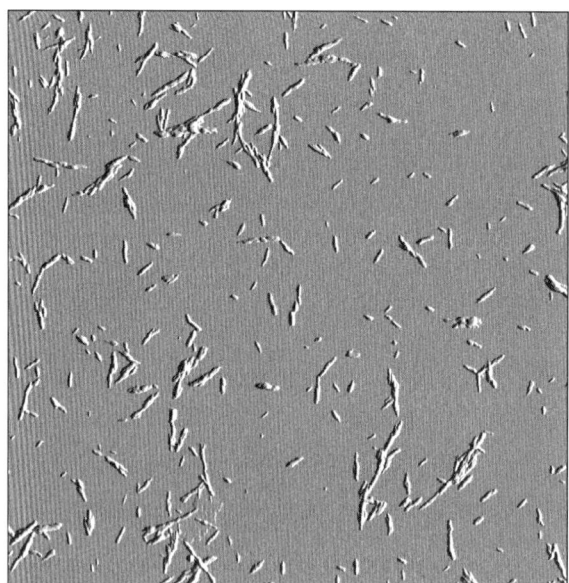

Figure 2. AFM amplitude image of cellulose nanocrystals (scan size: 5 μm^2).
(Reprinted with permission from ref 12. Copyright 2007 American Chemical Society)

For cellular uptake studies, cellulose nanocrystals were labeled with FITC, one of the most widely used fluorophores in fluorescence methods (16). Suspensions of the labeled nanocrystals in deionized water appeared clear and yellow, as opposed to those of unlabeled nanocrystals, which were colorless and slightly opaque (Figure 3). As described elsewhere (12), the FITC content of the labeled cellulose nanocrystals was 0.03 mmol/g of cellulose, equivalent to 5 FITC moieties per 1000 anhydroglucose units. Assuming a nanocrystal diameter of 4.5 nm (17), a density of 1.6 g/cm^3 (18), and neglecting the end surfaces of the particles, 0.03 mmol/g of cellulose corresponds to a surface concentration of 0.037 FITC moieties per nm^2 or 1 FITC moiety per 27 nm^2.

The fluorescence intensity of the labeled cellulose nanocrystals at excitation and emission wavelengths of 485 and 530 nm, respectively, was assessed with a fluorescence microplate reader (Figure 4). At a concentration of 0.5 μg/mL, the fluorescence intensity of the sample was comparable to that of the control (0 μg/mL). However, at a concentration of 1 μg/mL the fluorescence was higher than that of the control ((20.4 ± 0.6) units vs. (16.4 ± 0.2) units), with statistical significance (P < 0.05), and was as high as 4290 units at 500 μg/mL.

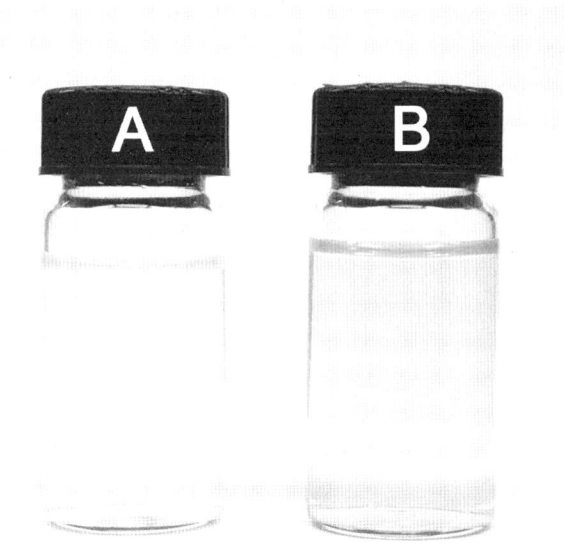

Figure 3. Aqueous suspensions of (A) cellulose nanocrystals (0.8 wt %) and (B) FITC-labeled cellulose nanocrystals (0.5 wt %). (Reprinted with permission from ref 12. Copyright 2007 American Chemical Society) (see page 1 of color insert)

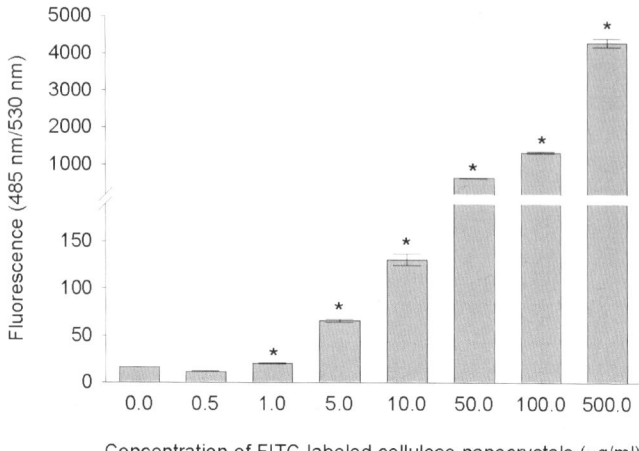

Figure 4. Fluorescence vs. concentration of FITC-labeled cellulose nanocrystals. Data are means ± SD of 4 determinations. Asterisk indicates significant difference with respect to the control (P < 0.05).

Cytotoxicity of Cellulose Nanocrystals

To determine the cytotoxicity of cellulose nanocrystals, an MTT assay was performed using HBMECs. MTT (3-[4,5-dimethyl-2-thiazol]-2,5-diphenyl-2H-tetrazolium bromide) is a tetrazolium salt, which gets metabolized within the mitochondria of viable cells. The absorbance of the produced formazan at 570 nm is proportional to the number of living cells in the sample. The viability of HBMECs, in percent of cell viability of the control, after exposure to different concentrations of cellulose nanocrystals for up to 72 h is shown in Figure 5. At each exposure time (24, 48, and 72 h), the cell viability for different concentrations of cellulose nanocrystals (10, 25, 50 μg/mL) stayed within the error margins of the cell viability of the control. Our cytotoxicity results indicate that cellulose nanocrystals are non-toxic to HBMECs over the concentration range examined.

Cellular Uptake of Cellulose Nanocrystals

To determine the cellular uptake of cellulose nanocrystals, the cells were exposed to fluorescently-labeled cellulose nanocrystals for a period of time, after which excess nanocrystals were removed and the remaining fluorescence was measured. The uptake of labeled cellulose nanocrystals by HBMECs at different exposure times of up to 24 h and a nanocrystal concentration of 50 μg/mL is shown in Figure 6. The cellular uptake was minimal even after 24 h of exposure. Limited uptake of cellulose nanocrystals by the cells could explain the lack of cytotoxicity observed for unlabeled cellulose nanocrystals. A low uptake of untargeted nanoparticles is desired in targeted drug delivery applications since it allows control over cellular uptake through targeting of specific uptake mechanisms.

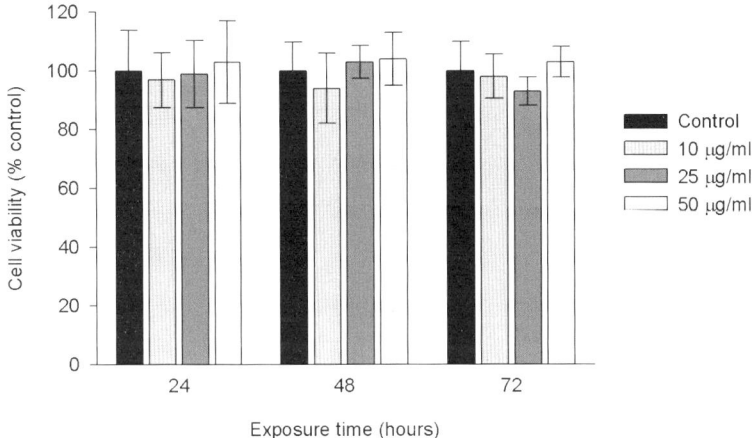

Figure 5. Cell viability of human brain microvascular endothelial cells after exposure to different concentrations of cellulose nanocrystals. Data are means ± SD of 4 determinations.

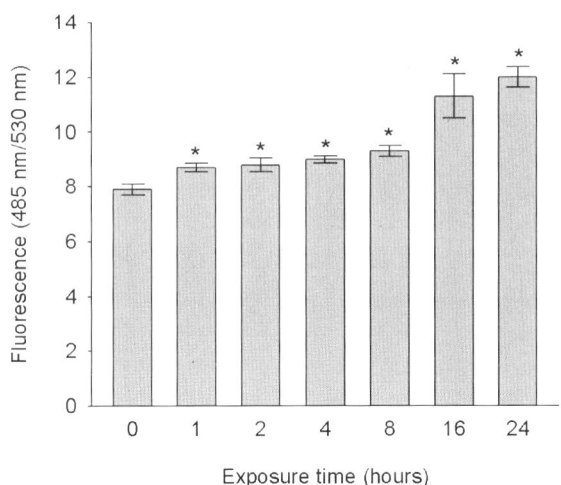

Figure 6. Uptake of FITC-labeled cellulose nanocrystals (50 μg/mL) by human brain microvascular endothelial cells. Data are means ± SD of 4 determinations. Asterisk indicates significant difference with respect to the control (P < 0.05).

Conclusions

In this study, the cytotoxicity of cellulose nanocrystals to HBMECs has been measured by MTT assay. Furthermore, the uptake of fluorescently labeled cellulose nanocrystals by HBMECs has been assessed by cell uptake experiments. From our results it can be concluded that cellulose nanocrystals are non-toxic to human brain microvascular endothelial cells in the concentration range studied, and that cellular uptake of untargeted cellulose nanocrystals is minimal. The lack of toxicity and untargeted uptake support the potential of cellulose nanocrystals as carriers in targeted drug delivery applications.

Acknowledgements

This project was supported by the National Research Initiative of the USDA Cooperative State Research, Education and Extension Service, grant number 2005-35504-16088. Moreover, the authors gratefully acknowledge partial funding of the Institute for Critical Technology and Applied Science (ICTAS) in this effort. Additional support from Omnova Solutions, Inc. and Tembec, Inc. is also acknowledged.

References

1. Gregoriadis, G.; Wills, E. J.; Swain, C. P.; Tavill, A. S. *Lancet* **1974**, *1*, 1313–1316.
2. Couvreur, P.; Vauthier, C. *Pharm. Res.* **2006**, *23*, 1417–1450.
3. Klemm, D.; Schumann, D.; Udhardt, U.; Marsch, S. *Prog. Polym. Sci.* **2001**, *26*, 1561–1603.
4. Owens, D. E., 3rd; Peppas, N. A. *Int. J. Pharm.* **2006**, *307*, 93–102.
5. Lemarchand, C.; Gref, R.; Couvreur, P. *European Journal of Pharmaceutics and Biopharmaceutics* **2004**, *58*, 327–341.
6. Rabinovich, M. L.; Melnik, M. S.; Boloboba, A. V. *Appl. Biochem. Microbiol.* **2002**, *38*, 305–321.
7. Lynd, L. R.; Weimer, P. J.; van Zyl, W. H.; Pretorius, I. S. *Microbiol. Mol. Biol. Rev.* **2002**, *66*, 506–+.
8. Christov, A.; Ottman, J. T.; Grammas, P. *Neurol. Res.* **2004**, *26*, 540–546.
9. Savoia, C.; Schiffrin, E. L. *Clin. Sci.* **2007**, *112*, 375–384.
10. Proescholdt, M. A.; Jacobson, S.; Tresser, N.; Oldfield, E. H.; Merrill, M. J. *J. Neuropathol. Exp. Neurol.* **2002**, *61*, 914–925.
11. Barone, F. C.; Kilgore, K. S. *Clin. Neurosci. Res.* **2006**, *6*, 329–356.
12. Dong, S.; Roman, M. *J. Am. Chem. Soc.* **2007**, *129*, 13810–13811.
13. Porath, J.; Fornstedt, N. *J. Chromatogr.* **1970**, *51*, 479–489.
14. Swoboda, G.; Hasselbach, W. *Z. Naturforsch., C: Biosci.* **1985**, *40*, 863–875.
15. Stins, M. F.; Gilles, F.; Kim, K. S. *J. Neuroimmunol.* **1997**, *76*, 81–90.

16. Schauenstein, K.; Schauenstein, E.; Wick, G. *J. Histochem. Cytochem.* **1978,** *26,* 277–283.
17. Beck-Candanedo, S.; Roman, M.; Gray, D. G. *Biomacromolecules* **2005,** *6,* 1048–1054.
18. Ganster, J.; Fink, H. P. In *Polymer Handbook;* Brandrup, J.; Immergut, E. H.; Grulke, E. A., Eds.; Wiley & Sons: New York, 1999; pp V/135–V/157.

Chapter 5

Enhanced dissolution of poorly soluble drugs from solid dispersions in carboxymethylcellulose acetate butyrate matrices

Michael C. Shelton[1], Jessica D. Posey-Dowty[1], Larry Lingerfelt[1], Shane K. Kirk[1], Sandra Klein[2], and Kevin J. Edgar[3]

[1]Eastman Chemical Company, Box 1972, Kingsport TN 37662, [2]Institute of Pharmaceutical Technology, 9 Marie Curie Street, Johann Wolfgang Goethe University, Frankfurt, 60439, Germany [3]Department of Wood Science and Forest Products, 230 Cheatham Hall, Virginia Tech, Blacksburg, Virginia 24601

This paper describes a novel cellulose ester for drug delivery, carboxymethyl cellulose acetate butyrate (CMCAB). We present data that highlights the ability of CMCAB to form amorphous matrices with poorly soluble drugs to enhance aqueous solubility. The ease of use, compatibility with numerous pharmaceutical actives, high glass transition temperature, and pH sensitivity make this new cellulose ester broadly promising for drug delivery systems. Formulation of various drugs with CMCAB affords slow, often zero-order release, pH-controlled release, stable solid blends, enhanced drug solubility, and enhanced stabilization in solution against drug recrystallization. We highlight the dramatic impact upon dissolution behavior of formulation into an amorphous polymer matrix, as compared with physical blends of drug and polymer.

Techniques, such as high throughput screening and computational chemistry, have revolutionized the pharmaceutical industry and greatly increased the number of drug candidates in recent years. However, the increase in candidates has not led to a concomitant increase in new marketed drugs[1]. This is due at least in part to the high proportion of modern drug candidates which are poorly soluble in water.[2] According to one estimate, more than one third of the drugs listed in the United States Pharmacopoeia are either poorly soluble or insoluble in water.[3] Poor water solubility can cripple drug bioavailability and thus may render otherwise potent drugs unsuitable for patient care.[4] Therefore, improving aqueous solubility for these poorly soluble compounds is an area of intense interest in pharmaceutical research. Drug delivery systems that enhance solubility can result in the commercialization of new drugs, and can enhance the performance of existing drugs by reducing dosage, cost, side-effects, and/or variability.

The bioavailability of a drug substance is influenced by two main factors, its aqueous solubility and the rate of its permeation through lipid bilayer membranes. The biopharmaceutics classification system (BCS), first introduced by Amidon, et al., is an accepted way of distinguishing between classes of drugs based on solubility and permeability.[5] This system divides drugs into four classes: Class 1 - high solubility-high permeability, Class 2 - low solubility-high permeability, Class 3 - high solubility-low permeability, and Class 4 - low solubility-low permeability.

Increasing the aqueous solubility of a BCS Class 2 drug will also increase its bioavailability, thereby enhancing therapeutic utility.[6,7] Numerous techniques have been developed to improve the solubility of BCS Class 2 drugs, including the use of solid dispersions,[8] nanoparticles[9], complexation by substituted cyclodextrins[10], formulation into liposomes[11], and manipulation of drug crystallinity[12]. Herein we report recent results on improvement of solubility of certain BCS Class 2 drugs by incorporating into solid dispersions with carboxymethylcellulose acetate butyrate (CMCAB, Figure 1), a recently developed cellulose ether ester.[13,14,15,16]

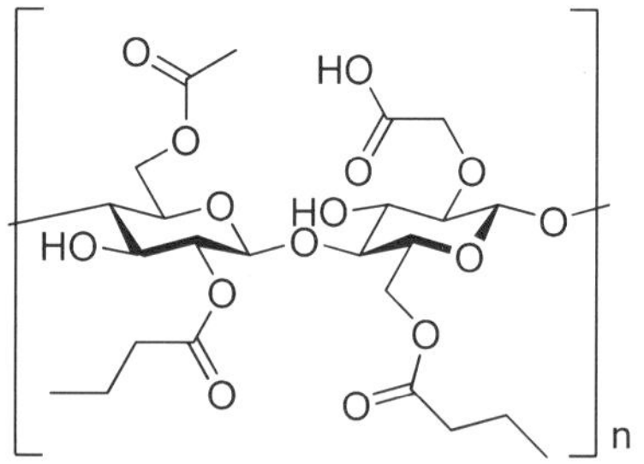

Figure 1. Structural depiction of CMCAB. Average DS for CMCAB: Ac= 0.4, Bu= 1.6, CH$_2$COOH= 0.3. Substituent locations represent only one possible dimer among many possibilities.

Previous studies from our laboratories[17] have shown that CMCAB has interesting properties for solid pharmaceutical formulations. It is a fundamentally hydrophobic polymer due to its high degree of substitution (DS) of butyryl and acetyl groups, giving it the potential to form miscible blends with hydrophobic actives. Its pendant carboxyl groups give it pH sensitivity and the opportunity for specific interactions with drug functional groups, such as amino moieties. At the same time, the rather low DS of the carboxymethyl groups (0.3) means that the polymer has low water solubility, even at pH above 7; rather, it swells at neutral and higher pH. The use of an insoluble polymer as a carrier in solid dispersions is not commonly practiced in drug delivery system development. However, one should not discount the potential benefit of using swellable polymers in conjunction with highly miscible drug substances.

In our earlier study we showed that these physical properties of CMCAB are reflected in performance; physical blends of CMCAB with several hydrophilic and hydrophobic drugs were shown to exhibit slow (often 24 h) release, in some cases zero-order with respect to time, as well as pH-controlled release. These properties along with the usually biocompatible and benign nature of cellulose esters, and the many productive uses of cellulose esters in pharmaceutical applications,[18] led us to think that CMCAB might have especially suitable properties for the solubility enhancement of poorly water soluble drugs by formation of amorphous matrices of drug in CMCAB.

By definition, a solid dispersion (or solid solution) is a drug substance entrapped (or "dissolved") in a carrier, usually a hydrophilic polymer. When the drug is miscible with the polymer (forming a molecular dispersion, often referred to as an amorphous matrix), drug crystallinity is disrupted to a greater or lesser extent, depending on whether there is any residual short range order in the dispersion. Disruption of the crystal lattice of a high-melting drug often increases the rate of dissolution and thus increases the bioavailability of the drug.[19] Polymers must possess certain properties to be attractive for these amorphous matrix formulations: they must be non-toxic, miscible with the (often hydrophobic) drug, and must have a trigger mechanism for drug release, such as pH sensitivity. The polymer-drug formulation must be stable in transport and storage for many months, so the polymer must possess high glass transition temperature (in order to inhibit drug migration through the matrix), and must not itself have a tendency to crystallize.[20] It is preferable that all crystallinity of the drug substance be suppressed by the polymer carrier since small pockets of crystallinity can act as seeds, resulting in additional crystal growth and ultimately reduced shelf-life of the drug formulation.[21] Another important characteristic that is often overlooked is the ability to stabilize the drug after dissolution and release; the solubility enhancement resulting from the amorphous form of the drug can lead to a super-saturated drug solution, from which precipitation will occur if the polymer is unable to stabilize the drug in solution. To date, research on enhancement of drug solubility through amorphous matrices has largely focused on polymers previously used in pharmaceutical formulations, such as polyethylene glycol (PEG) and poly(vinylpyrollidinone) PVP. The utility of PEG may be limited by its high water solubility and its tendency to crystallize,[22] while that of PVP may be limited somewhat by an inability to stabilize some actives in the solid phase.[23] Recently, Pfizer scientists have recognized the interesting properties of hydroxypropylmethylcellulose acetate succinate (HPMCAS) in amorphous matrix formulations.[24]

This study investigates the dissolution properties of blends of CMCAB (and control polymers) with several BCS Class 2 drugs: ibuprofen, glyburide, and griseofulvin, as well as fexofenadine, which is highly soluble in pure water, but is a Class 2 drug in solutions of significant ionic content, such as physiological fluids. The study compares results from physical drug/polymer blends with those from molecular dispersions of drug in polymer in which drug crystallinity has been suppressed.

Throughout this report, the terms solid dispersions, intimate blends, amorphous blends, amorphous matrices, and amorphous matrix formulations are used interchangeably. As used here, the term amorphous blend means largely amorphous, and does not necessarily indicate that a drug formulation has 100 % amorphous character (where possible, the measured percent crystallinity is noted).

Experimental

General Experimental Materials and Methods

Cryogenic grinding employed a SPEX liquid nitrogen grinder, to a 20 micron particle size target. CMCAB was from Eastman Chemical Company, and had DS (carboxymethyl) of 0.3, DS_{Ac} of 0.44, and DS_{Bu} of 1.63. Ibuprofen, glyburide, and griseofulvin were Sigma-Aldrich reagent grade; Fexofenadine HCl was from Apin Chemical, Ltd. Brief descriptions of each of these drugs are presented in Table I.

Table I. Drugs

Compound	M_w	pK_a^{25}	Solubility, mg/mL (37 °C , pH 6.8)	MP(°C)
Ibuprofen[a]	206	4.4	0.01 [26]	75-77
Glyburide[a]	494	5.3	0.031(X), 0.39(A)[c,27]	169-170
Griseofulvin[a]	352	Neutral	0.0056 [28]	220
Fexofenadine HCl[b]	538	4.4, 9.6	0.35[d]	195-197[e]

NOTE: [a]Sigma-Aldrich; [b]Apin Chemical, LTD.; [c](X) = Crystalline value, (A) = Amorphous value; [d]Measured in this study by the shake flask method in USP standard pH 6.8 phosphate buffer at 37.0 °C; [e]melting point of free base.

A number of polymers, plasticizers/additives, and solvents were used in this study. Polymers used include: carboxymethylcellulose acetate butyrate (CMCAB), hydroxypropylmethylcellulose acetate succinate (HPMCAS), hydroxypropylmethylcellulose (HPMC), and PVP. Plasticizers/additives used include: diethyl phthalate (DEP), triacetin, Vitamin E TPGS, sucrose acetate isobutyrate (SAIB), sodium dodecylsulfate (SDS), and Tween 80™. Solvents used include: acetone, dimethylsulfoxide (DMSO), methylene chloride, and acetonitrile.

Preparation of Solid Dispersions

The solid dispersions we investigated were prepared by solvent evaporation and co-precipitation using several techniques. Solvent evaporation was accomplished by rotary evaporation under reduced pressure, by spray drying, or by casting films at atmospheric pressure. General descriptions of these techniques are described below.

Active/Polymer/Additive Solution Preparation

A glass bottle was charged with polymer, drug, and/or plasticizer and an organic solvent or mixture of solvents (*e.g.,* acetone, methylene chloride, acetonitrile, and ethanol). The bottle was sealed tightly and mixed on mechanical rollers until a clear, homogeneous solution was obtained. The viscosity of the solution was controlled by adjusting the polymer level in the system. These solutions were combined and mixed again by mechanical rolling as appropriate for a particular experiment.

Solvent evaporation, reduced pressure method

The solvents were removed from the active/polymer/additive solutions under reduced pressure by rotary evaporation at 50 °C. The sample was then dried on a high vacuum line overnight to remove as much residual solvent as possible. The resulting solid dispersion was then removed from the round-bottomed flask using a spatula and further dried for at least 12 hours at 45 °C in a vacuum oven. The dried solid dispersion was pulverized to a particle size of approximately 20 microns using a SPEX® cryogenic grinder, at liquid nitrogen temperatures. The solid dispersion was stored in a desiccator until needed.

Solvent evaporation, film formation method

Solutions of polymer, drug, and additive were prepared and poured or cast into a glass or Teflon® sheet, and the solvent was allowed to slowly evaporate until a film was formed. Note that when casting films using CMCAB as the polymer, the preferred substrate is Teflon® due to the propensity of CMCAB to adhere to glass and to metal. The films were dried under ambient conditions, then removed from the substrate and dried overnight at 45 °C in a vacuum oven. The resulting dried solid dispersion films were pulverized to a particle size of approximately 20 microns using a SPEX® cryogenic grinder at liquid nitrogen temperatures. The samples were stored in a desiccator until needed.

Spray Drying

System Description: A Buchi Model B-290/B295 Mini Spray Dryer (lab-scale glass spray dryer) was employed with a N_2 atmosphere for safe operation with organic solvents. Typical process conditions are given below in Table II.

Table II. Spray Drying Conditions

Conditions	1	2
Inlet Temp °C	55	55
Outlet Temp °C	42	43
Fan setting (%)	100	100
Atomization Pressure	30	30
Feed Wt. (g)	200	226.5
Run Time (min.)	42	42
Pump Setting (%)	17	20
Feed Rate (g/min)	4.76	5.39
Yield	6.5	24.3

Co-precipitation

"Co-precipitation" is the general term used here to describe the combination of a solution or mixture containing a polymeric carrier (*e.g.*, a carboxyalkylcellulose ester), a pharmaceutically active agent, and optionally one or more other additives dissolved in an organic solvent with a non-solvent, typically aqueous, to produce a precipitate that is an intimate mixture of the non-volatile components of the organic solution/mixture. Co-precipitation methods used for the preparation of the compositions of this study are flake precipitation and powder precipitation. Flake precipitation is accomplished by adding a thin stream of the polymer/drug/solvent mixture to the aqueous non-solvent, typically with rapid mixing. There are a number of process variables, including temperature, rate of addition, mixing rate, concentration of solids in the organic mixture, pH of the non-solvent, organic solvent content in the precipitate mixture, and hardening time, that can be adjusted to modify the physical nature (*i.e.* morphology, particle size) of the co-precipitate, the composition of the co-precipitate, and likely the dissolution profile of the solid dispersion.

In flake co-precipitation, an appropriate organic solvent or mixture of solvents is added to a vessel (typically a glass bottle) containing the desired amount of the polymer carrier, and the vessel is mixed (typically on a roller or by stirring) until a clear or nearly clear solution is obtained. The pharmaceutical active, and any additives such as plasticizers, may be added to the same vessel and dissolved, or may be dissolved separately and the solutions combined. The combined solvents are added as a small stream to an excess of nonsolvent with rapid mixing. Typically a ratio of at least 1:3 organic to aqueous solution is appropriate to induce flake precipitation,. Once precipitation is complete, the precipitate is collected by filtration on a coarse fritted funnel, dried overnight at 45 °C in a vacuum oven, and pulverized to a particle size of approximately 20 µm (typically less than 200 µm) in a cryogenic grinder. The samples are stored in a desiccator or vacuum desiccator until needed.

Powder precipitation, a process commonly practiced in the field of cellulose ester chemistry, is accomplished by adding the aqueous non-solvent to the polymer/drug/solvent mixture with appropriate mixing and temperature. As with flake precipitation, a number of process variables, including but not limited to temperature, rate of addition, mixing rate, concentration of solids in the organic mixture, pH of the nonsolvent, organic solvent content in the precipitate mixture, and hardening time can be adjusted to modify the physical nature (i.e., morphology, particle size) of the co-precipitate, the composition of the co-precipitate, and likely the dissolution profile of the solid dispersion. The precipitate is collected, dried and ground as in flake precipitation.

Preparation of Paired Physical Blend and Solid Dispersion Samples

Blends of drugs, polymers, and in some cases plasticizers were prepared by mixing and grinding to *ca.* 20 μm particle size. To this mixture was added a finely ground blend of magnesium stearate (release aid) and carbon black (to affirm good mixing), each at a level of 0.1 wt % of the polymer/drug mixture. The resulting sample was split into two portions.

Solid dispersion – One portion of the sample was dissolved in acetone/ethanol (6/1 ratio by weight, respectively). The solution was cast on a Teflon-coated glass plate and air-dried. The film was removed and then dried under vacuum at 50 °C for 24 hours. The film was cryogenically ground, and the resulting powder packed into gelatin capsules, size 00.

Physical blend – The other portion of the sample was used as the control physical blend. It was either packed into gelatin capsules, size 00, or formed into tablets using a single punch tablet press set at 2500 psi.

Detailed Example: Preparation of Ibuprofen Homogeneous and Physical Blends

Homogenous blend: Ibuprofen/CMCAB blends were prepared by mixing CMCAB (2.9768 g, 9.48 mmol) with ibuprofen (1.1779 g, 5.71 mmol) and cryogenically grinding. Separately, magnesium stearate (10.489 g) was mixed with charcoal (Darco G-60, 1.262 g) until an even color was achieved. The ground ibuprofen/CMCAB blend was mixed with a portion of the carbon black/magnesium stearate blend (0.0307 g) until a uniform color was achieved. A portion of this sample was used to make tablets. Tablets were pressed using a single pill tablet press set at 2500 psi. The other portion was dissolved in a blend of acetone/ethanol (6/1 weight ratio) and then cast onto Teflon sheets to air dry. After drying in air overnight, the film was placed in a vacuum oven at 50 °C for 24 hours. The films were cryogenically ground. The resulting powders were added to gelatin capsules, size 00.

Physical blend: A similar procedure was used to form ibuprofen/ hydroxypropylmethylcellulose (HPMC) physical blends, except the following amounts were used: HPMC (4.6822 g), ibuprofen (1.1779 g, 5.71 mmol), and magnesium stearate/carbon black blend (0.0262 g). Tablets and powders of physical and intimate mixtures were made in the same ways as for CMCAB blends.

Samples were analyzed for ibuprofen by UV spectroscopy using a Hewlett-Packard UV-Vis Spectrophotometer calibrated with 9 standard solutions. The absorbance was measured at 274 nm with background corrections at 271 nm and 278 nm, affording a correlation coefficient of 0.99990.

Determination of percent crystallinity by X-ray diffraction

A Scintag PAD V diffractometer using Cu K-alpha X-ray was employed. Known weights of sample and corundum (Al_2O_3, diffraction standard) were mixed. Each mixture was pelletized with a hydraulic press and the XRD pattern of the pellet was measured from 5 to 45 degree scattering angle. A diffraction response factor, R, was calculated for each species according to the equation (R = $wc/ws*l_s/l_c$), where "wc" is the weight fraction of corundum, "ws" is the weight fraction of the species of interest, "l_c" is the net intensity of the major diffraction line of corrundum and "l_s" is the net intensity of the major diffraction line of the drug or in the case of the polymers, the net intensity of the maximum of the amorphous scattering curve.

The net intensity of the maximum of the amorphous scatter from the polymer, l_p, and the net intensity of the major diffraction line of the drug, l_d, were determined from the resulting scattering curve. The wt % crystalline drug was calculated using the equation:

$$\% \text{ crystalline drug} = \frac{100\ (l_d/R_d)}{l_d/R_d + l_p/R_p} \tag{1}$$

where "R_d" is the response factor for the drug and "R_p" is the response factor for the polymer.

General Dissolution Testing Protocol

Dissolution testing was performed in Simulated Intestinal Fluid (without pancreatin) pH 6.8 (SIF_{sp}, pH 6.8) as described by Galia, et al.[29] using a Varian VK7025 instrument, equipped with a Varian VK8000 Fraction Collector (USPII calibrated). Teflon-coated stir rods were used. A stir rate of 50 rpm was employed, except for fexofenadine and ibuprofen which were stirred at 100 rpm. Aliquots (5.0 mL) were removed at preprogrammed times and filtered through 10 micron filter tips. The vessel temperature was 37 °C. Weighed capsules or tablets were added to the vessels, and stainless steel capsule weights were used with gelatin capsules to prevent them from floating during the dissolution testing.

Unless stated otherwise, all samples were analyzed using an Agilent 1100 series HPLC equipped with a Zorbax Eclipse XDB-C8 4.6 x 150 mm column with a 5 μm particle size. The mobile phase was 55 % acetonitrile and 45 % ammonium acetate buffer (2.6 g ammonium acetate/1000 mL of water with the pH adjusted to pH 5.25 with glacial acetic acid). The flow rate was 1.5 mL/min, and retention times were typically 1.5 to 2.5 minutes. A UV detector was used, and commonly 5 signals were collected, 214 nm, 222 nm, 254 nm, 287 nm, 291 nm, and 325 nm. Typically 254 nm was the wavelength selected, but 291 nm was used for griseofulvin. Standard curves were prepared using USP standards.

Additional experimental details for most samples can be found in US Patent Application 2007/0178152.

Samples were analyzed for fexofenadine HCl by UV spectroscopy using a Hewlett-Packard UV-Vis Spectrophotometer calibrated with 3 standard solutions covering the ranges measured. The absorbance was measured at 235 nm with background corrections at 350 nm, affording a correlation coefficient of 0.99983.

Table III provides a description of the samples evaluated. It should be noted that the bulk density of samples often led to difficulty in producing formulations (capsules or tablets) with identical amounts of drug substance and/or additives between samples. As a result, data is often provided in both parts per million of drug substance released and percent of the drug substance released. Presenting both forms of data provides the reader some insight into possible limitations introduced to the systems based on the amount of drug substance available.

Table III. Description of Samples

	Polymer	Drug (wt %)	Blend Type	Method	% Xstal	Additive (wt %)
1	CMCAB	Ibuprofen (15)	Amorph	Film	4	DEP (5)
2	CMCAB	Ibuprofen (15)	Amorph	Film	2	Triacetin (6)
3	CMCAB	Ibuprofen (15)	Amorph	Film	2	SAIB (6)
4	CMCAB	Ibuprofen (30)	Amorph	Film	20	None
5	CMCAB	Ibuprofen (30)	Phys	NA	100	None
6	HPMC	Ibuprofen (30)	Amorph	Film	20	None
7	HPMC	Ibuprofen (30)	Phys	NA	1	None
8	CMCAB	Glyburide (9)	Amorph	Spray	3	TPGS (5)
9	CMCAB	Glyburide (8)	Amorph	Spray	0	None
10	CMCAB	Glyburide (11)	Amorph	CO	26	None
11	HPMCAS	Glyburide (10)	Amorph	CO	0	None
12	CMCAB	Griseofulvin (9)	Amorph	Film	0	None
13	CMCAB	Griseofulvin (9)	Amorph	CO	0	None
14	CMCAB	Griseofulvin (10)	Amorph	Roto	0	
15	HPMCAS	Griseofulvin (10)	Amorph	Film	3	None
16	HPMCAS	Griseofulvin (9)	Amorph	CO	0	None
17	PVP	Griseofulvin (8)	Amorph	Film	3	None
18	PVP	Griseofulvin (9)	Amorph	CO	0	None
19	PVP	Griseofulvin (10)	Amorph	Roto	0	None
20	CMCAB	Griseofulvin (10)	Amorph	Roto	0	SDS (5)
21	CMCAB	Griseofulvin (10)	Amorph	Roto	0	Tween 80 (5)
22	CMCAB	Griseofulvin (10)	Amorph	Roto	0	Tween 80 (1)
23	CMCAB	Fexofenadine (30)	Phys	NA	100	None
24	CMCAB	Fexofenadine (30)	Amorph	Film	20	None

Note: % Xstal is % Crystallinity measured by X-Ray diffraction. Film stands for cast film; Roto for rotovap; CO for Coprecipitation; Amorph for amorphous; Phys for physcial blend; Spray for spray dry; NA for not applicable.

Results and Discussion

Our previous work[13] showed that physical blends of CMCAB with several different drugs (for example, aspirin and ibuprofen) afforded slow release of the drug, often zero-order with respect to time. The process of compressing CMCAB and drug is far simpler than the osmotic pump system,[30,31] the most commonly employed system currently for achieving zero-order release with respect to time. The release was also pH-controlled, and much more rapid at small intestinal pH (6.8) than at gastric pH (1.2). Most intriguing were hints that physical blends with CMCAB might afford enhanced solubility for certain poorly soluble actives. We saw significantly enhanced solubility for both fluconazole and fexofenadine, when incorporated even into physical blends with CMCAB. It seemed likely, based on recent literature precedent that even greater solubility enhancement could be obtained if we could prepare molecular dispersions of drug in polymer. This led us to perform the current, more extensive study in which we compare the results of physical blends of drugs with CMCAB with molecular dispersions of drug in CMCAB, prepared by several different techniques. In addition, we compare the CMCAB/drug blends with blends of the same drugs and other polymers used in amorphous matrix drug delivery, including hydroxypropylmethylcellulose (HPMC), hydroxypropylmethylcellulose acetate succinate (HPMCAS), and poly(vinylpyrollidinone) (PVP). We wish to develop enhanced understanding of the structure-property relationship that governs miscibility of drugs with cellulose derivatives and other polymers, as well as the release of those drugs upon exposure to aqueous media at physiological pH levels. We also seek to develop understanding of the structure-property relationships governing stabilization of the amorphous form of the drug against crystallization in the solid phase, and in solution after drug release.

Solid Dispersions with CMCAB

Solid dispersions can be prepared by a number of methods including solvent evaporation (spray drying, rotary evaporation, or film casting), freeze drying (lyophilizing), and co-precipitation. The use of thermal (*e.g.*, extrusion) or mechanical *(e.g.*, roll-milling) energy to promote intimate mixing can also be useful techniques where drug and polymer are sufficiently stable and unreactive with one another. We describe here the performance of dispersions made by low-energy techniques such as co-precipitation, film casting, and rotary evaporation of the polymers noted above with several different drugs, including ibuprofen, glyburide, griseofulvin, and fexofenadine. Since CMCAB contains both carboxylic acid and hydroxyl groups, crosslinking with itself as well as reaction with many drug molecules could be a concern if high-temperature extrusion techniques were employed.

Ibuprofen Solid Dispersions

Ibuprofen solid dispersions with CMCAB and, in some cases plasticizer, were made according to the solvent evaporation film formation method described in the Experimental section. Analysis of these samples by X-ray diffraction indicates (Figures 2, 3) that ibuprofen crystallinity develops easily if the concentration of ibuprofen becomes greater than approximately 10 % or if the glass transition temperature of the blend drops below approximately 100 °C. Comparing plots of the percent amorphous ibuprofen to the ibuprofen concentration in the blend, with and without plasticizers (Figure 2), shows that plasticizers did not inhibit crystal growth. At ibuprofen concentration ≤ 10 %, plasticizers have little impact on drug crystallinity in the blend. With increasing concentration of ibuprofen in the formulation, crystallinity increased. Higher concentrations of ibuprofen increase the probability of crystal formation. Additionally, the T_g is reduced, thus enhancing molecular motion in the blend and promoting crystallization. Consistent with this interpretation, Figure 3 shows that high blend glass transition temperatures generally favor higher percentages of amorphous ibuprofen.

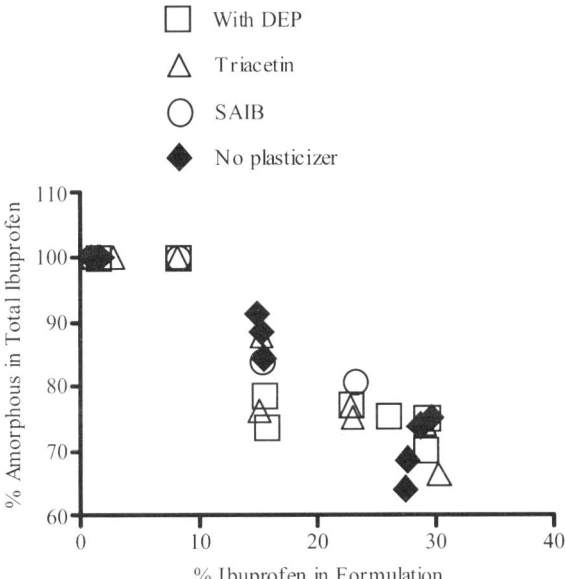

Figure 2. The Effect of Ibuprofen Concentration on Crystallinity
Ibuprofen Dissolution Studies

Ibuprofen is an important over-the-counter pain reliever and anti-inflammatory agent. It has poor aqueous solubility at stomach pH, and dissolves quickly at neutral and higher pH. We examined the impact of crystallinity on the apparent dissolution rates from solid dispersions in simulated intestinal fluid without pancreatin (SIF_{sp}) at 37 °C. Physical blends of ibuprofen with the polymers were used as controls.

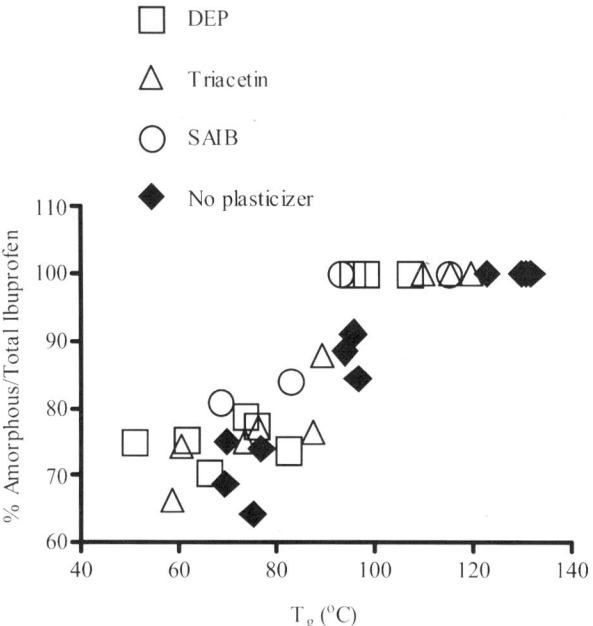

Figure 3. Percent crystalline drug as a function of glass transition temperature of CMCAB / ibuprofen / plasticizer blends

Formulating ibuprofen into an amorphous solid dispersion with CMCAB produced a remarkable acceleration in release rate, as shown in Figure 4. While a physical blend of ibuprofen with CMCAB afforded slow, nearly zero-order (with respect to time) release (24 hours to release 80 % of the ibuprofen), release of ibuprofen from an ibuprofen/CMCAB solid dispersion was rapid, reaching a plateau at approximately 80 % release within one hour. Note that both HPMC and CMCAB solid dispersions contained approximately 20 % crystalline and 80 % amorphous ibuprofen. It is interesting and perhaps instructive that the percent ibuprofen released is roughly equal to the percent amorphous ibuprofen in the blend. The term "loading" in Figure 4 refers to the total amount of drug substance, in this case ibuprofen, present in the delivery system and used for dissolution testing.

In contrast, as shown in Figure 5, the release profiles of ibuprofen/HPMC intimate blends and ibuprofen/HPMC physical blends are very similar. There is only a slight increase in dissolution rate for the intimate blend over the physical blend. It should be noted that in all HPMC/ibuprofen samples, HPMC did not fully disperse into the dissolution vessel, but instead remained a swollen gel throughout the experiment. Thus, it appeared that diffusion through the swollen gel was the rate limiting step with HPMC even when it contained largely amorphous ibuprofen.

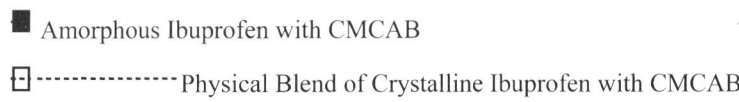

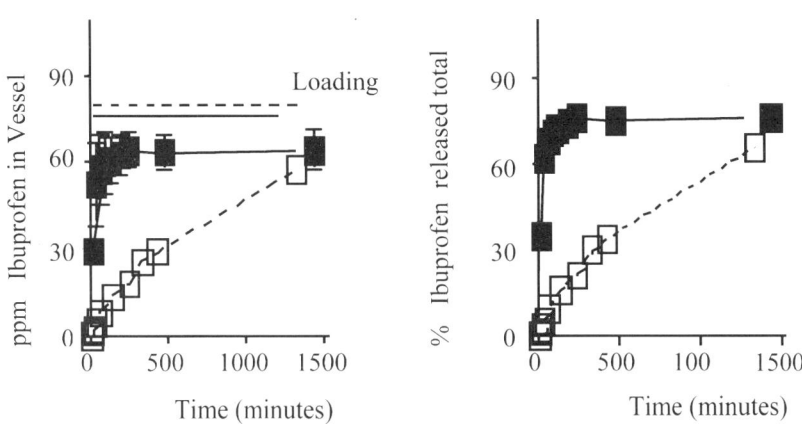

Figure 4. The dissolution rate of amorphous and crystalline blends of ibuprofen with CMCAB in pH 6.8 phosphate buffer at 37 °C.

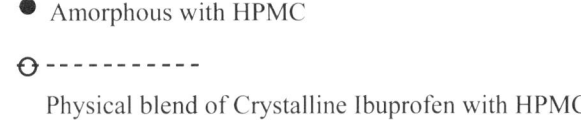

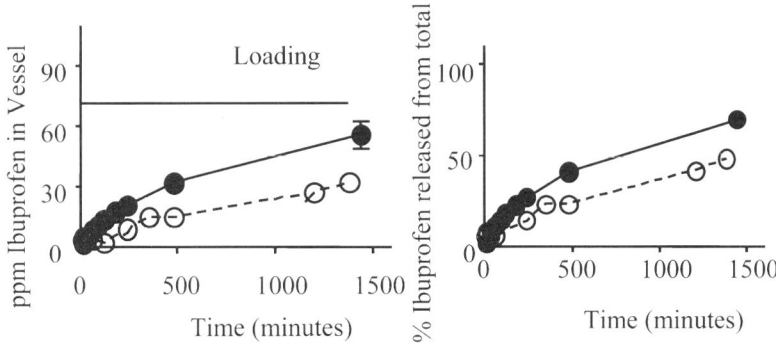

Figure 5. Ibuprofen release from amorphous and crystalline (physical) blends with HPMC in pH 6.8 phosphate buffer at 37 °C.

Glyburide Solid Dispersions

Glyburide is an important therapeutic agent for Type 2 diabetes mellitus, a malady of growing importance in the developed world.[32] It is weakly acidic (pK$_A$ 5.3) and poorly soluble in water (*ca.* 3 µg/mL), partly due to its high crystallinity (m.p. 172-174°C, log P value 4.8 at neutral pH). The poor solubility of glyburide translates directly into poor bioavailability, for example 14.7 % in dogs.[33] We prepared glyburide/CMCAB solid dispersions by two methods, co-precipitation and spray drying. The polyethylene glycol (Mw 1000) ester of α-tocopherol succinate (referred to hereafter by its commercial name, Vitamin E TPGS, Figure 6) can help to solubilize insoluble drugs and is a modestly potent inhibitor of P-glycoprotein-mediated efflux transport.[34] Since it is also a miscible plasticizer for CMCAB, we included Vitamin E TPGS as a functional additive/plasticizer in one of the spray dried samples. For comparison, one physical blend of glyburide and CMCAB was prepared, along with two glyburide/HPMCAS solid dispersions (prepared by co-precipitation). In all cases, DMSO was required as a co-solvent due to the limited solubility of glyburide in less polar solvents (*e.g.*, acetone). Complete removal of DMSO from the glyburide solid dispersions was extremely difficult. Additionally, the use of DMSO as a co-solvent limited the options for solid dispersion preparation (in particular, film casting under ambient conditions was not acceptable since complete removal of the DMSO was difficult). Glyburide and DMSO levels in the solid dispersions were determined by HPLC. The glyburide and DMSO levels in the solid dispersions varied depending on the method of preparation. Levels of residual DMSO ranged from < 0.1 wt % to 10.5 wt %, with higher levels observed in the spray dried samples. Experimental results on glyburide release are presented in Figures 7-8. Co-precipitated CMCAB solid dispersions contained a significant amount of crystalline glyburide (approximately 27 %), while similarly prepared solid dispersions in HPMCAS contained no detectable levels of crystalline glyburide. The total amount released from the solid dispersions was roughly equal to the percent amorphous glyburide in the blend, with the exception of the solid dispersion produced by spray drying. The cause of this phenomenon is not understood and could warrant future investigation. The amorphous blends with CMCAB or with HPMCAS afforded large increases in glyburide solubility, and supersaturated solutions which remained stable with respect to glyburide re-precipitation over the course of the experiment. This suggests that the cellulosic polymers not only stabilize amorphous glyburide in the solid state, but also help stabilize dissolved glyburide in supersaturated solutions. The mechanism of this solution stabilization of the amorphous drug by the polymer is not well understood. Residual DMSO leads to changes in glyburide dissolution profiles. Unfortunately, our data fails to demonstrate a consistent influence, be it positive or negative, on the dissolution rate. In some cases, high residual level of DMSO improved drug release and in others lower release rates were observed (not all data is presented in this report). Vitamin E TPGS had minimal impact on the amount of glyburide dissolved (Figure 7). The blends containing Vitamin E TPGS had higher amounts of crystalline glyburide, probably due to the enhanced solid state mobility of the plasticized polymer.

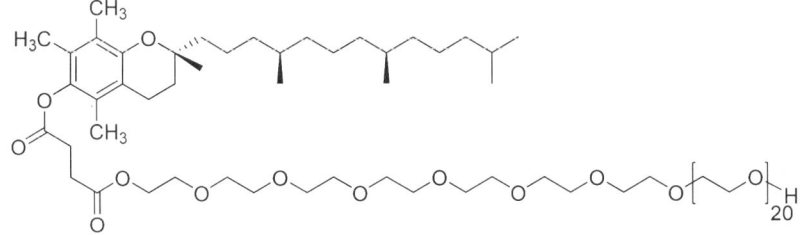

Figure 6. Structure of PEG α-Tocopherol Succinate (Vitamin E TPGS)

Spray Dried

■—With CMCAB   ▲—With Vit E TPGS/CMCAB   ⊖··Crystalline·

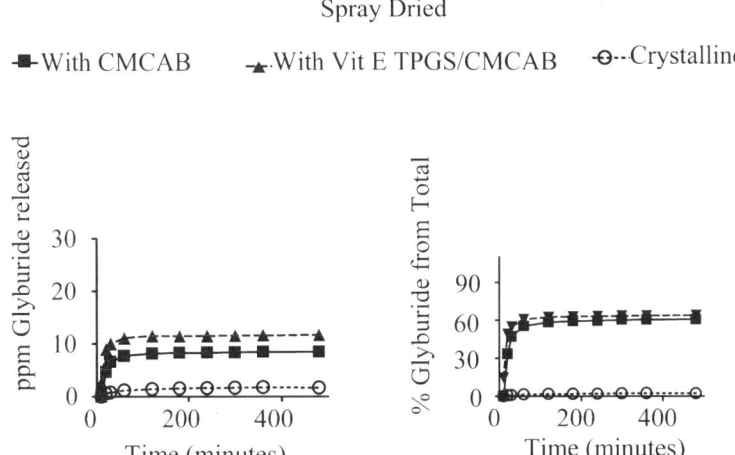

Figure 7. Dissolution of Crystalline and Amorphous Glyburide from Solids Dispersions with CMCAB at 37 °C in phosphate buffer at pH 6.8.

·⊖·····Crystalline·   ▲—High DMSO   △·Low DMSO·-

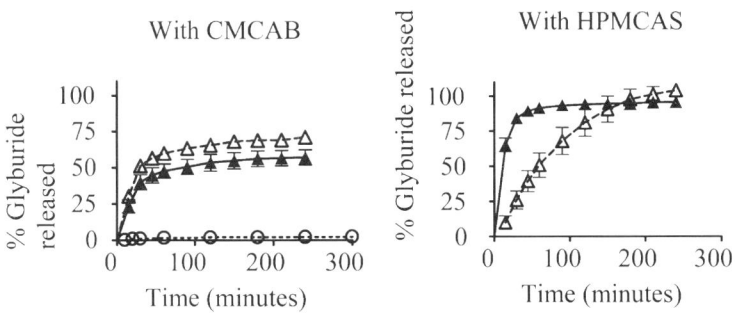

Figure 8. % Glyburide released from CMCAB and HPMCAS blends using two different methods of sample preparation, in pH 6.8 phosphate buffer at 37 °C.

Griseofulvin Solid Dispersion

Griseofulvin is an antifungal drug that has been in use since the 1960s, and is still therapeutically important.[35] It has very poor aqueous solubility (in the range of 12 μg/mL),[36] and a high melting point (220°C). It is somewhat unusual among marketed drugs in that it has neither amino nor carboxyl functionality. These characteristics make griseofulvin a particular challenge for drug delivery.

Completely amorphous blends of griseofulvin with CMCAB, HPMCAS, and PVP were prepared by solvent evaporation (rotary evaporation) or co-precipitation from blended solutions, griseofulvin in methylene chloride and polymer in either acetone (CMCAB, HPMCAS) or acetonitrile (PVP). In contrast, casting films from the same solutions gave a completely amorphous blend only with CMCAB (see Table IV). Co-precipitation of griseofulvin/PVP solid dispersions was not practical since the high water solubility of PVP dramatically reduced the yield from a co-precipitation process.

There are subtle differences in the griseofulvin release profiles based on the polymer choice and the method of solid dispersion preparation (see Figures 9 – 11). Initial release from solid dispersions in HPMCAS or PVP was rapid, but soon leveled off. Release of griseofulvin from solid dispersions in CMCAB followed a more linear profile. In particular, CMCAB-based solid dispersions of griseofulvin lacked the initial rapid release period demonstrated by the HPMCAS- and PVP-based solid dispersions of griseofulvin. One can imagine using such subtle differences in release profiles as a way to tailor the performance of the drug formulation.

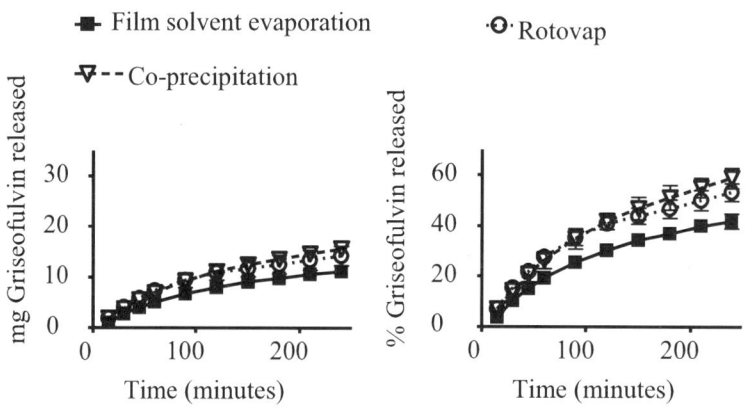

Figure 9. Griseofulvin release from CMCAB solid dispersions (37 °C, pH 6.8 phosphate buffer).

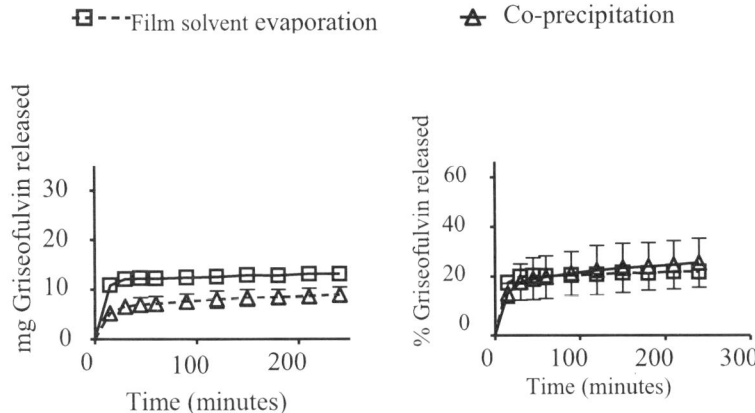

Figure 10. Griseofulvin release from HPMCAS solid dispersions vs. isolation methods (37 °C, pH 6.8 phosphate buffer).

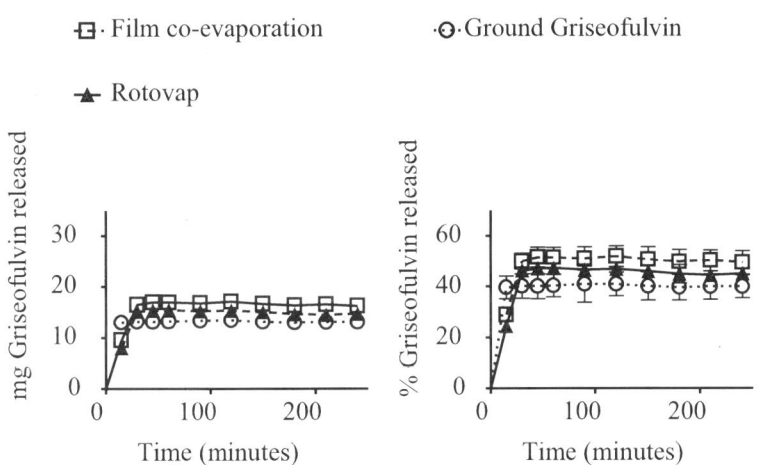

Figure 11. Release from PVP solid dispersions vs. pure griseofulvin (37 °C, pH 6.8 phosphate buffer).

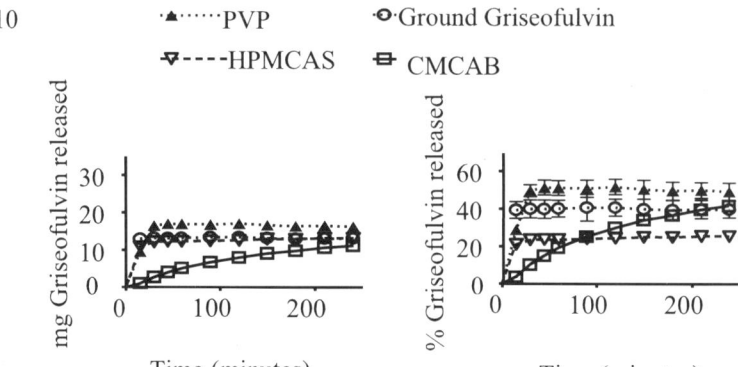

Figure 12. Dissolution from solid dispersion films vs. matrix polymer and pure griseofulvin control (37 °C, pH 6.8 phosphate buffer)

Incorporation of surfactants into CMCAB/griseofulvin solid dispersions improves the dissolution of griseofulvin (Figure 13). CMCAB is particularly compatible with surfactants, having both polar and non-polar groups itself. In the presence of Tween 80, concentrations of griseofulvin up to 30 mg/mL were obtained, in the range of 2500 times the solubility of crystalline griseofulvin.

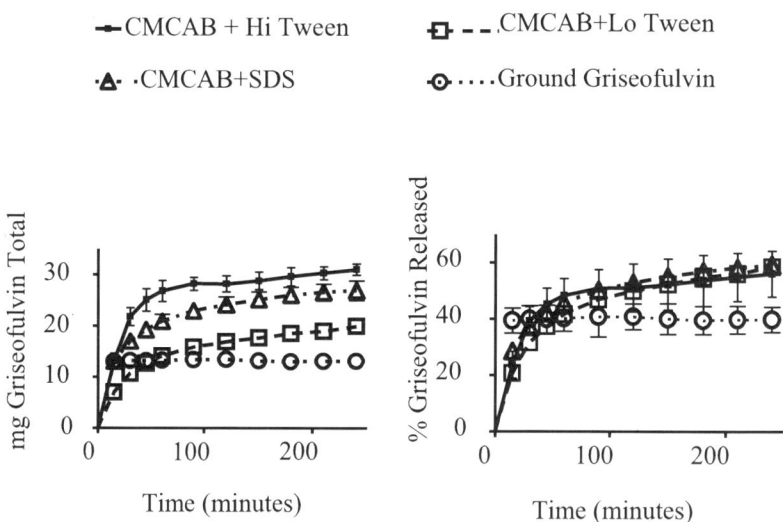

Figure 13. Dissolution of griseofulvin from CMCAB solid dispersions containing surfactants (37 °C, pH 6.8 phosphate buffer).

Fexofenadine Solid Dispersion

Fexofenadine HCl is a highly soluble, but poorly absorbable drug according to the BCS system. Even though the solubility of fexofenadine HCl in water is quite high, 1000 mg/L, its solubility in USP standard phosphate buffer (0.05M) at pH 6.8 is only 0.35 mg/L, meaning that fexofenadine HCl is sparingly soluble under these conditions. The reduction in fexofenadine HCl solubility in water with significant ionic content may be due to the zwitterionic nature of fexofenadine, in which the carboxylic acid can internally protonate the amine.

We found that only 50 % of the crystalline fexofenadine HCl dissolved at pH 6.8 from a physical blend with CMCAB, while 100 % of the fexofenadine HCl from an amorphous CMCAB/fexofenadine solid dispersion dissolved within the same time period, under the same conditions (see Figure 14). Amorphous blends of fexofenadine in CMCAB not only sharply enhance fexofenadine solubility, but afford enhanced solution stability (*i.e.*, little or no fexofenadine crystallization over time). Dissolution from the experimental physical blends gave fexofenadine concentrations well below the intrinsic solubility limit, (80 μg/mL vs. a 350 μg/mL limit). The reason for the observed lower solubility from these physical blends is unclear.

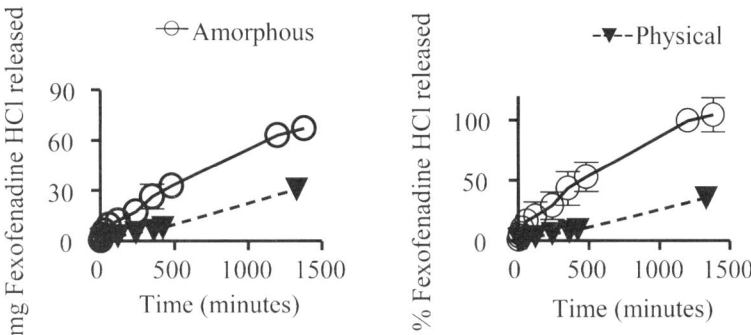

Figure 14. Dissolution of physical and amorphous blends of fexofenadine HCl and CMCAB, in SIF$_{sp}$ (pH 6.8, 37 °C).

Conclusion

CMCAB is a potentially valuable tool for pharmaceutical formulation scientists, and provides numerous advantages over other polymeric matrices commonly incorporated in pharmaceutical solid dispersions. CMCAB can be readily incorporated into pH-controlled, slow-release, often zero-order matrix formulations by the simple technique of physical blending with the drug. These formulations are effective even with drugs of low solubility. Amorphous matrices of CMCAB and drug can be made from solutions of drug and polymer, by a number of techniques including co-precipitation, spray-drying and film casting. This methodology is made easier by the broad solubility of CMCAB in organic solvents.[9] Differences in solubility between the polymer and drug in an amorphous matrix formulation often limit the number of processes applicable to preparation of the solid dispersion. Amorphous drugs in solid dispersions of CMCAB have greatly enhanced solubility as well as very fast release. The contrast between physical and molecular blends of drug and CMCAB is often remarkable. Additionally, functional additives such as TPGS and surfactants can readily be incorporated into the amorphous matrices. Caution should be exercised, since the added molecular mobility of the drug in the presence of the plasticizer could enhance drug crystallization in the solid state. CMCAB is observed to enhance the stability of the amorphous drug with respect to crystallization not only in the solid state, as expected, but also in solution. This result is of particular practical value, and there is little literature precedent that would guide understanding of the phenomenon and its mechanism. Studies of the mechanism of this solution stabilization, the structure-property relationships that govern it, and the impact of polymer aqueous solubility on stabilization are a particular focus for our current research. It is clear from the data in this and our previous publications that CMCAB is a highly promising polymer for drug delivery, and an important lead for design and synthesis of polysaccharide derivatives for enhanced delivery of poorly soluble drugs.

Acknowledgement

The authors wish to express their appreciation to Professor Jennifer Dressman, who provided valuable consultation during the selection of model drugs for this study.

References

1   DiMasi, J.A., *Clin. Pharm. & Ther.*, **2001**, *69*, 297-307.
2   Lipinski, C.A., *J. Pharm. Tox. Methods*, **2000**, *44*, 235-249.
3   Pace, S.N.; Pace, G.W.; Parikh, I.; Mishra, A.K. *Pharm. Tech.* **1999**, *23*, 116-134.
4   Horter, D.; Dressman, J.B., *Adv. Drug Del. Rev.*, **2001**, *46*, 75-87.
5   Yu, L.X.; Amidon, G.L.; Polli, J.E.; Zhao, H.; Mehta, M.U.; Conner, D.P.; Shah, V.P.; Lesko, L.J.; Chen, M.-L.; Lee, V.H.L.; Hussain, A.S., *Pharm. Res.* **2002**, *19*, 921-925.
6   Wu, C.-Y.; Benet, L.Z. *Pharm. Res.* **2005**, *22*, 11-23.
7   Custodio, J.M.; Wu, C.-Y.; Benet, L.Z. *Adv. Drug Del. Rev.* **2008**, *60*, 717-733.
8   Leuner, C.; Dressman, J. *Eur. J. Pharm. Biopharm.* **2000**, *50*, 47-60.
9   Merisko-Liversidge, E.; Liversidge, G.G.; Cooper, E.R. *Eur. J. Pharm. Sci.* **2003**, *18*, 113-120.
10 Szejtli, J. *Chem. Rev.* **1998**, *98*, 1743-1753.
11 Constantinides, P.P.; Han, J.; Davis, S.S. *Pharm. Res.* **2006**, *23*, 243-255.
12 Zhou, D.; Zhang, G.G.Z.; Law, D.; Grant, D.J.W.; Schmitt, E.A. *Mol. Pharm.* **2008**, *5*, 927-936.

13 Allen, J.M.; Wilson, A.K.; Lucas, P.L.; Curtis, L.G. *U.S. Patent 5,668,273* **1997**.

14 Allen, J.M.; Wilson, A.K.; Lucas, P.L.; Curtis, L.G. *U.S. Patent 5,792,856* **1998**.

15 Posey-Dowty, J.D.; Wilson, A.K.; Curtis, L.G.; Swan, P.M.; Seo, K.S. *U.S. Patent 5,994,530* **1999**.

16 Posey-Dowty, J.D.; Seo, K.S.; Walker, K.R.; Wilson, A.K. *Surface Coatings Intl. Part B: Coat. Trans.*, **2002**, *85*, 203-208.

17 Posey-Dowty, J.D.; Watterson, T.L.; Wilson, A.K.; Edgar, K.J.; Shelton, M.C.; Lingerfelt, L.R., Jr. *Cellulose*, **2007**, *14*, 73-83.

18 Edgar, K.J. *Cellulose*, **2007**, *14*, 49-64.

19 Hancock, B.C.; Parks, M. *Pharmaceutical Research*, **2000**, 17(4), 397-404.

20 Konno, H.; Taylor, L.S. *Pharm. Res.* **2007**, *25*, 969-978.

21 Florence, A.T., Atwood, D., **1998**. Physicochemical Principles of Pharmacy, 3rd ed. MacMillan Press, London.

22 Gines, J.M.; Arias, M.J.; Moyano, J.R.; Sanchez-Soto, P.J. *Int. J. Pharm.* **1996**, *143*, 247-253.

23 Marsac, P.J.; Konno, H.; Rumondor, A.C.F.; Taylor, L.S. *Pharm. Res.* **2007**, *25*, 647-656.

24 Curatolo, W.J.; Nightingale, J.A.S.; Shanker, R.M.; Sutton, S.C. *U.S. Patent 6,548,555*, **2003**.

25 Wilson and Gisvold's Textbook of Organic Medicinal and Pharmaceutical Chemistry; Block, J. H. and Beale, J. M., Eds; Lippincott Williams and Wilkins: Philadelphia, PA, 2004; pp 948-956.

26 Kasim, N.A.; Whitehouse, M.; Ramachandran, C.; Bermejo, M.; Lenneräs H. Hussain, A.S.; Junginger, H.E.; Stavchansky, S.A.; Midha, K.K.; Shah, V.P.; and Amidon, G.L. **2004**, *Mol. Pharm.* 1(1)85-96.

27 Salem, M.S.; Najib, N.M.; Hassan, M.A.; and Suleiman, M.S. *Acta Pharmaceutica Hungarica 1997*, 67,13-17.

28 Fujioka, Y.; Kadano, K.; Fujie, Y.; Metsugi, Y.; Ogawara, K.; Higakik K.; Kimura, T. *J. Controlled Release* **2007**, 119(2), 222-8.

29 Nicolaides, E.; Galia, E.; Efthymiopoulos, C.; Dressman, J.; Reppas, C. *Pharm. Res.* **1999**, *16*, 1876-1882.

30 Theeuwes, F., *Drug Dev. Ind. Pharm.* **1983**, *9*, 1331-1357.

31 Makhija, S.N.; Vavia, P.R., *J. Contr. Rel.* **2003**, *89*, 5-18.

32 WHO Fact Sheet 312, http//www.who.int/mediacentre/factsheets/fs312/en, Geneva, Switzerland, **2006**.

33 Savolainen, J.; Jarvinen, K.; Taipale, H.; Jarho, P.; Loftsson, T.; Jarvinen, T., *Pharm. Res.* **1998**, *15*, 1696-1701.

34 Collnot, E.-M.; Baldes, C.; Wempe, M.F.; Kappl, R.; Hüttermann, J.; Hyatt, J.A.; Edgar, K.J.; Schaefer, U.F.; Lehr, C.-M. *Mol. Pharm.* **2007**, *4*, 465-474.

35 Bennett, M.L.; Fleischer, A.B., Jr.; Loveless, J.W.; Feldman, S.R. *Pediatric Derm.* **2000**, *17*, 304-309.

36 Trotta, M.; Gallarate, M.; Carlotti, M.E.; Morel, S. *Int. J. Pharm.* **2003**, *254*, 235-242.

Chapter 6

Modification of cellulose in ionic liquids towards biomedical applications

T. Liebert, J. Wotschadlo, M. Gericke, S. Köhler, P. Laudeley, T. Heinze[*]

Centre of Excellence for Polysaccharide Research, Friedrich Schiller University of Jena, Humboldtstraße 10, D-07743 Jena, Germany, *Member of the European Polysaccharide Network of Excellence, EPNOE (www.epnoe.eu)

Ionic liquids (ILs) are valuable media for the defined regeneration and the homogeneous functionalization of cellulose towards new materials for biomedical applications. It was possible to prepare highly porous cellulose membranes from cellulose dissolved in 1-ethyl-3-methylimidazolium acetate (EMIMAc). These membranes are useful for the separation of lysozyme from protein mixtures. Furthermore, the homogeneous functionalization of cellulose in the ILs 1-butyl-3-methylimidazolium chloride (BMIMCl) and EMIMAc in the presence of co-solvents yielded defined cellulose sulfates and trimethylsilyl celluloses. The sulfates prepared in this manner are suitable for manufacture of capsular polyelectrolyte complexes for the encapsulation of proteins such as glucose oxidase, which is still active in the capsule. Using the silylated derivatives pure cellulose particles in the nanoscale range were accessible. They are potential carrier materials or contrasting agents.

The polysaccharide cellulose and its derivatives are well-suited for a number of biomedical applications because of their biocompatibility and their tendency to form defined supramolecular architectures. To exploit these features, tailoring of the structures on both the molecular and supramolecular levels is necessary. For the majority of these defined shaping and chemical modification methods, homogeneous conditions are indispensable. Commercially applied methods for cellulose dissolution such as the viscose and the cupro process or the application of *N*-methylmorpholine-*N*-oxide are of limited utility because the systems applied contain water and the regeneration is usually carried out in aqueous media. In the case of chemical modification, this leads to side reactions such as hydrolysis. Thus, numerous non-aqueous cellulose solvents were developed over the last three decades.

N,*N*-dimethyl acetamide (DMA)/LiCl is considered as the most valuable, non-aqueous solvent system for homogeneous chemical modification of cellulose up to now, especially for tailored esterification reactions (1-3). It was demonstrated that esterification in this solvent leads to a broad variety of derivatives not accessible via heterogeneous methods (3,4). Nevertheless, the solvent is expensive and a recycling of this two-component medium is rather complicated hindering large scale utilization. The same is true for the fairly new solvents dimethyl sulfoxide (DMSO)/tetrabutylammonium fluoride trihydrate (5-7) and the isolated organo-soluble cellulose intermediates, *e.g.* cellulose formates (8), which are reasonable only for lab scale experiments. Therefore, alternative media for the dissolution of cellulose are necessary. One exciting development in this context is the utilization of ionic liquids (ILs).

The first report on salt-like cellulose solvents was published by Greanacher (9) who applied pyridinium salts. In 2002 the use of ILs as cellulose solvents was revived (10,11). It was shown that the most promising ILs for the modification of cellulose are 1-alkyl-3-methylimidazolium salts. Most of these substances melt below 100°C. Some of them are liquid at room temperature. These water-free liquids consist completely of ions. In addition to the non-aqueous conditions both for the regeneration and the chemical conversion, ILs exhibit a huge structural diversity (12,13). Therefore, tailoring of the solvent properties is possible.

With this paper we intend to discuss opportunities created by ILs for the preparation of materials suitable for biomedical applications, including the preparation of membranes and nanoparticles of cellulose useful for the isolation of proteins and as carrier materials, as well as the homogeneous sulfation of cellulose using ILs leading to low substituted, water soluble cellulose sulfates for polyelectrolyte complex (PEC) formation.

Results and Discussion

Preparation of cellulose membranes from ILs for protein separation

Among the first applications of ionic liquids (ILs) in the field of polysaccharide research was the regeneration of cellulose mainly for the production of fibers (11,14). In contrast, the film formation from ILs is scarcely studied although their usefulness for the preparation of biocompatible membranes or loaded membranes with biological activity was demonstrated (15-17). Cellulose membranes are valuable materials for medical and biotechnological applications. Fractionation of biological fluids, such as blood, is of increasing interest. Cellulose membranes are already well established in the field of dialysis. Most of the membranes used are produced via the viscose process or the cupro (aqueous ammoniacal solution of copper (II) hydroxide) process (18). They are usually homogeneous materials with a defined pore size of about 10 nm (see Figure 1a). It is known that the fractionation characteristics and the biocompatibility of such cellulosic materials are strongly influenced by the surface properties (19). In case of ILs these surface properties can be largely affected by the precipitation step because a coagulation of the cellulose in a broad variety of solvents and solvent mixtures is possible even under completely anhydrous conditions. The formation of very porous structures with a large surface should be possible using this approach. It was intended to use the membranes for the selective isolation of lysozyme, an antibacterial protein, from protein mixtures. Lysozyme has a binding unit comparable to that of cellulase (20) and should bind efficiently to cellulose chains if a large surface is provided. Only a small retention is observed using commercial cellulose-based dialysis membranes, perhaps due to their smooth surfaces (21).

In a first set of experiments cellulose was regenerated from ILs by precipitation in various organic liquids. Membranes with a highly porous structure were obtained using 1-ethyl-3-methylimidazolium acetate (EMIMAc) as solvent and ethanol as precipitating agent. The films were stable and can be dried without undesired changes. A scanning electron microscopy (SEM) image of a cellulose membrane prepared in this manner is shown in Figure 1 (right). A clear difference is visible between the surface topographies of films obtained from the viscose process and those obtained by regeneration from the IL. The surface roughness of IL-regenerated films is in the range of 70 to 80 nm as revealed by atomic force microscopy (AFM) (Figure 2).

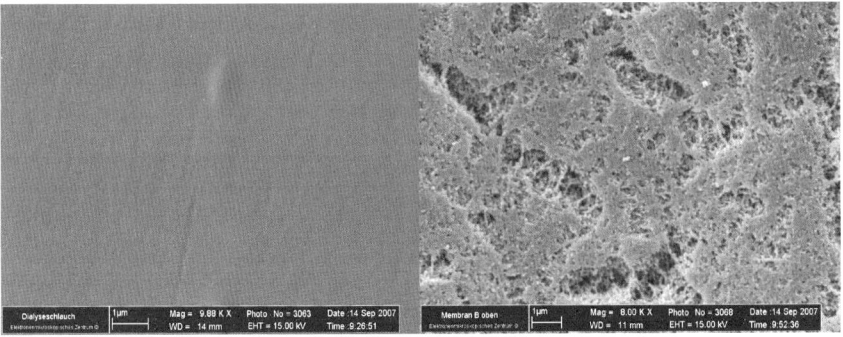

Figure 1. Scanning electron microscopy images of cellulose membranes prepared via the viscose process (left) and prepared by dissolution of cellulose in EMIMAc and precipitation with ethanol (right)

118

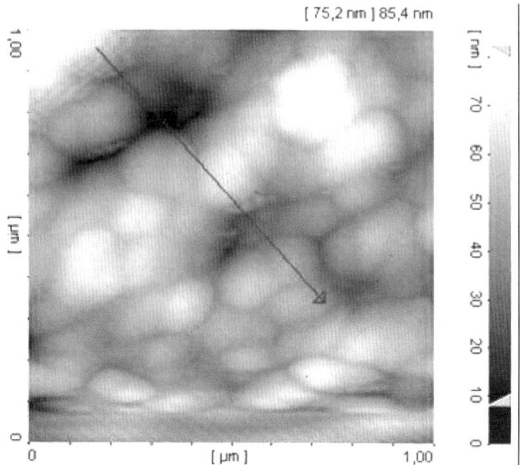

[75,2 nm] 85,4 nm

Figure 2. Atomic force microscopy image of a cellulose membranes obtained from a EMIMAc solution by precipitation with ethanol

Specific binding of proteins such as lysozyme, ovalbumin and bovine serum albumin (BSA) onto these porous layers was investigated. Figure 3 shows cellulose membranes prepared from EMIMAc after treatment with protein solutions, washing and staining with Coomassie blue. Only in the case of lysozyme was staining of the membrane observed, indicating protein attachment (Samples 1 and 2 in Figure 3). This reveals that lysozyme can bind to the material whereas the other proteins (BSA, ovalbumin) do not show adhesion and could be separated by dialysis on these materials according to molecular mass only.

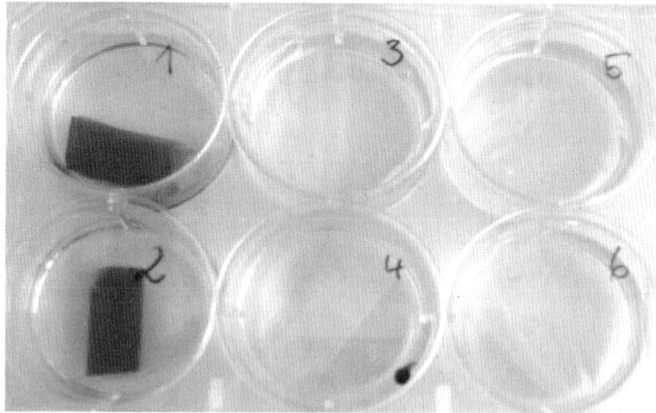

Figure 3. Cellulose membranes prepared from EMIMAc after treatment with protein solutions, washing and staining of the membranes with Coomassie blue: Sample 1 and 2 lysozyme; sample 3 and 4 ovalbumin and sample 5 and 6 bovine serum albumin

As an evidence for the fact that this loading of cellulose is due to a specific binding on readily accessible cellulose chains, experiments with cellulose acetate (degree of substitution, DS=2.5) based membranes were carried out. It was observed that the lysozyme is no longer fixed to the membrane surface and can be found in the washing solution, which might be due to the modification of the cellulose polymer preventing specific recognition by the protein. Surprisingly, these more hydrophobic surfaces retain ovalbumin. It is not yet clear if this binding is due to a recognition effect or if it is caused by hydrophobic-hydrophobic interaction between the polymers.

These experiments on the preparation of readily accessible cellulose membranes from ILs illustrate that the new materials can be a suitable tool for defined protein absorption. The phenomenon of selective binding is now being studied for protein isolation from protein mixtures obtained for example from fermentation media, which could permit easy protein fractionation, lead to new analytical methods based on fractionation, or yield loaded materials with antibacterial properties.

Sulfation of cellulose in ILs and preparation of polyelectrolyte complexes (PEC)

Besides the defined regeneration of cellulose under non-aqueous conditions, ILs offer the opportunity of homogeneous functionalization reactions, including sulfation of cellulose. Water soluble cellulose sulfates (CSs) are the material of choice for several medical and biotechnological applications because of their beneficial properties such as anticlotting behaviour (20). In combination with polycations such as poly(dimethyldiallylammonium chloride) (PolyDADMAC) the polyanion CS forms capsular shaped polyelectrolyte complexes (PEC) that possess great potential for the encapsulation and immunoisolation of biological matter (23). This is a valuable approach for the encapsulation of cells and proteins. Cells, which are included into PEC, are well protected against detection and attack by the immune system (24) and therefore are expected to be suitable for xenotransplantation, for example in diabetes therapy. It was shown that, besides other cell types, insulin producing porcine islet cells can be entrapped in CS/PolyDADMAC symplex capsules. Neither cell growth nor glucose-dependent insulin production are influenced by the encapsulation (25). Enzymes entrapped into PEC capsules show higher process stability and reusability in comparison to non-encapsulated enzymes and can easily be removed and recycled after reaction (26, 27).

The preparation of PEC requires CS with complete water solubility. Furthermore, the average degree of substitution (DS) is limited to a maximum of about 0.65 (28). At higher DS values the stability of the PEC is significantly reduced due to the larger density of negative charges attached to the polymer chain. Suitable CS with low DS and water solubility has to be prepared under homogeneous conditions to guarantee a statistical (even) distribution of substituents along the polymer chain. Moreover, the sulfation should not be accompanied by significant degradation of the polymer during the reaction, to assure a sufficient viscosity of the CS solutions for the PEC preparation.

During the past three decades several pathways for the synthesis of CS have been developed (2). Heterogeneous sulfation is not feasible for commercial applications because of strong chain degradation and poor water solubility of the resulting CS at low DS. CS with DS of about 0.3 can be prepared homogeneously in N_2O_4/N,N-dimethyl formamide (DMF) (29) but the toxic nature of the system prohibits the application of this synthetic route in non-laboratory scale, especially for medicinal use. Several other cellulose solvents including N-methylmorpholine N-oxide or LiCl/ N,N-dimethyl acetamide (DMA) have also been tested for the homogeneous sulfation of cellulose but showed coagulation of the reaction medium yielding CS only partial soluble in water (30). Sulfation *via* cellulose derivatives, in particular cellulose acetate and trimethylsilyl cellulose (TMSC), and subsequent cleavage of the acetate or TMS function is a valuable route for the preparation of CS (32-33). The major disadvantages here are the requirement of larger amounts of chemicals and the effort necessary both for the reaction and purification process. Consequently, there is still a strong demand for new, simple methods for the preparation of water-soluble CSs with low DS. Thus, homogeneous sulfation of cellulose dissolved in ILs was investigated using SO_3-pyridine, SO_3-DMF and chlorosulfonic acid as reagents. Because of the necessity for mild reaction conditions, the room temperature liquid EMIMAc appeared to be an excellent solvent for the sulfation of cellulose. Surprisingly, during the reaction with different SO_3-complexes precipitation occurred, although CS and the SO_3-complexes are well soluble in EMIMAc. The resulting product had a rather low sulfur content (below 1%) and was water-insoluble. FT-IR spectra did not show signals at about 1250 and 815 cm^{-1} corresponding to a sulfate group attached to cellulose. In contrast, a signal at 1734 cm^{-1} was observed, which is characteristic for a carbonyl moiety. Additionally, a ^{13}C NMR spectrum (Figure 4) of the isolated product dissolved in DMSO-d_6/LiCl revealed peaks at 170.8 and 21.2 ppm. These findings indicate the formation of cellulose acetate instead of CS (Figure 5).

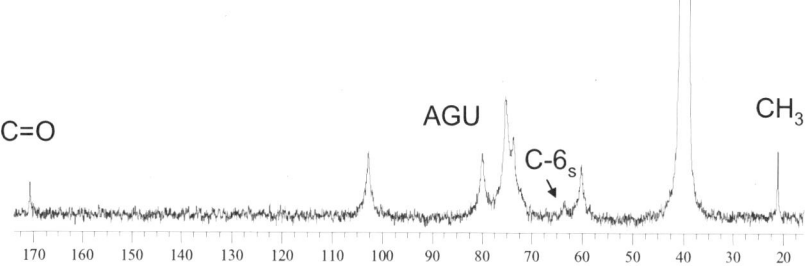

Figure 4. ^{13}C NMR spectrum of the reaction product obtained by conversion of cellulose with SO_3-DMF in EMIMAc

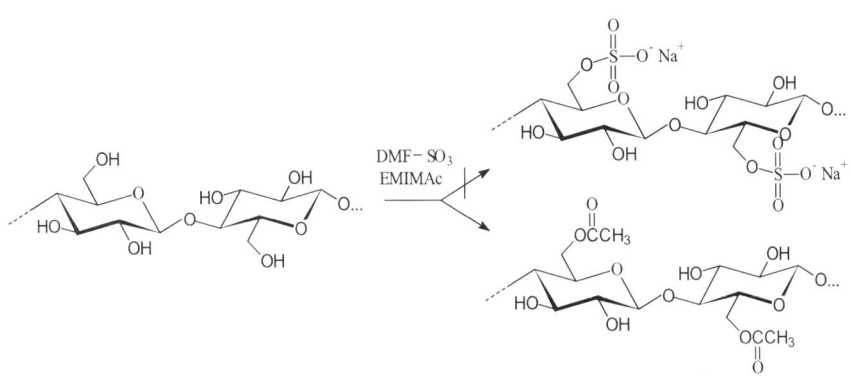

Figure 5. Schematic plot of the expected (upper part) and the found reaction (lower part) of cellulose in EMIMAc with SO₃ -DMF

Recently, it has been reported that the acetyl ion in EMIMAc might undergo several side reactions during the acylation, tritylation and tosylation of cellulose (34). According to the spectral data, a similar side reaction can be assumed for the sulfation of cellulose in EMIMAc. Consequently, 1,3-dialkyl imidazolium based ILs bearing acetate as counter ion are not proper solvents for the homogeneous cellulose sulfation.

In a next set of experiments sulfation of cellulose in 1-butyl-3-methylimidazolium chloride (BMIMCl) was attempted. When cellulose/BMIMCl solutions were cooled down to room temperature, they slowly solidified at about 45-50°C. Sulfation experiments in such highly viscous pasty mixtures at 25°C yielded rather high DS values of up to 0.89 but the resulting CS were not water soluble. This poor water solubility is most likely due to a non-uniform distribution of the sulfate groups along and between the cellulose chains caused by the heterogeneous conditions. Homogeneous sulfation with SO₃-DMF at 60°C resulted in black reaction mixtures and complete degradation of the cellulose. Water soluble CS was prepared at elevated temperatures using SO₃-pyridine. Nevertheless, significant decomposition of the polymer chain was concluded from the resulting low viscosity (1.47 mPa·s) of a 1 wt.-% aqueous solution of CS. It is known that after dissolution of cellulose, cellulose/IL solutions can be diluted with dipolar aprotic solvents, *e.g.* DMF or dimethyl sulfoxide (DMSO), without precipitation of cellulose (35). In contrast to cellulose/BMIMCl solutions, their mixtures with DMF and DMSO remained liquid upon cooling to room temperature. Consequently, reactions were carried out in the presence of co-solvents to reduce the solution viscosity and guarantee a better mixing with the reagents, and to permit lower reaction temperature while maintaining liquidity. Sulfation reactions were carried out in solutions containing 5 g cellulose, 45 g BMIMCl and 70 ml DMF. Cellulose was dissolved in BMIMCl at 80°C. Co-solvent was added and the mixture was stirred while cooling down to room temperature. Sulfation with SO₃-complexes at different molar ratios of anhydroglucose unit (AGU) to sulfating agent (see Table 1) proceeded completely homogeneously over the whole reaction time. Isolation of CS was possible by addition of 70 ml of 1 M NaOH followed by 160 ml water and precipitation of this clear solution into 1 l isopropanol/water (90:10).

In the preliminary experiments CSs were prepared from microcrystalline cellulose (MC; $[\eta]_{Cuen}= 128$ cm^3/g; MW$_{Cuen}$ = 61600 g/mol determined according to ref. 36) and had rather low solution viscosities. Therefore sulfation of spruce sulfite pulp (SSP; $[\eta]_{Cuen}$ 435 cm^3/g; MW$_{Cuen}$ = 254700 g/mol) cellulose in BMIMCl/DMF was investigated. The results are summarized in Table I.

The DS can be simply controlled by variation of the amount of sulfating reagent. SO$_3$-pyridine and chlorosulfonic acid showed comparable reactivity while SO$_3$-DMF was slightly less reactive. The sulfation of cellulose in BMIMCl/DMF proceeded very fast. It is usually completed within 30 min. Prolongation to 4 h did not change the DS values significantly while after 24 h a slight decrease of 20% was observed (data not shown). Above a DS of about 0.25 the CSs prepared in BMIMCl/DMF dissolved completely in water and therefore fulfilled one precondition for the preparation of PEC capsules.

Table I. DS values and water solubility of cellulose sulfates obtained by sulfation of spruce sulfite pulp dissolved in BMIMCl/DMF for 2 h at 25°C

No	Reaction conditions		Product	
	Sulfating agent (SA)	SA/AGU[a]	DS	Water solubility[b]
1	SO$_3$-pyridine	0.7	0.14	-
2	SO$_3$-pyridine	0.8	0.25	+
3	SO$_3$-pyridine	0.9	0.48	+
4	SO$_3$-pyridine	1.1	0.58	+
5	SO$_3$-pyridine	1.2	0.62	+
6	SO$_3$-pyridine	1.4	0.81	+
7	SO$_3$-DMF	1.0	0.34	-
8	SO$_3$-DMF	1.4	0.64	+
9	SO$_3$-DMF	1.5	0.78	+
10	HSO$_3$Cl	1.0[c]	0.49	+
11	SO$_3$-pyridine	1.1[d]	0.41	+

[a] Anhydroglucose unit, [b] + Soluble, - Insoluble, [c] Reaction time 3 h, [d] Recycled IL used

The high molecular weight of the CS and the formation of aggregates even in aqueous solution complicated size exclusion chromatographic (SEC) experiments (37). Initial results suggest that unlike several other methods described for the preparation of CS, the sulfation of cellulose in BMIMCl/DMF at 25°C does not lead to significant polymer degradation. For sample 6 a molecular weight of 310,000 g/mol was determined. The aqueous solutions (2 wt.-% in physiological saline solution) obtained from the CS prepared in BMIMCl/DMF using SO$_3$-pyridine had rather high viscosities, ranging from 93.1 mPa·s (sample 6) to 374.6 mPa·s (sample 3), which supports the assumption that only slight polymer degradation occurs during the sulfation.

A ^{13}C NMR spectrum of CS with DS 0.48 dissolved in D$_2$O and an assignment of the peaks is given in Figure 6. The signal at 67.3 ppm corresponds to sulfation at position 6. Peaks in the region of 82 ppm that would correspond to sulfated positions 2 or 3 are missing in the spectrum and no splitting of the C-1-signal can be observed. It demonstrates that at low DS (0.48), sulfation of cellulose in BMIMCl/DMF is regioselective, yielding the 6-O-sulfated product.

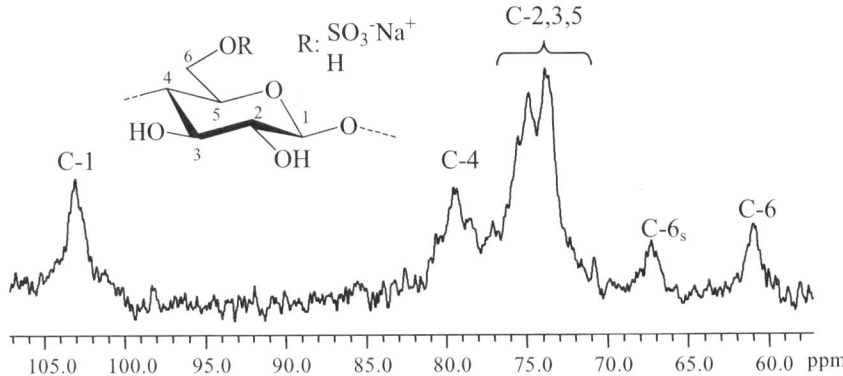

Figure 6. ^{13}C NMR spectrum of sample 3 (in D_2O) obtained in BMIMCl/DMF with SO_3-DMF

A crucial question concerning the efficiency of this sulfation procedure was the recycling of BMIMCl used as medium. After precipitation of the CS and filtration, the remaining solvent mixture was treated in vacuum to remove volatile components, mainly isopropanol. Removal of inorganic salts from this crude IL was carried out by precipitation with chloroform. Subsequently volatile compounds like water and DMF were removed by distillation at 80°C and reduced pressure ($< 10^{-2}$ mbar). Finally the ILs were freeze dried and obtained as solids with melting points of about 68°C, which is in the range of the one given for the starting BMIMCl. The recovery rate was 82%. The ^1H NMR spectrum of BMIMCl purified according to the procedure described above is represented in Figure 7.

The spectrum is identical to that of the starting IL and all signals can explicitly be assigned to the corresponding IL protons. This clearly shows that BMIMCl is stable under the sulfation conditions, with no side reactions leading to decomposition of the IL within the limits of detection of ^1H NMR spectroscopy (less than 1%).

The recycled ILs readily dissolved cellulose. Several sulfation reactions were repeated with same molar ratios but utilizing recycled IL in order to evaluate their reusability. As an example, sample 11 is given in Table I. The CS was water soluble and had a DS of 0.41, which is comparable to sample 4 prepared in commercially available BMIMCl. These results confirm that the IL used for the cellulose sulfation can be reused.

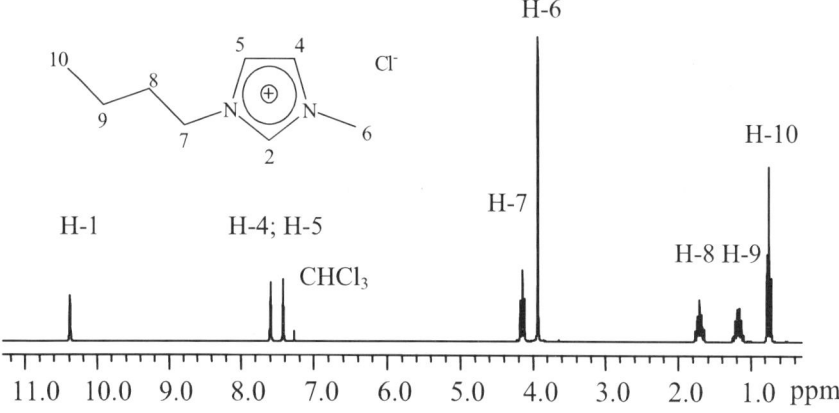

Figure 7. ^1H NMR spectrum of recycled BMIMCl (in chloroform-d$_1$)

The CSs prepared in this study were tested for their potential application as PEC-forming polyelectrolytes. Therefore, aqueous CS solutions (4 wt.-%) were dropped from a syringe into a solution of PolyDADMAC in water (4 wt.-%). All solutions contained 0.9% NaCl in order to mimic physiological conditions needed for the encapsulation of whole cells. Spherical PEC with a diameter of about 2-3 mm were formed from high molecular weight CS prepared from SSP (Figure 8), which can be easily handled without destruction.

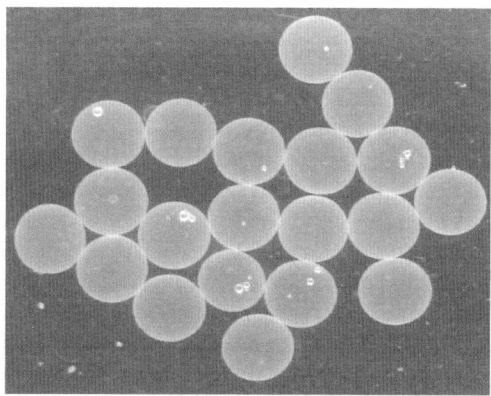

Figure 8. Capsules obtained from an aqueous solution (4 wt.-%) of sample 3 (DS 0.48) dropped into physiological saline solution (0.9% NaCl) containing 4 wt.-% PolyDADMAC

In contrast, CS prepared from MC did not yield capsular shaped membranes due to the low molecular weight of the starting cellulose and the resulting low solution viscosities of the corresponding CS solution.

In order to evaluate the feasibility of enzyme encapsulation within PEC capsules prepared from high MW-CS, glucose oxidase (GOD) was used. GOD is an enzyme which catalyzes the oxidation of β-D-glucose by molecular oxygen to δ-gluconolactone, producing gluconic acid (from lactone hydrolysis) and hydrogen peroxide. This enzyme has considerable commercial importance in the removal of D-glucose and oxygen from food and in the formation of gluconic acid (38). However, the most important application of GOD is as a biosensor for the quantitative determination of D-glucose, for example in body fluids or food products (39). For all these processes immobilization and encapsulation of GOD is desired in order to enhance reusability and operational stability. GOD containing PEC-capsules were prepared simply by dissolving the enzyme (5 mg/ml) in an aqueous CS solution (4 wt.-% in 0.9% NaCl solution) and dropping this solution into the precipitation bath containing a PolyDADMAC solution (4 wt.-% in 0.9% NaCl solution). After 30 min the formed capsules were removed, washed and stored in salin solution. The capsules had diameters of about 2-3 mm and were slightly yellow, supporting the presence of GOD. Additionally, the enzyme was detected by staining with Coomassie blue. Moreover, the activity of the protein and the suitability of the PEC capsule were verified with a color reaction using the horse radish peroxidase (HPO)/ 2,2'-azino-bis(3-ethylbenzthiazoline-6-sulphonic acid) (ABTS) system (40, Figure 9).

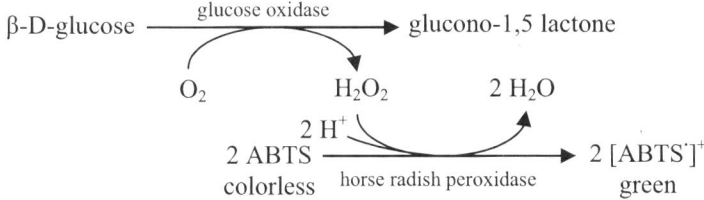

Figure 9. Mechanism for the conversion of horse radish peroxidase/ABTS in with glucose in the presence of glucose oxidase capsules, ABTS: 2,2'-azino-bis(3-ethylbenzthiazoline-6-sulphonic acid)

For this purpose, a O_2-saturated β-D-glucose/ABTS solution containing HPO was treated with a GOD-containing PEC capsule. Within 1 min a change in color was observed (Figure 10) confirming both the activity of the protein and the suitability of the PEC complex for the encapsulation. The PEC membrane is able to retain the protein (no protein leaking was observed by the Bradford method) but allows permeation of small molecules such as glucose and H_2O_2 as can be concluded from this experiment.

Figure 10. O₂-saturated β-D-glucose/2,2'-azino-bis(3-ethylbenzthiazoline-6-sulphonic acid) solution containing horse radish peroxidase before (a) and 1 min after (b) the addition of glucose oxidase capsules (see page 1 of color insert)

Preparation of cellulose particles via trimethylsilylation of cellulose

Particles of cellulose are valuable tools for a variety of biomedical applications including contrasting agents, carriers, and materials for immobilization, and in new separation methods for bioactive molecules (41). The preparation of defined small cellulose spheres is especially challenging because of the high tendency of cellulose for non-specific aggregation during regeneration steps caused by hydrogen bond formation between the chains. A reasonable path is to detour by means of shaping of hydrophobic cellulose derivatives, followed by regeneration of the cellulose while retaining the formed architecture. In this regard TMSC should be a favourable starting material for the preparation of pure cellulose particles. The solubility of the derivatives can be adjusted *via* the DS and the regeneration of pure cellulose nanostructures might even be possible by gaseous HCl as shown for Langmuir Blodgett layers (42). Therefore, we investigated the preparation of TMSC in ILs and the preparation of cellulose nanospheres *via* TMSC samples applying a dialysis process.

The conversion of celluloses with 1,1,1,3,3,3-hexamethyldisilazane (HMDS) was studied applying the ILs BMIMCl, EMIMAc and 1-ethyl-3-methylimidazolium chloride (EMIMCl) as reaction media. No additional base or catalyst was used. During the reaction TMSCs with higher DS precipitated. The silylation proceeded within one hour. It was found that the reaction is most efficient in EMIMAc. TMSC with a DS of 2.67 can be obtained at a molar ratio HMDS/AGU of 3/1, at 80°C. As mentioned, HMDS is insoluble in the ILs applied and the TMSC may precipitate during the synthesis. To assure a homogeneous conversion the addition of the co-solvent chloroform was studied because TMSCs with a DS > 2.0 are soluble in chloroform (43). Moreover, the use of co-solvents guarantees a better mixing of HMDS with the cellulose solution. Table II summarizes the results of the synthesis of TMSCs using 3 ml chloroform added to a reaction mixture of 0.5 g cellulose dissolved in 4.5 g IL (44).

The beneficial effect of the co-solvent was obvious for the reaction of HMDS with cellulose dissolved in BMIMCl at lower molar ratios HMDS/AGU. Thus, silylation in BMIMCl with HMDS/AGU ratios of 3/1 and 5/1 was successful in the presence of chloroform (samples 12a and 14a). However, in the system EMIMAc/cellulose the addition of chloroform decreased the silylation DS. For the reactions discussed commercial ILs were used. BMIMCl (BASF quality, purity ≥ 95 %) contains 0.2 - 0.4 wt.-% 1-methyl-1*H*-imidazole (MI), 0.1 - 0.2 wt.-% n-butyl chloride and 0.1 wt.-% water according to the manufacturer. EMIMAc (BASF quality, purity ≥ 90 %) contains 0.5 wt.-% MI, 1.0 wt.-% EMIMCl, 1.0 wt.-% potassium acetate, 0.5 wt.-% methanol and 0.5 wt.-% water according to the manufacturer. Interestingly, no TMSC was obtained if pure BMIMCl (dry, purity ≥ 99 %) was used as reaction medium.

Table II. Influence of chloroform on the efficiency of the trimethylsilylation of cellulose with 1,1,1,3,3,3-hexamethyldisilazane (HMDS) in ionic liquids (1 h, 80°C).

Reaction Conditions		Products			
Molar Ratio	Solvent	No.	DS_{Si}^{b}	No.	DS_{Si}^{b}
HMDS/AGU^{a}			with CHCl$_3$		without CHCl$_3$
3	BMIMCl	12a	1.94^{c}	12b	0
3	EMIMAc	13a	2.19	13b	2.67
5	BMIMCl	14a	1.71^{c}	14b	0
5	EMIMAc	15a	2.28	15b	2.73
8	BMIMCl	16a	0.43	16b	1.85
8	EMIMAc	17a	2.89	17b	2.85

[a] Anhydroglucose unit, [b] Degree of substitution determined by ^{1}H NMR spectroscopy after peracetylation, [c] Determined by elemental analysis

Therefore, it was important to understand the possible effects of added MI, perhaps involving base catalysis. Towards this end, comparable reactions with defined amounts of MI were carried out (see Table III).

Table III. Influence of 1-methyl-1*H*-imidazole (MI) on the degree of substitution of trimethylsilyl cellulose (1h, 80°C).

Reaction conditions			Products	
Molar Ratio		Solvent	No.	DS_{Si}^{c}
HMDS^{a}/AGU^{b}	MI/AGU^{b}			
8	0	BMIMCl	18	0.02
8	0.07	BMIMCl	17	1.85
8	0.2	BMIMCl	20	0.84
3	0.09	EMIMAc	21	2.67
3	3	EMIMAc	22	2.11

[a] 1,1,1,3,3,3-Hexamethyldisilazane, [b] Anhydroglucose unit

[c] Degree of substitution determined by ^{1}H NMR spectroscopy after peracetylation

It was found that, in BMIMCl, catalytic amounts of MI were necessary to start the reaction and increase the efficiency of the silylation. The activation of HMDS with MI can be explained by the occurrence of a pentacoordinated complex formed as intermediate (45).

In contrast, in EMIMAc an excess of MI 3/3/1 (MI/HMDS/AGU) seems to reduce the efficiency of the etherification. These findings show that knowledge about the content of substances with a potential catalytic effect in commercial ILs is an important prerequisite for the reproducibility of such conversions. The ILs applied for the silylation can be recycled in a comparable manner as discussed for the sulfation reaction. It was possible to reuse these recycled solvents for the silylation. ^{13}C NMR spectra confirmed the purity of all the TMSC synthesized. Only signals for the substituent (0.0 - 2.0 ppm) and the AGU (103.0 - 60.8 ppm) were found.

The formation of nanoparticles was carried out by a dialysis process. It is based on the slow exchange of an organic solvent against a non-solvent. Dialysis of TMSC dissolved in tetrahydrofuran (THF) or DMA, respectively, against water was performed for 4 d. The produced particles formed aqueous suspensions and were analyzed regarding their chemical structure, size and shape. Surprisingly, the silyl ether groups of the TMSC were completely split off during the dialysis process as clearly indicated by the corresponding FT-IR spectra (Figure 11). The typical signals of TMSC disappeared and the characteristic signals for cellulose increased in the spectra. Cellulose particles free of TMS groups were obtained, i.e. all hydrophobic and non-polar silyl moieties were removed from the surface and the interior of the particles. No separate splitting of the TMS functions is necessary making this process an easy and efficient access to pure cellulose particles. The particle size is in the range from 176 to 3165 nm (measured by light scattering, Table IV).

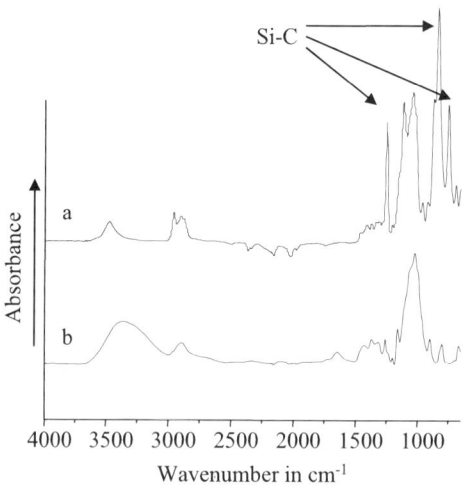

Figure 11. FT-IR spectra of a) trimethylsilyl cellulose (TMSC; DS 1.94) and b) cellulose particles formed via dialysis of TMSC.

Figure 12 presents a SEM image of cellulose particles with particle size of 265 nm formed from TMSC with DS 1.94 (sample 12a).

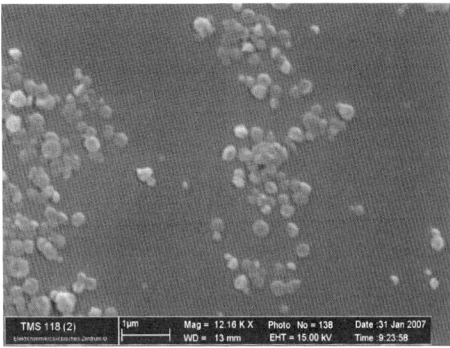

Figure 12. Scanning electron microscope image of particles formed via dialysis of trimethylsilyl cellulose with a degree of substitution of 1.94

It is obvious that the particles are well separated and exhibit a desirable spherical shape. On the contrary, particles prepared from TMSC samples of higher or lower DS showed no particle formation or yielded aggregates (Table IV). This means that the balance of silyl groups and the OH groups, with its concomitant influence on polymer polarity, plays a large role in the formation, size and shape of the resulting particles.

Table IV. Size of cellulose particles formed via trimethylsilyl celluloses of different degree of substitution (DS).

Sample DS	Solvent	Particle	Size^a (nm)	PDI
0.88	DMA	-	-	-
1.85	DMA	+	176	0.11
2.19	THF	-	-	-
1.94	DMA	+	265	0.28
2.26	THF	+	1150	0.76
2.28	THF	-	-	-
2.85	THF	+	3165	0.46

^{a} Determined with light scattering (+ means particle formation, - means no particles were obtained)

Viscosity measurements of the nanoparticle suspensions revealed viscosity in the range of water. Particle suspensions from sample 12 had a viscosity of 0.948 mPa·s at 25 °C. Therefore parenteral application as drug carrier or contrasting agent is possible.

Conclusion

It was shown that ionic liquids (ILs) are valuable tools both for the defined regeneration of cellulose and for the homogeneous functionalization of the polysaccharide. Tailoring of the molecular and the supramolecular features of cellulose and its derivatives can be achieved making the products promising new materials for biomedical applications. Thus, it was possible to prepare highly porous membranes from cellulose in IL for the separation of lysozyme from protein solutions. Using homogeneous functionalization of cellulose in ILs/co-solvent mixtures, defined cellulose sulfates (CSs) and silylated cellulose were accessible. The CSs prepared in this manner are suitable for the preparation of capsular polyelectrolyte complexes (PEC) for the encapsulation of biological material. In this paper we showed the encapsulation of glucose oxidase, which is still active in the capsule. The silylated derivatives can be applied as precursors for manufacturing of pure cellulose particles in the nanoscale range suitable as carrier materials or contrasting agents. Further investigation is focused on the preparation of bioactive membranes, loading of the PEC complexes with cells and loading of the cellulose particles with proteins and other bioactive molecules.

Acknowledgement

The authors would like to thank the "Fachagentur Nachwachsende Rohstoffe e.V." (project 22021905) for the financial support.

References

1. McCormick, C.L.; Lichatowich, D.K. *J. Polym. Sci., Polym. Lett. Ed.* **1979**, 17, 479-484.
2. Heinze, T.; Liebert, T.; Koschella, A. In *Esterification of Polysaccharides*; Springer Verlag: Heidelberg, 2006; Vol. 1 (ISBN 3-540-32103-9).
3. Heinze, T; Liebert, T. *Prog. Polym. Sci.* **2001**, 26, 1689-1762.
4. Liebert, T.; Heinze T. *Biomacromoleculse* **2005**, 6, 333-340.
5. Heinze, T.; Dicke, R.; Koschella, A.; Kull, A.H.; Klohr, E.-A.; Koch, W. *Macromol. Chem. Phys.* **2000**, 201, 627-631.
6. Hussain, M.A., Liebert, T.; Heinze, T. *Macromol. Rapid Commun.* **2004**, 25, 916-920.
7. Köhler, S.; Heinze, T. *Cellulose* **2007**, 14, 489-495.
8. Liebert, T.; Heinze, T.; Klemm, D. *J. Macromol. Sci.-Pure Appl. Chem.* **1996**, 33, 613-626.
9. Graenacher C. U.S. Patent 1,946,176, 1934.
10. Swatloski, R.P.; Spear, S.K.; Holbrey, J.D.; Rogers, R.D. *J. Am. Chem. Soc.* **2002** 124, 4974-4975.
11. Swatlowski, R.P.; Rogers, R.D.; Holbrey, J.D. WO 03/029329, 2003.

131

12. Wasserscheid P.; Welton, T. In *Ionic Liquids in Synthesis*, Wiley VCH: Weinheim, 2003.
13. Liebert, T.; Heinze, T. *BioResources* **2008**, 3, 576-601.
14. Kosan, B.; Michels, C.; Meister, F. *Cellulose* **2008**, 15, 59-66.
15. Murugesan, S.; Wiencek, J.M.; Ren, R.X.; Linhardt, R.J. *Carbohydr. Polym.* **2006**, 63, 268-271.
16. Turner, M.B.; Spear, S.K.; Holbrey, J.D.; Rogers R.D. *Biomacromolecules* **2004**, 5, 1379-1384.
17. Turner, M.B.; Spear, S.K.; Holbrey, J.D.; Daly, D.T.; Rogers R.D. *Biomacromolecules* **2005**, 6, 2497-2502
18. Staude E. In *Membranen und Membranprozesse*, Wiley-VCH, Weinheim 1992.
19. Wan, Y.; Lub, J.; Cui, Z. *Separ. Purific. Technol.* **2006**, 48, 133-142.
20. Davies, G.J.; Dodson, G.G.; Hubbard R.E.; Tolley S.P.; Dauter Z.; Wilson K.S.; Hjort C.; Mikkelsen J.M.; Rasmussen G., Schulein M. *Nature* **1993**, 365, 362-364.
21. Bonomini, M.; Pavone, B.; Sirolli, V.; Del Buono, F.; Di Cesare, M.; Del Boccio, P.; Amoroso, L.; Di Ilio, C.; Sacchetta, P.; Federici, G.; Urbani, A. *J. Proteome Res.* **2006**, 5, 2666-2674.
22. Wang, Z.M.; Li, L.; Zheng, B.S.; Normakhamatov, N.; Guoa, S.Y. *Int. J. Biol. Macromol.* **2007**, 41, 376-382.
23. Dautzenberg, H.; Schuldt, U.; Grasnick, G.; Karle, P.; Müller, P.; Löhr, M.; Pelegrin, M.; Piechaczyk, M.; Rombs, K.V.; Gunzburg, W.H.; Salmons, B.; Saller, R.M. *Ann. N Y Acad. Sci.* **1999**, 875, 46-63.
24. Löhr, M.; Müller, P.; Karle, P.; Stange, J.; Mitzner, S.; Jesnowski, R.; Nizze, H.; Hebe, B.; Liebe, S.; Salmons, B.; Günzburg, W.H. *Gen Ther.* **1998**, 5, 1070-1078.
25. Schaffellner, S.; Stadlbauer, V.; Stiegler, P.; Hauser, O.; Halwachs, G.; Lackner, C.; Iberer, F.; Tscheliessnigg, K.H. *Transplantation Proceedings* **2005**, 37, 248-252.
26. Mansfeld, J. ; Förster, M.; Schellenberger, A.; Dautzenberg, H.; *Enzyme Microb. Technol.* **1991**, 13, 240-244.
27. A. Vikartovská, A.; M. Bučko, M.; D. Mislovičcová, Potoprstý, D. V.; Lacík, I.; Gemeiner, P. *Enzyme Microb. Technol.* **2007**, 13, 748-755.
28. Dautzenberg, H.; Lukanoff, B.; Eckert, U.; Tiersch, B.; Schuldt, U.; *Ber. Bunsenges. Phys. Chem.* **1993**, 100, 1045-1053.
29. Wagenknecht, W.; Nehls, I.; Philipp, B. *Carbohydr. Res.* **1993**, 240, 245-252.
30. Wagenknecht, W.; Phillip, B.; Keck, M. *Acta Polym.* **1985**, 36, 697-698.
31. Lukanoff, B.; Dautzenberg, H. *Das Papier* **1994**, 48, 287-296.
32. Wagenknecht, W.; Nehls, I.; Stein, A.; Klemm, D.; Philipp, B. *Acta Polym.* **1992**, 43, 266-269.
33. Wagenknecht, W. *Papier* **1996**, 12, 712-720.
34. Köhler, S.; Liebert, T.; Schöbitz, M.; Schaller, J.; Meister, F.; Günther, W.; Heinze, T.. *Macromol. Rapid Commun.* **2007**, 28, 2311-2317.
35. Heinze, T.; Schwikal, K.; Barthel, S. *Macromol. Biosci.* **2005**, 5, 520-525.
36. Evans, R. ; Wallis, A.F.A. *J. Appl. Polym. Sci.*; **1989**, 37 2331-2340.

37. Gericke, M.; Liebert, T.; Heinze, T. *Macromol. Biosci.*; **2008** to be submitted.
38. Ahmad, A.; Akhtar, Md.S.; Bhakuni, V.; *Biochemistry* **2001**, *40*, 1945-1955
39. Rauf, S.; Ihsan, A.; Akhtar, K.; Ghauri, M.A.; Rahman, M.; Anwar, M.A.; Khalid, A.M.; *J. Biotechnol.* **2006**, 121, 351-360
40. Ukeda, H.; Fujita, Y.; Ohira, M.; Sawamura, M. *J. Agric. Food Chem.* **1996**, 44, 3858-3863.
41. Dong, S.; Hirani, A. A.; Lee, Y. W.; Roman, M.; Abstracts of Papers, Proceedings of the 233rd ACS National Meeting, Chicago, IL, United States, March 25–29, **2007**.
42. Schaub, M.; Wenz, G.; Wegner, G.; Stein, A.; Klemm, D. *Adv. Mater.* **1993**, 5, 919-922.
43. Stein, A.; Ph.D. thesis, *Friedrich Schiller University of Jena*, 1991.
44. Köhler, S.; Liebert, T.; Heinze, T. *J. Polym. Sci., Part A: Polym. Chem.* **2008**, 46, 4070-4080.
45. Nouvel, C; Dubois, P.; Dellacherie, E.; Six, J.-L. *Biomacromolecules* **2003**, 4, 1443-1450.

Chapter 7

Cellulose composites prepared using ionic liquids (ILs) - Blood Compatibility to Batteries

Tae-Joon Park [a], Saravanababu Murugesan [a1], and Robert J. Linhardt [a, b, c]

Department of Chemical and Biological Engineering [a] Chemistry and Chemical Biology [b], and, Biology [c], Center for Biotechnology and Interdisciplinary Studies
Rensselaer Polytechnic Institute, Biotech Center 4005,
110 8th Street, Troy, NY 12180-3590, USA
[1] Currently at Development Engineering, Process Research and Development, Bristol-Myers Squibb, New Brunswick, NJ 08903
* To whom correspondence should be addressed:
Biotechnology 4005, Rensselaer Polytechnic Institute, Troy, NY 12180 (USA)
Fax: (+ 1) 518-276-3405, E-mail: linhar@rpi.edu

Introduction

Cellulose and cellulose composites

The application of biomaterials in the bioengineering of medical devices can be considered only if these are safe for patients. The biocompatibility and blood compatibility of biomaterials are often among the most relevant issues associated with safety. Biocompatibility and blood compatibility are important in pharmaceutical and imaging applications because the lack of these properties often trigger undesirable side effects and can prevent a biomaterial from performing its intended function. Biomaterials can also contain biologically derived components that enhance their properties. In their first generation, biomaterials were inert structural materials but more recent biomaterials actively interact with tissues and organs [1, 2]. Biomaterials have been applied in many areas, including structural transplants, drug delivery transplants, and tissue engineering.

Cellulose, a linear polysaccharide of up to 15,000 D-glucose residues linked by β-(1→4)-glycosidic bonds, is biocompatible and has excellent thermal, mechanical properties [3,4]. It is one of the most abundant renewable organic polymers and is considered easily biodegradable, thus less contaminating to the environment. Cellulose derivatives and composites offer an excellent biocompatibility, and are considered as promising materials for biochemical engineering for economic and scientific reasons [5]. Chemically modified celluloses are found in a wide range of biomaterials, particularly in the preparation of dialysis membranes and implantable sponges [6, 7]

Cellulosic derivatives such as cellulose acetate, cellulose propionate and cellulose acetate-butyrate, when cast as membranes, have been reported as useful supports for immobilizing various enzymes such as catalase, alcohol oxidase and glucose oxidase [8]. These supports gave better activity and storage stability for the enzymes. Graft copolymers of cellulose have been utilized as coupling supports for cells, acid and alkaline phosphatase, glucose oxidase, trypsin, ascorbate oxidase and various other biomolecules [9-13]. Ikeda and coworkers reported the successful immobilization of several biocatalysts such as -galactosidase, α-chymotrypsin, invertase, urease, lipase and *Saccharomyces cereviciae* using a gel formation of cellulose acetate and metal alkoxide [14]. In this work, the biocatalysts were physically entrapped among the gel networks and distributed throughout the gel fiber. The immobilized biocatalyst offered stable activity for repeated cycles, and for an elongated time period. Further, it was possible to perform continuous reaction in a column reactor packed with this immobilized biocatalyst.

Cellulose is highly crystalline as a result of an extensive hydrogen bonding network, making it insoluble in most conventional organic and aqueous solvents and limiting its utility. *N*-methylmorpholine-*N*-oxide, CdO/ethylenediamin [15] LiCl/N,N-dimethylacetamide and near supercritical water [16] are some more unconventional solvents capable of dissolving unmodified cellulose. Chemical modification of undissolved and unmodified cellulose, while routinely accomplished when modifying cellulose with small molecules, represents a major challenge when modifying cellulose with other biopolymers (e.g., heparinization) [17, 18].

Heparin

Heparin, a linear, poly-disperse, negatively-charged, and highly sulfated polysaccharide, is widely used as an injectable anticoagulant for acute coronary syndromes (e.g., myocardial infarction, arterial fibrillation, deep-vein thrombosis and pulmonary embolism) [19, 20]. Heparin is usually administered intravenously since subcutaneously or orally administered heparin has low bioavailability and negligible activity [21]. Heparin has a short biological half-life of approximately 1 h, making it necessary to administer by continuous infusion. This can be inconvenient and is quite costly. In addition, heparin can cause serious side effects including hemorrhage, extensive bleeding, and difficulty in breathing. Efforts to eliminate the portions of the heparin polysaccharide which are not necessary for anti-thrombotic activity have been fruitful, resulting in compounds such as low molecular weight heparins and a synthetic pentasaccharide, Arixtra® (fondaparinux sodium). Despite these advances, low molecular weight heparin and synthetic heparin oligosaccharides are expensive to produce and still display adverse side-effects. Hence it is important to strive to avoid the systemic administration of soluble heparin. The bonding of heparin to devices, rather than the traditional delivery of soluble heparin, promises reduced costs and diminished side-effects, such as uncontrolled bleeding. Heparinized devices, include currently used macrodevices such as kidney dialyzers in extracorporeal circuits, indwelling catheters and stents [22, 23] or implantable nanodevices or nanomachines under development for applications such as drug delivery systems.

Heparinized materials were originally developed for the affinity chromatographic purification of heparin-binding proteins [24, 25]. More recently, there have been many reports of using heparinization to enhance the surface properties of various polymeric materials for medical applications [17, 22, 26-28]. Our laboratory reported the preparation of various types of such cellulose-heparin composites using a new class of solvents called room temperature ionic liquids (RTILs) [17]. These composites, based on their biological properties and morphology, were proposed in various applications including as blood compatible, hollow fiber, nanoporous membranes for kidney dialyzers, eliminating the requirement for systemic heparinization. The following chapters discuss in detail the significance of the ILs and the use of these ILs in preparing various cellulose derivatives including the cellulose-heparin composite materials.

Ionic Liquids (ILs)

Ionic liquids (ILs) or organic liquid salts are a new class of solvents having potential applications in the chemical and biotechnology industries. ILs, unlike molecular solvents, are completely composed of ions, and are considered excellent substitute solvents for the environmentally harmful organic solvents used in conventional chemical processes. Moreover, ILs are useful in overcoming the lack of solubility of polysaccharides in conventional organic solvents [29, 30]. Salts that are liquid between -50°C and 300°C have great utility. RTILs are a family of ILs referred to as "molten salts" that remain liquid at room temperature. ILs are "designer solvents" that can be prepared with specific physical and chemical properties [31]. Because their properties such as density, viscosity, melting point, polarity and refractive index, can be tuned by varying the structure of the cation and anion, designer ILs often have attractive physical and chemical properties. A few other useful properties include low volatility due to very high boiling point, thermal stability, non-flammability, odorlessness, and a recyclable nature [29-33]. Cations and anions (Figure 1) commonly employed in ILs [29, 30] result in a wider range of specific desirable physical and chemical properties than are commonly encountered in organic solvents.

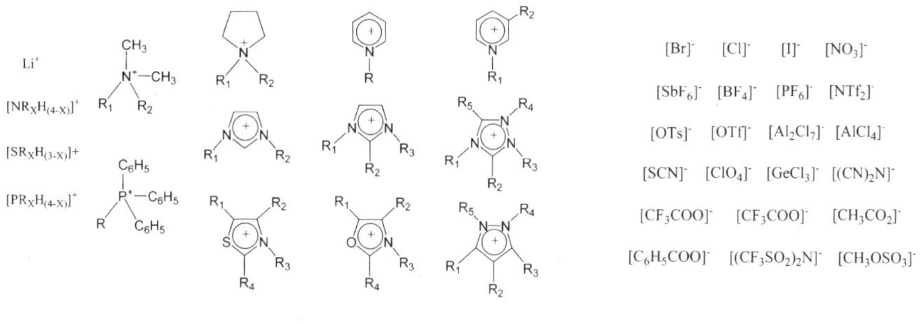

(a)   (b)

Figure 1. Structures of a few (a) cations, and (b) anions that can constitute an ILs.

Moreover, the water miscibility of ILs can be altered by modifying the IL cation or anion structure or both based on process requirements [29]. For example, $[PF_6]^-$, and $[(CF_3SO_2)_2N]^-$ generally reduce water solubility, while $[CH_3COO]^-$, $[CF_3COO]^-$, $[NO_3]^-$, Br^-, I^-, Cl^-, $[Al_2Cl_7]^-$ and $[AlCl_4]^-$ generally increase water solubility. ILs also play an important role in electrochemical applications because they do not easily crystallize and can solubilize either polar or non-polar solutes [29, 34-46].

ILs can often be quantitatively recovered because of their nearly zero vapor pressure. This low volatility and lack of propensity to crystallize make it possible to exploit ILs under a wide range of temperatures, from sub zero to > 100 °C. Most of the ILs are recoverable and reusable by liquid-liquid extraction, affording products in a separate liquid layer, such as supercritical CO_2 or water. The recoverability and reusability of ILs enable them to diminish the adverse environmental impact of various chemical and biotechnological processes.

Solubility of Polysaccharides in ILs

Polysaccharides are highly complex, chiral organic compounds that are challenging to modify and are insoluble in most common organic solvents. Only certain polysaccharides can be dissolved and/or modified in selected solvents (e.g., water, pyridine, formamide, dimethylformamide and dimethylsulfoxide) [29, 47-49]. Thus, it is important to investigate new solvent systems capable of dissolving polysaccharides. The introduction of ILs has led to a dramatic increase in the number of publications aimed at overcoming the lack of suitable solvent systems in the field of polysaccharide chemistry.

Solubility of cellulose in ILs

Unmodified or modified (e.g., cellulose acetate) carbohydrate polymers are suitable for use as a matrix for a variety of applications. Commonly used carbohydrates include cellulose, starch, xylose, inulin, chitin, chitosan, glycogen, pectin, carrageenan, fucan, and fucoidin [18]. Cellulose is particularly important as it is synthesized by a great diversity of living organisms and has excellent thermal, mechanical, and biocompatibility properties [4, 50]. Chemically modified celluloses have been applied for the preparation of dialysis membranes and implantable sponges.

There is a need for additional environmentally benign solvents for use in cellulose processing [51, 52]. Recently, IL solvents were introduced to dissolve cellulose. N-ethylpyridinium chloride was one of the first organic molten salts introduced. It was not considered commercially useful because of its relatively high melting point of 118 °C. N-ethylpyridinium chloride was suggested as a cellulose solvent by Graenacher in 1934 [53, 54]. Extensive studies by Rogers and coworkers [55-61] led to the effective use of ILs as solvents for cellulose. In 2002, Swatloski and coworkers [58] tested the ability of ILs having 1-n-butyl, 3-methylimidazolium [bmIm] cations with various anions to dissolve cellulose. The most effective anion was found to be the chloride salt. The maximum solubility of cellulose in [bmIm][Cl] IL (25% (w/w)) was achieved by microwave irradiation of the mixture, affording a transparent solution. In 2003, Ren et al, [62] synthesized 1-allyl 3-methylimidazolium chloride ([amIm][Cl]) and showed that it dissolves cellulose, creating a solution in which a homogeneous acetylation reaction [63] can be carried out. Heinze et al., [64] in 2005 tested the ability of 1-n-butyl-3-methylimidazolium chloride [bmIm][Cl], 3-methyl-n-butyl-pyridinium chloride [bmPy][Cl], and benzyldimethyl(tetradecyl)ammonium chloride (BDTAC) to dissolve cellulose. This group reported that [bmPy][Cl] is more effective than [bmIm][Cl] and that BDTAC showed the lowest ability to dissolve cellulose. In 2007, Erdmenger, et al. [53] studied the effect of alkyl chain length, from C_2-C_{10}, on the ability of 1-alkyl-3-methylimidazolium chlorides to dissolve cellulose. An odd-even effect was observed for the small (less than C_6) alkyl chains. ILs with even-numbered alkyl chains were more able to dissolve cellulose than those with odd-numbered alkyl chains. The results of the solubility studies on various types of cellulose with a variety of ILs are presented in Table 1.

Table 1. Solubility of cellulose in ILs

Type of Cellulose	Ionic liquid	method	Temp. (°C)	Solubility (% w/w)	Ref.
Pulp	[bmIm][Cl]	Heat	100	10	[58]
Pulp	[bmIm][Cl]	Heat	70	3	[58]
Pulp	[bmIm][Cl]	Heat +sonication	80	5	[58]
Pulp	[bmIm][Cl]	Microwave		25	[58]
Pulp	[C$_6$Im][Cl]	Heat	100	5	[58]
Pulp	[C$_8$Im][Cl]	Heat	100	Slightly soluble	[58]
Pulp	[bmIm]SCN	Microwave		5~7	[58]
Pulp	[bmIm]Br	Microwave		5~7	[58]
microcrystalline	[bmIm][Cl]	Heat	110	4.8	[65]
microcrystalline	[bm$_2$Im][Cl]	Heat	110	4.5	[65]
microcrystalline	[bmIm][Cl]	Microwave	115	4	[65]
microcrystalline	[bm$_2$Im][Cl]	Microwave	115	4	[65]
Paper filter	[bmIm][Cl]	Heat	110	5.2	[65]
Paper filter	[bm$_2$Im][Cl]	Heat	110	4.5	[65]
Paper filter	[bmIm][Cl]	Microwave	115	4	[65]
Paper filter	[bm$_2$Im][Cl]	Microwave	115	4	[65]
Avicel® PH-101	[bmIm][Cl]	Heat	100	20	[53]
Avicel® PH-101	[emIm][Cl]	Heat	100	10	[53]
Avicel® PH-101	[C$_7$Im][Cl]	Heat	100	5	[53]
Avicel	[bmIm][Cl]	Heat	83	18	[64]
Avicel	[bmPy][Cl]	Heat	105	39	[64]
Avicel	BDTAC	Heat	62	5	[64]
Spruce sulfite pulp	[bmIm][Cl]	Heat	83	13	[64]
Spruce sulfite pulp	[bmPy][Cl]	Heat	105	37	[64]
Spruce sulfite pulp	BDTAC	Heat	62	2	[64]
Cotton linters	[bmIm][Cl]	Heat	83	10	[64]
Cotton linters	[bmPy][Cl]	Heat	105	12	[64]
Cotton linters	BDTAC	Heat	62	1	[64]

*Pulp: high molecular weight dissolving pulp (degree of polymerization 100)

Solubility of Heparin in ILs

Heparin, like cellulose, is soluble in only a very few conventional organic solvents. Heparin dissolves in water, dimethyl sulfoxide, dimethyl formamide and formamide. Therefore, many studies have been conducted to test the solubility of various heparin-like glycosaminoglycans (GAGs) in novel solvents [17, 29, 30]. Our laboratory suggested the use of ILs for the dissolution of glycosaminoglycans. We studied the synthesis and properties of ILs having benzoate as the anion and a cation comprised of alkyl substituted imidazolium, pyridinium or phosphonium moieties [66]. The aim of this work was to find ILs that could dissolve the sodium or imidazolium salts of heparin.

The synthesis of benzoate-based ILs was easily performed in ethanol, resulting in protonation of ethanol followed by nucleophilic displacement by halide [67], or by using microwave irradiation (Figure 2). These two methods result in products with slightly different properties due to the presence of trace amounts of different by-products. Four different ILs ([emIm][[ba], [bmIm][ba] [bmIm][PF_6], and [bmIm][BF_4]) were used to study GAG solubility [66]. Dissolution data in mg/ml are presented in Table 2. A total of eight GAGs, four with sodium and four with imidazolium counterions, were tested in these dissolution studies. The IL [emIm][[ba] showed the best dissolution of GAGs, and the imidazolium salts of GAGs dissolved better than their corresponding sodium salts.

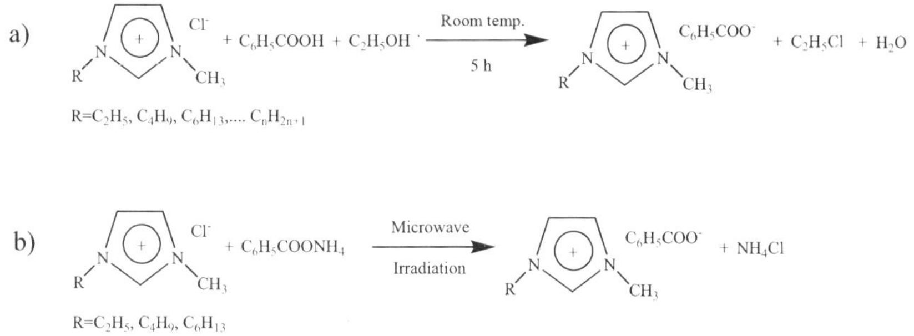

Figure 2. Synthesis of ILs (a) in ethanol (b) with microwave irradiation

Therefore, in further studies of the application of ILs in carbohydrate chemistry, [emIm][[ba] was selected as the desired solvent. This IL was first examined as a solvent for peracetylation and perbenzoylation of monosaccharides [68]. Following our initial success, this IL was used in the glycosylation of unprotected donors with protected acceptors [69]; the fabrication of blood compatible composite membranes [17]; and the preparation of biopolymer fibers by electrospinning [31].

Table 2. Dissolution data (mg/ml) of GAGs in ILs measured by carbazole assay

GAG	Counterion	[emIm][ba]	[bmIm][ba]	[bmIm][PF$_6$]	[bmIm][BF$_4$]
Heparin	Imidazolium	7.00	7.00	0	0
	Sodium	0	0	0	0
Heparan Sulfate	Imidazolium	3	2.83	0.21	0.07
	Sodium	0	0	0.07	0
Chondroitin Sulfate	Imidazolium	9.87	5.67	0.03	0.19
	Sodium	1.30	0.67	0.51	0.03
Hyaluronic Acid	Imidazolium	10.00	10.00	3.38	2.60
	Sodium	8.60	5.00	1.00	0.15

Cellulose Composites using ILs

We define biomaterials for our purposes as nondrug materials that may be synthetic, natural or modified natural products, useful for diagnosis, treatment, enhancement, or replacement of any tissue, organ, or function in an organism. Biomaterials can be used in many applications including artificial transplants, drug delivery implants, tissue engineering and other biomedical applications. The design of biomaterials at the molecular level is the subject of study in the field of biomedical engineering. The release of drugs from polymer coated stents or drug delivery systems for proteins such as human growth hormone represent some of the important current medical applications of biomaterials.

Covalent immobilization of heparin to the surface can reduce clot formation on biomaterials. All kidney dialysis systems require heparin to maintain blood flow in patients on extracorporeal therapy. Cellulose dialysis membranes are difficult to modify chemically with drugs such as heparin to improve their blood compatibility. Our laboratory used ILs to prepare blood-compatible cellulose-heparin nanoporous nanocomposite membranes for use in kidney dialysis [17]. Next, we expanded our efforts to examine nanocomposite heparin-cellulose fibers prepared by electrospinning as biomaterials for the preparation of vascular grafts [31]. In addition, we devised a foldable and bendable postage-stamp-sized cellulose-carbon nanotube (CNT) nanocomposite battery that can be readily cut to size [70]. This flexible sheet of integrated paper battery is made of cellulose with CNTs and lithium electrodes. This implantable battery technology may one day demonstrate the importance of cellulose nanocomposites prepared using IL for a variety of *in vivo* biomedical applications. Each of these novel cellulose composites will be described in detail in the following sections. Figure 3 shows the schematic representation of the strategy for preparing various cellulose-heparin composites using ILs.

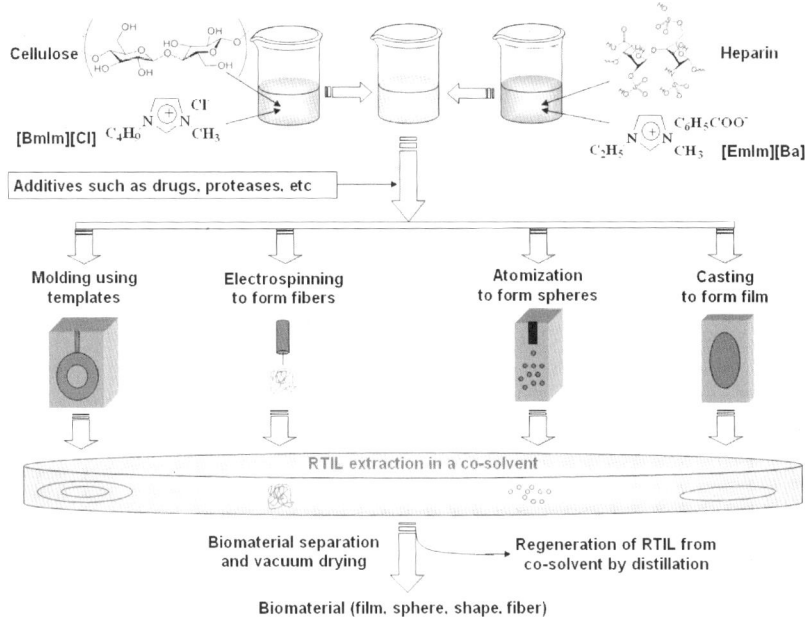

Figure 3. Schematic representation for the preparation of heparin-cellulose composites

Blood compatible membranes and fibers

The surfaces of extracorporeal and prosthetic medical devices that come directly into contact with blood or body fluids and tissues must be biocompatible. Such devices should not trigger blood clotting, nor should they induce inflammatory responses when brought into contact with tissues. Blood compatibility is a major factor in biocompatibility because thrombogenesis is induced by surfaces of medical devices that are not blood compatible. Thrombus formation on the surface of an extracorporeal medical device or an implanted biomedical device can result in heart attack, stroke or pulmonary embolism.

Kidney dialysis membranes must have excellent compatibility with blood, and should not adsorb plasma proteins, promote platelet adhesion or activate the complement and coagulation cascades. Soluble heparin can improve the blood compatibility of cellulose-only membranes during kidney dialysis. Unfortunately, it is difficult to immobilize heparin onto cellulose membranes without compromising membrane function. Composites of cellulose and heparin are difficult to prepare due to the limited solubility of the polysaccharides. Therefore our laboratory has primarily focused on the design of biomaterials with enhanced blood compatibility to reduce thrombogenesis.

A novel cellulose-heparin based biocomposite was first prepared by solvent casting from ILs in which heparin and cellulose were homogenously dispersed

142

[17]. One IL [emIm][ba] was used to dissolve heparin, and a second IL [bmIm][Cl] was used to dissolve cellulose. These two ILs were combined to afford a solution of both cellulose and heparin that was then solvent cast into a cellulose-heparin-composite film after extracting the ILs with ethanol

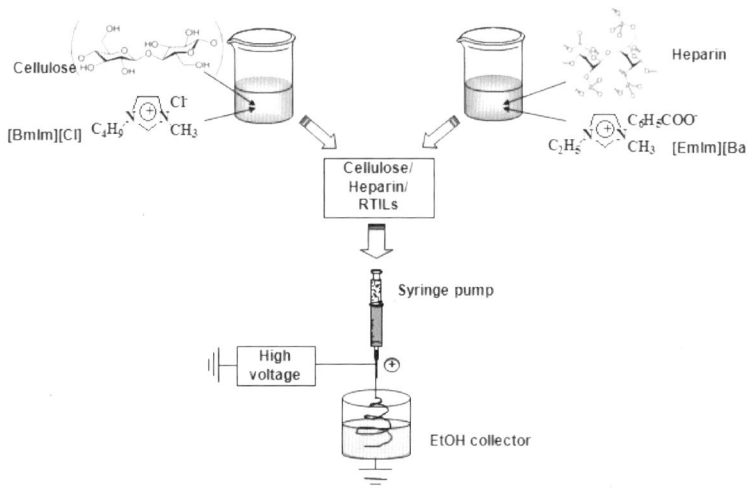

Figure 4. Schematic Representation of Electrospinning from IL solutions

Electrospinning refers to a derivative of the electrospray process as both use high voltage to form a liquid jet. It is a useful tool for the preparation of micron to nanometer sized fibers of various polymers from volatile solvents [31]. Electrospun fibers can be applicable in the fabrication of nanofiber-reinforced composites, membranes, biosensors, electronic and optical devices and as enzyme and catalytic supports [71]. The IL solution of cellulose and imidazolium heparin could also be electrospun into biopolymer fibers. Typically electrospinning requires a volatile solvent. In a novel approach heparin and cellulose were electrospun from non-volatile IL by using ethanol as co-solvent to remove the IL (Figure 4).

Surface characterization of two different types of biocomposites (film and fibers) was performed using field emission scanning electron microscopy (FESEM) (Figure 5). The blood compatibility of these materials was evaluated by activated partial thromboplastin time (APTT) and thromboelastography (TEG). These assays are routinely used to measure anticoagulant activity [17, 22, 72]. The film cast cellulose-heparin composite membrane was also evaluated by equilibrium dialysis using urea and serum albumin to evaluate its utility as a dialysis membrane. The blood compatibility assay demonstrated that heparin was accessible and functional in the composite, while the equilibrium dialysis method showed the membrane to be stable (since no heparin was extracted from the composite). Furthermore, the high voltage involved in electrospinning did not reduce anticoagulant activity of the heparin present in the electrospun fibers. These biocomposite films might be useful in eliminating the requirement for systemic heparinization required for dialysis, providing a potentially valuable blood compatible biomaterial for the fabrication of hollow fiber, nanoporous membranes for use in kidney dialysis.

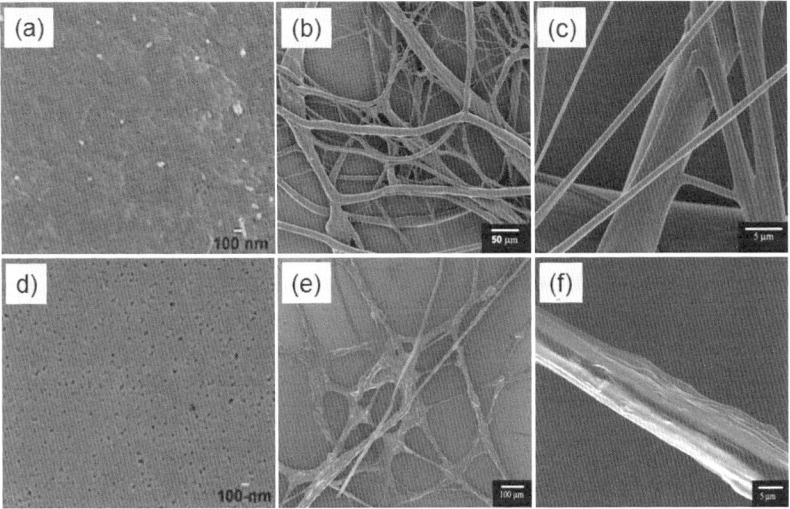

Figure 5. FESEM images: (a) Cellulose film, (b,c) cellulose fibers, (d) cellulose-heparin film, (e, f) cellulose-heparin composite fibers.

Foldable battery/supercapacitor

Supercapacitors, referred to as "double-layer capacitors", "gold capacitors", "ultracapacitors", "power cache" or "electrochemical capacitors", are capacitors that can store energy within the electrochemical double-layer in the electrode/electrolyte interface [73-75]. The first supercapacitor patent was filed in 1957 [76], and the first supercapacitor reached the market in 1969 [77]. Supercapacitors have recently become increasingly important energy storage devices. In terms of design and manufacture, a simple supercapacitor requires two electrodes, a separator and an electrolyte [78]. Typical energy storage devices are compared in Figure 6 [73]. The power density of conventional capacitors is not as low as that of batteries and fuel cells so that supercapacitors may improve the power density of batteries. Moreover, supercapacitors also improve on the low energy density of conventional capacitors. Hence, it is important for supercapacitors to fill a complementary role with conventional capacitors (e.g., electrolytic capacitors or metalized film capacitors) and batteries.

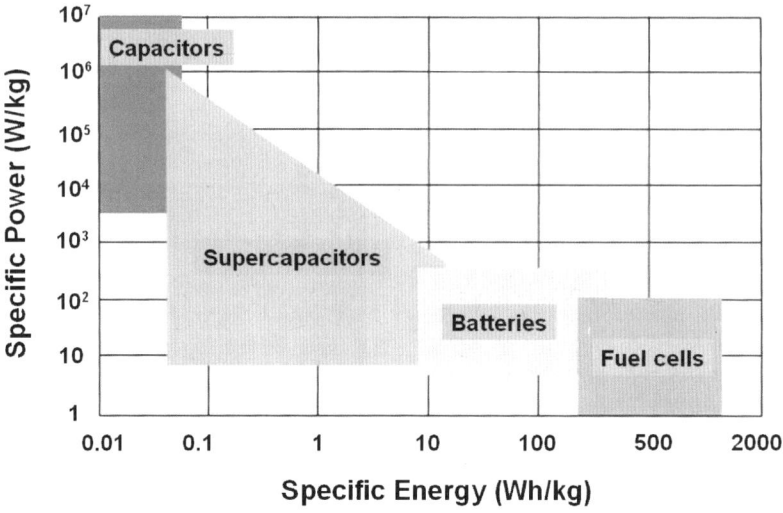

Figure 6. The comparison of typical energy storage devices [73].

Figure 7 shows a schematic representation of a supercapacitor consisting of porous separator and the electrolytic solution. A porous separator separates electrodes and it contains the electrolyte [75].

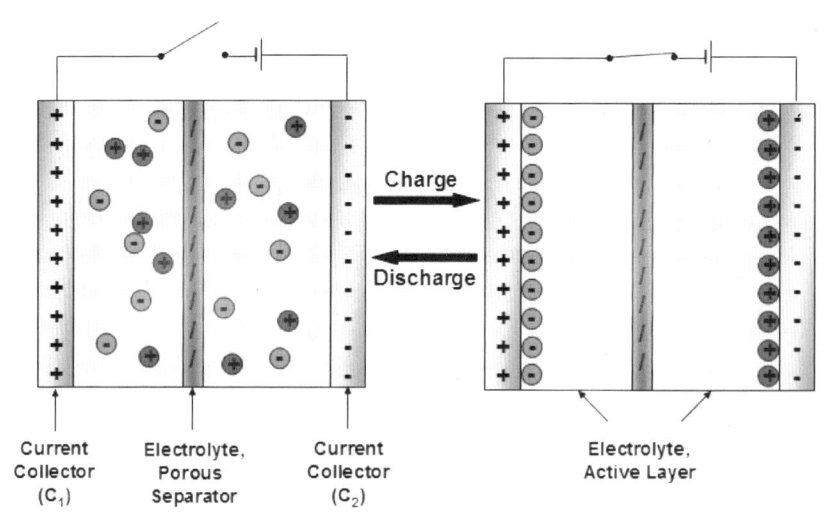

Figure 7. Principle of a supercapacitor.

Supercapacitors can be classified by several standards depending upon the electrode materials; the electrolyte used; or the cell design. The performance characteristics of supercapacitors are defined by the structural and electrochemical properties of electrode materials [70, 78-81]. The electrode requirements in supercapacitors are less important than those in batteries and the power density of supercacitors prevails over energy density [78]. In terms of electrode materials, three main materials are used in super capacitors, activated carbon, metal oxides and polymeric materials [75, 82]. Activated carbon is commonly used for electrode materials in electrochemical capacitors because of its large specific surface area (up to 2500 m^2/g), low cost and commercial availability [75, 83, 84]. High-surface-area activated carbon electrodes can be used to optimize the performance (capacitance and overall conductivity) of supercapacitors [78].

The choice of electrolyte (aqueous vs. nonaqueous) is another criterion used to categorize different supercapacitors. For example, aqueous electrolytes have several advantages, such as easy purification and drying processes during production, and higher conductance. However, organic electrolytes are more frequently used because they achieve higher voltage, even though organic (nonaqueous) electrolytes have the disadvantage of higher specific resistance [73]. Moreover, the cell voltage can be limited by the water content of the aqueous electrolytes. The ionic nature of ILs makes these an excellent choice as an electrolyte in supercapacitors [70, 85-88].

Cell design is the other important factor for energy storage devices. Our laboratory recently designed integrated structures for supercapacitor and battery using cellulose and ILs (Figure 8 and 9). This resulted in foldable, bendable, and cutable postage-stamp-sized supercapacitors [70]. Prior to our efforts, flexible energy storage devices had been designed based on separated thin-electrode and spacer layers [89]. Unfortunately, performance and handling of these devices were less than optimal because of the multiple interfaces between these layers, and the corresponding interfacial resistance. The fabrication of electrode-spacer-electrolyte-integrated carbon nanotube (CNT)-cellulose composites provided excellent flexibility and porosity to the system. Cellulose was used as an inexpensive insulating separator with excellent biocompatibility and adjustable porosity. CNTs, widely used as electrodes in electrochemical devices [90-113], also support the properties of extreme flexibility of our integrated system. The insolubility of cellulose, the major challenge to fabricate CNT-integrated cellulose composites, was overcome by using IL [bmIm][Cl]. Moreover, IL allows the integrated molecular assembly of all three components (cellulose, CNT, and IL) used in supercapacitors [87, 88].

Aqueous bodily fluids could replace the non-aqueous ILs as electrolytes in these novel supercapacitors. The CNT-cellulose-IL nanocomposite afforded higher operating voltage ($\approx$2.3 V) compared to aqueous KOH electrolyte ($\approx$ 0.9 V) and the resulting supercapacitors operated through a wider range of temperatures (195-450 K). These flexible nanocomposite sheets could also be configured into other energy-storage devices including Li-ion batteries and supercapacitor-battery hybrids. Hence, this robust integrated thin film composite structure showed excellent ability to function over a large range of temperatures and mechanical deformation and allows good electrochemical performance.

146

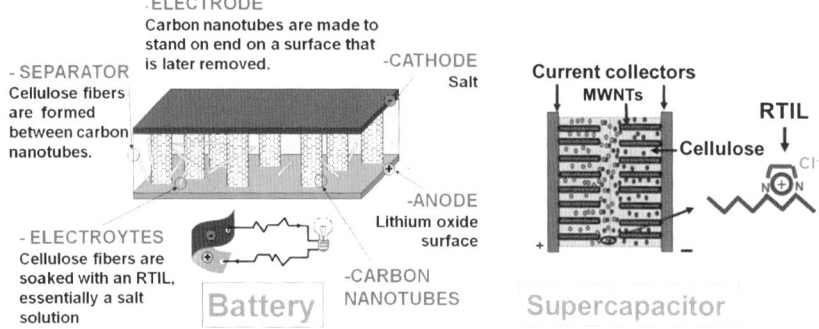

Figure 8. Schematic of the supercapacitor and battery assembled by using nanocomposite film (CNT-cellulose-IL). In the supercapacitor, Ti/Au thin film deposited on the exposed multi-walled nanotubes (MWNTs) acts as current collector.

Since our integrated structures include only three indispensable components (electrodes, spacer, and electrolyte of the supercapacitor), these foldable integrated energy storage devices (able to be cut to size) can be embedded into various functional devices including smart cards, displays, and pacemakers as implantable medical devices.

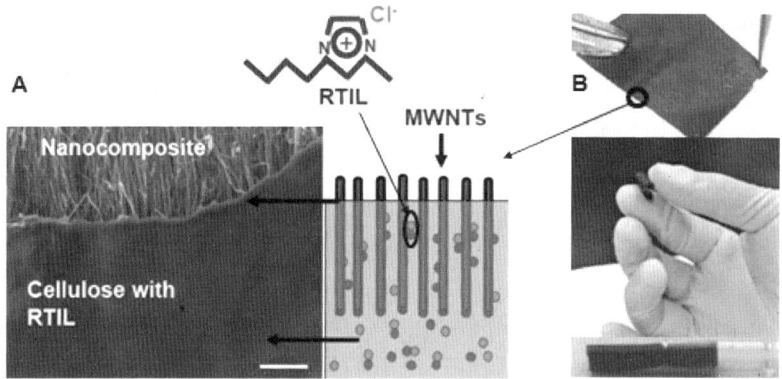

Figure 9. A postage-stamp sized foldable, bendable energy storage nanocomposite (CNT-cellulose-IL) as thin as paper; (a) Cross-sectional SEM image of the nanocomposite paper showing MWNTs protruding from the cellulose–IL ([bmIm][Cl]) thin films. (Scale bar, 2 μm.) The schematic displays the partial exposure of MWNT. (b) Photographs of the nanocomposite units demonstrating mechanical flexibility. Flat sheet (top), partially rolled (middle), and completely rolled up inside a capillary (bottom) are shown.

A variety of cellulose composites - bioactive films, CNT fibers, sensor platforms

A variety of cellulose-based composite materials can be formed, consisting of two or more diverse components in one or more phases (*e.g.*, rigid and flexible tubes, sheets, membranes, fibers or formed and contoured shapes). Murakami *et al.*, (2007) [114] prepared composites of cellulose with polymerized IL. The IL 1-4(4-acryl-oyloxybutyl)-3-methylimidazolium bromide was polymerized *in situ* in the presence of cellulose. The authors examined the compatibility of cellulose with the polymerized IL. Egorov *et al.*, (2007) [65] described the preparation of cellulose films using RTILs, [bmIm][Cl] and [bm$_2$Im][Cl] incorporating entrapped analytical reagents suitable for metal-ion detection. They found that two reagents, 1-(2-pyridylazo)-2-naphthol (PAN) and 1-(2-thiazolylazo)-2-naphthol (TAN), were effectively immobilized. These two films, PAN-cellulose and TAN-cellulose, were stable in aqueous media over a broad pH range. Cellulose composites modified with organic reagents are useful for quantitative determination of transition metal cations.

Zhang et al., (2007) [115] prepared regenerated cellulose fibers reinforced with CNTs by means of the ionic liquid [amIm][Cl]. The aim of this study was the improvement of mechanical properties and thermal stability of cellulose induced by the incorporation of CNTs. It is very difficult to disperse CNTs uniformly in the polymer matrix due to their tendency to form agglomerates by van der Waals attraction during the mixing process. To overcome this dispersion problem, CNTs were ground and well dispersed into an ionic liquid [amIm][Cl] IL that can also dissolve cellulose. Dry-jet wet-spinning was then used to fabricate regenerated cellulose and cellulose/CNT composite fibers (CNT loading of 1, 3, 4, 5, 6, 8, and 9 wt%), resulting in cellulose-CNT composites with increased tensile strength and storage modulus.

Rogers and coworkers have extensively explored the use of ionic liquids to dissolve cellulose and reconstitute cellulose composites. Turner, *et al.* (2004) [57] focused on the use of [bmIm][Cl] to encapsulate active biomacromolecules such as *Rhus vernificera laccase*, into cellulose membranes. A hydrophobic ionic liquid [bmIm][Tf$_2$N] coating resulted in higher enzyme activity, as this coating protected the enzyme from the high, inactivating concentrations of Cl$^-$ present in [bmIm][Cl]-cellulose environments. Turner et al., (2005) [61] also reported cellulose-polyamine composite films and beads using [bmIm][Cl] as solid support matrices for biocatalyst immobilization. They prepared the surface functionalized cellulose composites with primary amine functional groups necessary for chemical bonding between the enzyme and the support resulting in enhanced stability. The use of these materials as sensing devices was proposed to detect polyphenols, aromatic amines and aminophenols. Such materials can also be used as solid support materials for enzyme-catalyzed transformations. Poplin, *et al.* [59] (2007) developed sensor platforms based on encapsulating a probe molecule within the cellulose matrix. The IL [bmIm][Cl] was used to co-dissolve cellulose and the hydrophobic dye/metal complexant 1-(2-pyridylazo)-2-naphthol (PAN) which is a good extractant and indicator for metal ions.

Poplin and coworkers [59] demonstrated that their biodegradable sensor was versatile as it could be easily reconstituted prior to use.

Discussion and Conclusion

The fabrication of cellulose composites using ILs has been extensively studied. A variety of novel cellulose based materials prepared by dissolving cellulose in ILs have been described. They have promise in medical applications in the forms of biofilms, biofibers, templates or foldable energy storage devices. The use of these biomaterials for pharmaceutical applications should be considered only if these cellulose composites are safe for patients. Biocompatibility and blood compatibility, the relevant safety issues for patients, are very important in biomedical engineering since the lack of biocompatibility inevitably results in device failure. Heparin-cellulose based biomaterials show excellent blood compatibility and these might be useful in various applications for extracorporeal and implantable medical devices. The properties of ILs make these important novel solvent systems for cellulose-based biomaterials. Many ILs have been synthesized to have specific desirable physical and chemical properties, such as affinity for glycosaminoglycans, viscosity, melting point, and/or electrochemical properties that are tunable by varying the structure of their cation and anion. The use of ILs to dissolve cellulose is very attractive because cellulose is generally not easily dissolved and is of great utility in energy storage devices. We conclude that cellulose and heparin-cellulose composites prepared using ILs will provide further opportunities to develop biomaterial systems useful in a wide variety of biomedical devices.

References

1. Amass, W.; Amass, A.; Tighe, B. **1998**, *Polym. Intern 47*, 89-144
2. Langer, R.; Tirrell, D. A. *Nature* **2004**, *428*, 487-492
3. Mannhalter, C. *Sensor Actuat B-Chem* **1993**, *11*, 273-279
4. Brown, R. M. *J Polymscipol Chem* **2004**, *42*, 487-495
5. Gemeiner P., Stefca V. and Bales V. *Enzyme Microb Tech* **1993**, *15*, 551–566
6. Yan, L.; Ishihara, K. *J Polym Sci Part A Polym Chem* **2008**, *46*, 3306-3313
7. Hoenich N. A. *Int J Artif Organs* **2007**, *30*, 964-970
8. Murtinho, D.; Lagoa, A. R.; Garcia, F. A. P.; Gil, M. H. *Cellulose* **1998**, *5*, 299-308
9. *The chemistry and technology of cellulosic copolymers*; Hebeish, A.; Guthrie, J. T. Springer-Verlag: New York, 1982
10. Beddows, C. G.; Gil, M. H.; Guthrie, J. T. *Biotech Bioeng* **1986**, *28*, 51-57.
11. Beddows, C. G.; Gil, M. H.; Guthrie, J. T. *Polymer Bulletin* **1984**, 11, 1-6.
12. Alves da Silva, M.; Gil, M. H. *Graft copolymers as supports for the immobilization of biological compounds, Analytical Uses of Immobilised Compounds for Detection, Medical and Industrial Uses*; Guilbault, G. G. and Mascini, M. eds.; 1988; pp 177-185.

13. Gil, M. H.; Alegret, S.; Alves da Silva, M.; Alegria, A. C.; Piedade, A. P. *The use of modiÆed cellulose for biosensors, Cellulosics: Materials for Selective Separations and Other Technologies*; Kennedy, J. F., Philips, G. O. and William, P. A. Eds.; Ellis Horwood Ld.: New York, 1993; pp 163-171

14. Ikeda, Y.; Kurokawa, Y.; Nakane, K.; Ogata, N. *Cellulose* **2002**, *9*, 369–379

15. Rosenau, T.; Potthast A.; Adorjan I.; Hofinger A.; Sixta H.; Firgo H.; Kosma P. *Cellulose* **2002**, *9*, 283-291

16. Strlič, M.; Kolar, J. *J. Biochem. Biophys. Methods* **2003**, *56*, 265–279.

17. Murugesan, S.; Mousa, S.; Vijayaraghavan, A.; Ajayan, P. M.; Linhardt, R. J. *J Biomed Mater Res B Appl Biomater* **2006**, *79*, 298-304

18. Linhardt, R. J.; Murugesan S.; Park, T. J. *3230.3019 US1*

19. Linhardt, R. J., Gunay, N. S. *Sem Thromb Hem* **1999**, *25*, 5-16

20. Linhardt, R. J. *Chem. Indus* **1991**, *2*, 45-50

21. Langer, R.; Linhardt, R. J.; Cooney, C. L.; Klein, M.; Tapper, D.; Hoffberg, S. M.; Larsen, A. *Science* **1982**, *217*, 261-263

22. Murugesan, S.; Park, T. J.; Yang, H. C.; Mousa, S.; Linhardt, R. J. *Langmuir* **2006**, *22*, 3461-3463

23. Pierce, G.; Baird, J. A. *International patent WO 97/49434* **1997**

24. Jackson, R. L.; Busch, S. J.; Cardin, A. D. *Physiol Rev* **1991**, *71*, 481–539

25. Danishefsky, I.; Tzeng, F.; Ahrens, M.; Klein, S. *Thromb Res* **1976**, *8*, 131-140

26. Keuren, J. F.; Wielders, J. H.; Driessen, A.; Verhoeven, M.; Hendricks, M.; Lindhout, T. *Arterioscler Thromb Vasc Biol* **2004**, *24*, 613-617

27. Oliveira, G. B.; Carvalho, L. B.; Silva, M. P. C. *Biomaterials* **2003**, *24*, 4777-4783

28. Li, Y.; Neoh, K. G.; Cen, L.; and Kang, E. T. *Biotech Bioeng* **2003**, *84*, 305-313

29. Murugesan, S.; Linhardt, R. J. *Curr Org Synth* **2005**, *2*, 437-451.

30. El Seoud, O. A.; Koschella, A.; Fidale, LC.; Dorn, S.; Heinze, T. *Bio macromolecules* **2007**, *8*, 2629-2647

31. Viswanathan, G.; Murugesan, S.; Pushparaj, V.; Nalamasu, O.; Ajayan, P. M.; Linhardt, R. J.; *Biomacromolecules* **2006**, *7*, 415-418

32. Huddleston, J. G.; Visser, A.E.; Reichert, W. M.; Willauer, H.D.; Broker, G. A.; Rogers, R. D. *Green Chem* **2001**, *3*, 156-164

33. Ranke, J.; Stolte, S.; Störmann, R.; Arning, J.; Jastorff, B. *Chem Rev.* **2007** *107*, 2183-2206.

34. Hagiwara, R.; Ito, Y. *J Fluorine Chem* **2000**, *105*, 221-227

35. Welton, T. *Chem Rev* **1999**, *99*, 2071-2083

36. Ding, K.; Zhao, M.; Wang, Q. *Russ J Electrochem* **2007**, *43*, 1082-1090

37. Tsunashima, Katsuhiko.; Sugiya, Masashi. *Electrochemistry* **2007**, *75*, 734-736

38. Wang, Q.; Tang, H.; Me, Q.; Tan, L.; Zhang, Y.; Li, B.; Yao, S. *Electrochim Acta* **2007**, *52*, 6630-6637

39. Aliaga, C.; Santos, C. S.; Baldelli, S. *Phys Chem Chem Phys* **2007**, *9*, 3683-3700

40. O'Toole, S.; Pentlavalli, S.; Doherty, A. P. *J Phys Chem B* **2007**, *111*, 9281-9287

41. Cheng, H.; Zhu, C.; Huang, B.; Lu, M.; Yang, Y. *Electrochim Acta* **2007**, *52*, 5789-5794

42. Oyama, T.; Okajima, T.; Ohsaka, T. *J Electrochem Soc* **2007**, *154*, D322-D327

43. Doherty, A. P.; Koshechko, V.; Titov, V.; Mishura, A. *J Electroanal Chem* **2007**, *602*, 91-95

44. Paul, A.; Samanta, A. *J Phys Chem B* **2007**, *111*, 1957-1962

45. Silvester, D. S.; Compton, R. G. *Z Phys Chem* **2006**, *220*, 1247-1274

46. Wishart J. F.; Castner E. W. *J Phys Chem B* **2007**, *111*, 4639-4640

47. Chheda, J. N.; Dumesic, J. A. *Catal Today* **2007**, *123*, 59-70

48. Nadkarni, V. D.; Pervin, A.; Linhardt, R. J. *Anal Biochem* **1994**, 59-67

49. Hodosi, G.; Podanyi, B.; Kuszmann, J. *Carbohyd Res* **1992**, *230*, 327-342

50. Novoselov, N. P.; Sashina, E. S.; Kuz'mina, O. G.; Troshenkova, S. V. *Russ J Gen Chem* **2007**, *77*, 1395-1405

51. Zhu, S.D.; Wu, Y. X.; Chen, Q. M.; Yu, Z. N.; Wang, C. W.; Jin, S. W.; Ding, Y. G.; Wu, G. *Green Chem* **2006**, *8*, 325-327

52. Ye, J.; Zhao, X. F.; Xiong, *J. Prog Chem* **2007**, *19*, 478-484

53. Erdmenger, T.; Haensch, C.; Hoogenboom, R.; Schubert, U. S. *Macromol Biosci* **2007**, *7*, 440-445

54. Graenacher, C. *U.S. Patent 1,943, 176, 1934*

55. Remsing, R. C.; Swatloski, R. P.; Rogers, R. D.; Moyna, G. *Chem Commun* **2006**, *12*, 1271-1273.

56. Moulthrop, J. S.; Swatloski, R. P.; Moyna, G.; Rogers, R. D. *Chem Commun* **2005**, *12*, 1557-1559

57. Turner, M. B.; Spear, S. K.; Holbrey, J. D.; Rogers, R. D. *Biomacromolecules* **2004**, *5*, 1379-1384.

58. Swatloski, R. P.; Spear, S. K.; Holbrey, J. D.; Rogers, R. D. *J Am Chem Soc.* **2002**, *124*, 4974-4975.

59. Poplin, J. H.; Swatloski, R. P.; Holbrey, J. D.; Spear, S. K.; Metlen, A.; Gratzel, M.; Nazeeruddin, M. K.; Rogers, R. D. *Chem Commun* **2007**, 2025-2027

60. Swatloski, R. P.; Holbrey, J. D.; Weston, J. L.; Rogers, R. D. *Chim Oggi* **2006**, *24*, 31-35

61. Turner, M. B.; Spear, S. K.; Holbrey, J. D.; Daly, D. T.; Rogers, R. D. *Biomacromolecules* **2005**, *6*, 2497-2502

62. Ren, Q.; Wu, J.; Zhang, J.; He, J. S. *Acta Polym Sin* **2003**, 448-451

63. Wu, J.; Zhang, J.; Zhang, H.; He, J. S.; Ren, Q.; Guo, M. *Biomacromolecules* **2004**, *5*, 266-268

64. Heinze, T.; Schwikal, K.; Barthel, S. *Macromol Biosci* **2005**, *5*, 520-525

65. Egorov, V. M.; Smirnova, S. V.; Formanovsky, A. A.; Pletnev, I. V.; Zolotov, Y. A. *Anal Bioanal Chem* **2007**, 387, 2263-2269

66. Murugesan, S.; Wiencek, J. M.; Ren, R. X.; Linhardt, R. J. *Carbohyd Polym* **2006**, *63*, 268-271

67. Ren, R. X.; Wu, J. X. *Org Lett* **2001**, *3*, 3727-3728

68. Murugesan, S.; Karst, N.; Islam, T.; Wiencek, J. M.; Linhardt, R. J. *Synlett* **2003**, 1283-1286

69. Park, T. J.; Weiwer, M.; Yuan, X. J.; Baytas, S. N.; Munoz, E. M.; Murugesan, S.; Linhardt, R. J. *Carbohyd Res* **2007**, *342*, 614-620

70. Pushparaj, V. L.; Shaijumon, M.M.; Kumar, A.; Murugesan, S.; Ci, L.; Vajtai, R.; Linhardt, R. J.; Nalamasu, O.; Ajayan, P. M. *Proc Natl Acad Sci USA* **2007**, *104*, 13574-13577

71. Li, D.; Xia, Y. N. *Adv Mater* **2004**, *16*, 1151-1170

72. Denizli, A. J. *Appl Polym Sci* **1999**, *74*, 655-662

73. Kötz, R.; Carlen, M. *Electrochimica Acta* **2000**, *45*, 2483-2498

74. Frackowiak, E.; Beguin, F. *Carbon* **2001**, *39*, 937-950

75. Frackowiak E . Phys Chem Chem Phys **2007**, *9*, 1774-1785

76. Becker, H. E. *US Patent 2800616 (to General Electric)* **1957**

77. Boos, D.I. *US Patent 3536963 (to Standard Oil, SOHIO)* **1970**

78. Arico A. S.; Bruce, P.; Scrosati, B.; Tarascon, J. M.; Van Schalkwijk, W. *Nat Mater* **2005**, *4*, 366-377

79. Burke, A. *J Power Sources* **2000**, *91*, 37-50

80. Tarascon, J. M.; Armand, M. *Nature* **2001**, *414*, 359-367

81. Dresselhaus, M. S.; Thomas, I. L. *Nature* **2001**, *414*, 332-337

82. Sarangapani, S.; Tilak, B. V.; Chen, C. P. *J Electrochem Soc* **1996**, *143*, 3791-3799

83. Pandolfo A. G, Hollenkamp A. F. *J Power Sources* **2006**, *157*, 11-27

84. Frackowiak, E.; Lota, G.; Machnikowski, J.; Vix-Guterl, C.; Beguin, F. *Electrochim Acta* **2006**, *51*, 2209-2214

85. Frackowiak, E.; Lota, G.; Pernak, J. *Appl Phys Lett* **2005**, *86*, Art. No. 164104

86. Ania, C. O.; Pernak, J.; Stefaniak, F.; Raymundo-Pinero, E.; Beguin, F. *Carbon* **2006**, *44*, 3126-3130

87. Xu, B.; Wu, F.; Chen, R. J.; Cao, G. P.; Chen, S.; Wang, G. Q.; Yang, Y. S. *J Power Sources* **2006**, *158*, 773-778

88. Balducci, A.; Dugas, R.; Taberna, P. L.; Simon, P.; Plee, D.; Mastragostino, M.; Passerini, S. *J Power Sources* **2007**, *165*, 922-927

89. Sugimoto, W.; Yokoshima, K.; Ohuchi, K.; Murakami, Y.; Takasu, Y. *J Electrochem Soc* **2006**, *153*, A255-A260

90. Frackowiak, E.; Metenier, K.; Bertagna, V.; Beguin, F.; *Appl Phys Lett* **2000**, *77*, 2421-2423

91. Kim, C.; Ngoc, B. T. N.; Yang, K. S.; Kojima, M.; Kim, Y. A.; Kim, Y. J.; Endo, M.; Yang, S. C. *Adv Mater* **2007**, *19*, 2341-2346

92. Chao, Y. J.; Yuan, X. X.; Ma, Z. F.; *Rare Metal Mat Eng* **2007**, *36*, 1110-1114

93. Lota, G.; Lota, K.; Frackowiak, E. *Electrochem Commun* **2007**, *9*, 1828-1832

94. Du, C. S.; Pan, N. *Nanotechnology* **2006**, *17*, 5314-5318

95. Xu, B.; Wu, F.; Wang, F.; Chen, S.; Cao, G. P.; Yang, Y. S. *Chinese J Chem* **2006**, *24* 1505-1508

96. Du, C. S.; Pan, N. *J Power Sources* **2006**, *160*, 1487-1494

97. Futaba, D. N.; Hata, K.; Yamada, T.; Hiraoka, T.; Hayamizu, Y.; Kakudate, Y.; Tanaike, O.; Hatori, H.; Yumura, M.; Iijima, S. *Nat Mater* **2006**, *5*, 987-994

98. Kim, Y. J.; Kim, Y. A.; Chino, T.; Suezaki, H.; Endo, M.; Dresselhaus, M. S. *Small* **2006**, *2*, 339-345

99. Endo, M.; Kim, Y. J.; Chino, T.; Shinya, O.; Matsuzawa, Y.; Suezaki, H.; Tantrakarn, K.; Dresselhaus, M. S. *Appl Phys A-Mater* **2006**, *82*, 559-565

100. Shiraishi, S.; Kibe, M.; Yokoyama, T.; Kurihara, H.; Patel, N.; Oya, A.; Kaburagi, Y.; Hishiyama, Y. *Appl Phys A-Mater* **2006**, *82*, 585-591

101. Kim, Y. T.; Ito, Y.; Tadai, K.; Mitani, T.; Kim, U. S.; Kim, H. S.; Cho, B. W. *Appl Phys Lett* **2005**, *87*, Art. No. 234106

102. Du, C. S.; Yeh, J.; Pan, N. *Nanotechnology* **2005**, *16*, 350-353

103. Lee, C. Y.; Tsai, H. M.; Chuang, H. J.; Li, S. Y.; Lin, P.; Tseng, T. Y. *J Electrochem Soc* **2005**, *152*, A716-A720

104. Zhou, C. F.; Kumar, S.; Doyle, C. D.; Tour, J. M. *Chem Mater* **2005**, *17*, 1997-2002

105. Frackowiak, E.; Jurewicz, K.; Beguin, F. *Pol J Chem* **2004**, *78*, 1345-1356

106. Pico, F.; Rojo, J. M.; Sanjuan, M. L.; Anson, A.; Benito, A. M.; Callejas, M. A.; Maser, W. K.; Martinez, M. T. *J Electrochem Soc* **2004**, *151*, A831-A837

107. Yoon, B. J.; Jeong, S. H.; Lee, K. H.; Kim, H. S.; Park, C. G.; Han, J. H. *Chem Phys Lett* **2004**, *388*, 170-174

108. Chatterjee, A. K.; Sharon, M.; Banerjee, R.; Neumann-Spallart, M. *Electrochim Acta* **2003**, *48*, 3439-3446

109. Park, J. H.; Ko, J. M.; Park, O. O. *J Electrochem Soc* **2003**, *150*, A864-A867

110. Frackowiak, E.; Beguin, F.; *Carbon* **2002**, *40*, 1775-1787

111. Frackowiak, E.; Jurewicz, K.; Szostak, K.; Delpeux, S.; Beguin, F. *Fuel Process Technol* **2002**, *77*, 213-219

112. Jurewicz, K.; Delpeux, S.; Bertagna, V.; Beguin, F.; Frackowiak, E. *Chem Phys Lett* **2001**, *347*, 36-40

113. Frackowiak, E.; Jurewicz, K.; Delpeux, S.; Beguin, F. *J Power Sources* **2001**, *97*, 822-825

114. Murakami, M. A.; Kaneko, Y.; Kadokawa, J. I. *Carbohyd Polym* **2007**, *69*, 378-381

115. Zhang, H.; Wang, Z. G.; Zhang, Z. N.; Wu, J.; Zhang, J.; He, H. S. *Adv Mater* **2007**, *19*, 698-704

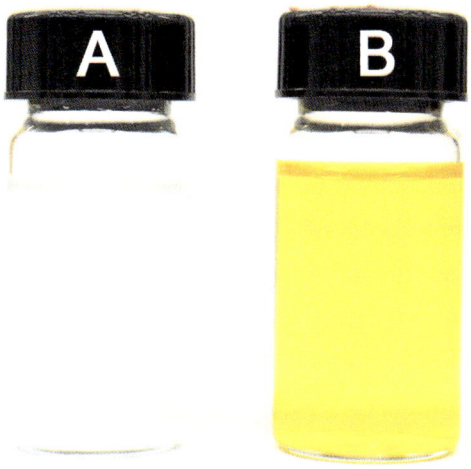

Figure 4.3. Aqueous suspensions of (A) cellulose nanocrystals (0.8 wt %) and (B) FITC-labeled cellulose nanocrystals (0.5 wt %). (Reprinted with permission from ref 12. Copyright 2007 American Chemical Society)

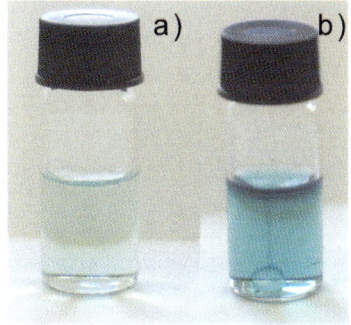

Figure 6.10. O$_2$-saturated β-D-glucose/2,2'-azino-bis(3-ethylbenzthiazoline-6-sulphonic acid) solution containing horse radish peroxidase before (a) and 1 min after (b) the addition of glucose oxidase capsules

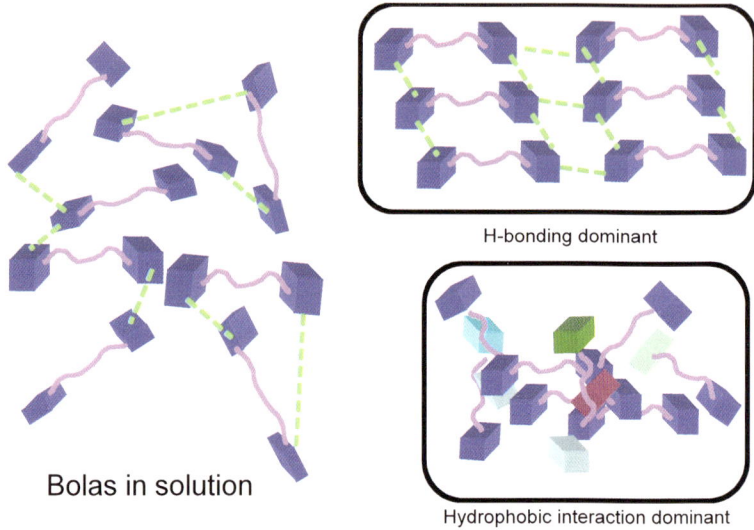

H-bonding dominant

Bolas in solution

Hydrophobic interaction dominant

Figure 14.6 - Schematic of bolaform self assembly in solution

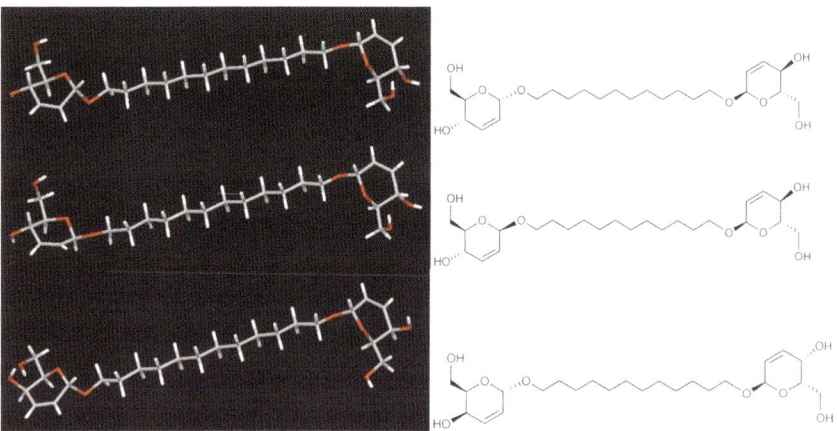

*Figure 14.7 - X-ray crystal structures of compounds **3d** (major and minor diastereomers) and **3e**.*

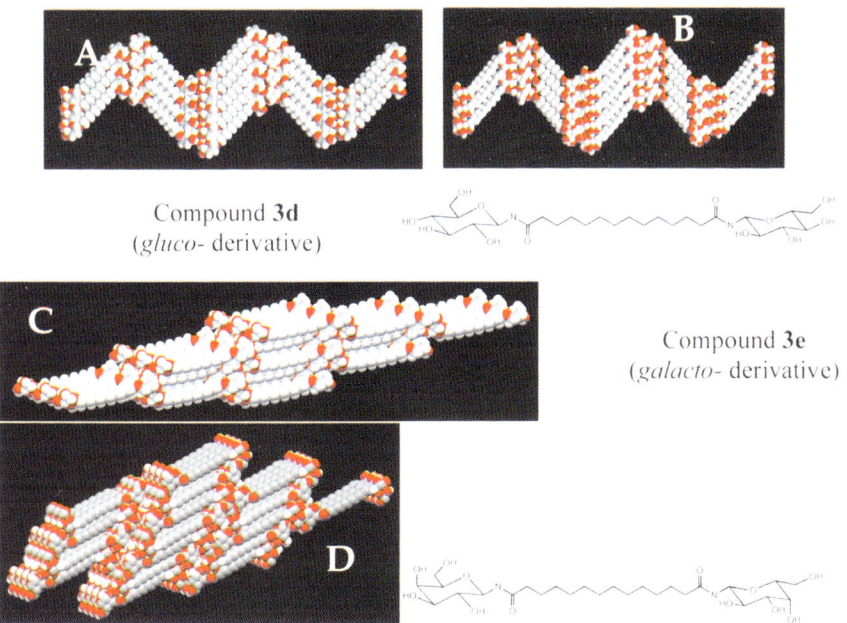

Compound **3d**
(*gluco*- derivative)

Compound **3e**
(*galacto*- derivative)

Figure 14.8 - Comparative X-ray structures of glycal bolaforms **3d** and **3e** with
similar carbohydrate bolaforms

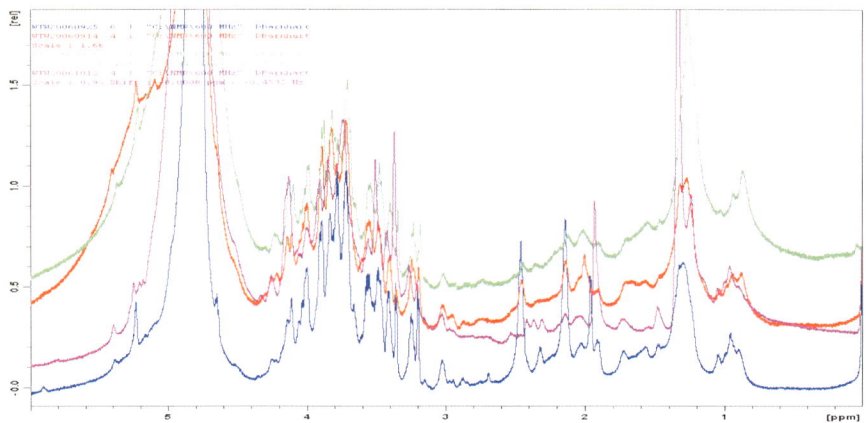

Figure 15.3. 600 MHz ^{1}H HR-MAS NMR spectra of WT (D), GH9C1-1 (C), GH9C3-1 (B), and GH9C2-2 (A) Arabidopsis thaliana root tissues scaled by carbohydrate region intensities. Arrows indicate significant differences in peak intensities in the acetate region of the spectra.

Chapter 8

Green Composites Prepared from Cellulose Nanoparticles

Jacob D. Goodrich [1,2] and William T. Winter [1]

[1] SUNY College of Environmental Science and Forestry, Cellulose Research
Institute and Department of Chemistry, Syracuse, NY 13210
[2] Eastman Chemical Company, Kingsport, TN 37662

Nanoparticles were prepared from bacterial cellulose. The particles were topochemically modified with poly(caprolactone), (PCL), via a ring-opening polymerization on the surface hydroxyl groups. The reaction is a facile, solvent-free system, containing only cellulose as the initiator, an organic acid as the catalyst, and ε-caprolactone as the monomer. Graft weight % was calculated by gravimetric methods and confirmed with NMR, while chemical attachment of PCL was confirmed by spectroscopic methods including Fourier Transform Infrared (FTIR), and [13]C Cross Polarization Magic-Angle Spinning NMR (CP-MAS). The grafting reaction was performed as a route to improve compatibility between the cellulose nanoparticles and a PCL matrix in thermoplastic composites. Nanocomposites were formed by extensive mixing of the continuous and filler phases in a twin-screw extruder to create good particle dispersions. The thermal and viscoelastic properties of the composites were investigated by differential scanning calorimetry (DSC), and dynamic mechanical analysis (DMA). Thermal analysis data does not indicate a significant enhancement in composite particle-matrix interactions with surface grafting. The tensile modulus of the PCL matrix was improved by reinforcement with all filler types. The grafted cellulose-PCL filled composites offered only marginal improvements in tensile modulus relative to unmodified cellulose nanoparticle filled composites. All cellulose nanoparticle based composites closely matched the mechanical properties of similarly prepared nanoclay composites.

153

Background

The past several years have seen increasing efforts to create composite materials incorporating natural fibers as the reinforcement phase. *(1,2)* Cellulose fiber reinforced materials possess the advantages of renewability, lower cost, and lower density over conventional reinforcement phases such as glass fiber and carbon fibers. The elastic modulus of the crystalline domains of cellulose approaches 150 GPa, approximately 2 times larger than that of glass fiber, while the crystals exhibit a density close to 1.6 g/cm^3, significantly lower than that of glass fibers. *(3,4)*

More recently, cellulose nanoparticles have been studied as reinforcement agents in composites. *(5-7)* These are prepared from cellulose fibers, with relative ease, by a process involving acid hydrolysis and mechanical shearing. The nanoparticles retain the crystalline properties and associated stiffness that are observed with cellulose fibers or microfibers, and possess the advantage of higher specific surface areas ($\sim$ 250 m^2/g). *(8)* The higher specific surface area enables increased particle-matrix interactions, often providing better composite performance.

A problem encountered with using cellulose in composite applications, is the poor compatibility of the hydrophilic particles with a hydrophobic matrix. Composite performance is strongly a function of the interaction between the particle and matrix phases. The guiding principle of 'like dissolves like' when estimating solubility, applies also to composites, where 'like' particle surface and matrix phases typically provide improved composite interactions. Such interactions provide better fiber-matrix load transfer, thereby enhancing the overall strength of the material. To improve compatibility and enhance particle-matrix bonding, a topochemical approach was used.

The conventional approach to topochemical modification of cellulose involves attaching single molecule organic functionalities to the hydroxyl groups through ester or ether linkages. *(9-11)* Typically, these functional groups increase the hydrophobic character of the particle surface. An alternate method that has been recently explored is a 'grafting from' approach, where polymer chains are grown directly from the surface of the particles. *(12,13)* Such work has been done using cellulose fibers, silica, and carbon whiskers as the substrates. *(13-15)* For monomers such as ε-caprolactone, a ring-opening polymerization (ROP) has been used to create grafts from particle surfaces. *(13,14,16-17)* For the ROP of ε-caprolactone, an alcohol can be used as the initiator, and the polymerization can proceed in the presence of a catalyst. Several catalyst systems have been developed including; stannous 2-ethylhexanoate (Sn(Oct)2), enzymatic catalysts, and organic acid catalysts such as tartaric acid. *(14,16-20)* For the organometallic type catalysts, toxic remnants are often found attached to the chain ends, and depending upon the application, they often need to be removed. *(21)* The organocatalytic approach has been shown to be an effective, mild, nontoxic reaction system free from metals and solvents, and was selected as the reaction system for our grafting reactions based upon these successes in previous findings. *(14)*

The grafted cellulose nanoparticles were melt processed with PCL to form thermoplastic nanocomposites. PCL was selected as the matrix plastic for the composites based upon its known biodegradability and mild melt processing conditions, and also based on the hypothesis that the surface grafted PCL oligomers would enhance the interaction of the cellulose nanoparticles with this material. The thermal and mechanical properties of the nanocomposites were studied and are summarized here. The aim of this paper is to study and characterize thermoplastic composites based upon grafted cellulose nanoparticles.

Materials and Methods

Raw Materials

Bacterial cellulose (Primacel™) was provided by CP Kelco Company (SanDiego, CA). Montmorillonite nanoclay (MMT) was purchased from Ward Scientific (Rochester, NY), and was used as received. The PCL used as the matrix phase in the composites was granulated, 80 K MW PCL, provided by Solvay Chemical Co. (Bridgeport, N.J.). The measured T_g and T_m of the PCL were –65 °C and +56 °C, respectively. All other chemicals were of reagent grade or higher and used as received.

Cellulose Nanoparticle Preparation

Bacterial cellulose as received was cleaned with a bleaching process using a sodium chlorite/glacial acetic acid mixture and washed with water until a neutral pH was reached. The material was hydrolyzed in 65% (w/v) sulfuric acid for 2 hours at 40°C after which the mixture was quenched with ice water, and the cellulose was separated from the bulk acid by vacuum filtration. *(22)* The hydrolyzed cellulose was rinsed repeatedly with distilled water and filtered until the material reached a neutral pH. The hydrolyzed cellulose was resuspended in distilled water using a Waring blender. The suspension was mechanically disrupted with an APV Gaulin 1000 homogenizer operating at 8000 psi with 10 to 15 recycling passes. The resulting bacterial cellulose nanocrystal aqueous suspension was exchanged in *t*-butanol and freeze-dried to minimize agglomeration. The solid material was kept dry in a dessicator prior to its use in the grafting reactions.

Grafting

The ROP of ε-caprolactone was performed according to a procedure developed by Hafren and Cordova. *(14)* ε-Caprolactone and tartaric acid were added to glass pressure tubes sealed with Teflon screw caps. The mole ratio of monomer to catalyst was selected based upon the results of a previous report, suggesting an optimum 10:1 mole ratio to obtain higher MW grafts. *(14)* The molecular weight of the grafts was previously found to decrease with increasing molar ratios above this limit, due to self-polymerization. *(14)* The mixture containing ε-caprolactone and tartaric acid was heated to 120 °C, and known amounts of bacterial cellulose nanocrystals (BCNC) were added to the reaction tubes, and allowed to react under various time intervals (1-6 h). After the reactions, the unreacted monomer and un-grafted polymer were washed from the fiber with several rinses of dichloromethane. Residual ungrafted polymer was further removed by Soxhlet extraction in dichloromethane for 24 hours. Following this treatment, the tartaric acid was removed by soxhlet extraction in water for 24 hours. The grafted cellulose (hereafter referred to as cellulose-PCL) was then vacuum dried in an oven at 60 °C for 24 h prior to analysis.

Nanocomposite Formation

Nanocomposites were formed by premixing the nanoparticles with granulated PCL in a sealed plastic bag with measured filler loadings ranging from 0 to 10 wt% of fiber. Five-gram batches of sample were fed into a DACA twin-screw bench-top extruder and extensively mixed under high pressure and shear rate. The processing temperature used was 150 °C. The melt temperature of the PCL, and the thermal degradation temperature of the fillers dictated the selection of the thermal processing temperatures. The chosen temperature was significantly above the PCL melting temperature (56 °C) and below the temperature at which any filler degradation would normally occur (180 °C). A screw speed of 125 rpm was used, and the material was mixed for 30 min in a recirculating mode prior to being released from the die. A recirculating mode was used to improve particle dispersions in the matrix phase. The melt extruded from the die was collected on water-cooled tape roller in approximate dimensions of 8 mm in width by 0.5 mm in thickness. Melt processing was selected over solvent casting for composite preparation to overcome sedimentation effects present in solvent casting techniques, and to mimic processing techniques that are more practical on a manufacturing scale.

Thermal Analysis

The thermal properties of the nanocomposites were measured by DSC. The instrument used for analysis was a TA Instruments DSC 2920, calibrated with an indium standard. Samples sizes varied from 10 to 12 mg and were heated in standard flat-bottom aluminum pans at 10 °C min^{-1} against a reference aluminum pan. Typical test methods consisted of an initial heating scan at 10 °C min^{-1} to 150 °C, a cooling cycle from 150 °C to –100 °C at 10 °C min^{-1}, and a second heating cycle from –100 °C to 150 °C at 10 °C min^{-1}. The first heating cycle was used to reset the thermal history of the composite samples. The cooling and second heating curves were used for transition analysis. The melting and crystallization temperatures were taken as the peak temperatures in the transitions. The glass transition temperature (T_g) was taken as the inflection point at the baseline change in heat capacity, corresponding to the glassy-to-rubbery transition. Enthalpies of melting and crystallization were measured by integration.

Mechanical Analysis

DMA was used to measure the viscoelastic properties of the polymer nanocomposites formed. Measurements were performed on a TA Instruments DMA 2980. Sample strips, with dimensions approximately 12 x 8 x 0.5 mm were analyzed in a film tension clamp. Data was collected using a single frequency temperature ramp profile. The temperature was ramped from 120 °C to 50 °C at a rate of 5 °C/min. The oscillation frequency was fixed at 1.0 Hz. The oscillation amplitude was set to 10 μm, and a static preload force of 0.01N was used in autotension mode with a setting of 125%. The static preload force was set to prevent sample buckling during the initial phase of testing. The force track autotension applies a static force proportional to the modulus of the material to reduce creep as the sample softens at higher temperatures.

Results

Microscopy

Transmission electron microscopy (TEM) was used to confirm the presence of cellulose nanoparticles after exposing bacterial cellulose fibers to treatments of acid hydrolysis, neutralization, and homogenization. Figure 1 is a TEM micrograph demonstrating the isolated cellulose nanoparticles. The image is very similar to bacterial cellulose nanoparticles (BCNC) produced by the same method and reported elsewhere. *(22)* The particles are approximately 10 – 20 nm in width and vary in length up to several micrometers.

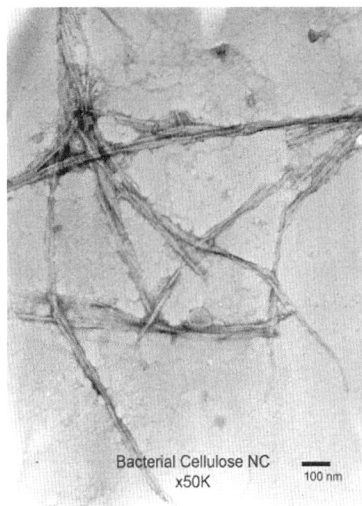

Figure 1. TEM micrograph of BCNC stained with uranyl acetate

Reaction Characteristics

The ROP of ε-caprolactone from whole cellulose fibers using organic acid catalysis has been described in a communication by Hafren and Cordova. *(14)* Our interests were in applying this reaction system to cellulose nanoparticles to produce surface decorated particles with improved surface properties, suitable for use as reinforcing agents in thermoplastic nanocomposites. In addition to replicating some of the work previously described for the cellulose fibers, we have determined the grafting weight % as a function of reaction time and molar ratio of monomer to cellulose. We have also found through NMR analysis, that the grafting occurs primarily at the C6 hydroxyl of the anhydroglucose residues.

The organic acid catalyzed ROP of ε-caprolactone from the hydroxyl groups of cellulose, is thought to proceed via a monomer activation mechanism. *(23)* The organic acid acts as a proton donor, to activate the monomer. The nucleophilic alcohol then reacts with the proton-activated monomer to form the ring-opened monoester adduct. *(23)* Polymerization proceeds through the terminal hydroxyl group as it reacts with subsequent units of the proton-activated monomer.

After purification and drying of the cellulose-PCL nanoparticles, a significant weight change was observed. This weight change is described as the graft weight %, and is calculated from the following equation: *(15)*

Graft weight % = [(Final grafted particle wt (g) – initial particle wt (g))/ initial particle wt (g)] x 100

Graft weight % changed significantly with reaction time and the ratio of monomer to glycosyl units. The graft weight % reached values as high as 50% under conditions of a 4 h reaction time and a monomer to glycosyl unit ratio of 40:1. As shown in Figure 2, graft weight % reaches a plateau after a 3 h reaction time at 120 °C, with a fixed ratio of 10:1 for monomer to glycosyl units. This reaction time is consistent with previous studies done with the organocatalytic ROP of ε-caprolactone from benzyl alcohol, where reaction completion was observed after 4 h. *(23)*

Since little increase in grafting weight % was observed after 4 h, all subsequent grafting reactions were carried out for 4 h reaction times to ensure that the maximum graft weight % was reached. Figure 3 shows the dependence of graft weight % on the mole ratio of monomer to glycosyl units. Graft weight % also reaches a maximum at a mole ratio of 40:1 monomer : glycosyl units. This condition was used in the bulk preparation of the cellulose-PCL due to these favorable levels of grafting achieved.

Previous studies on grafting PCL to cellulose fibers reported a maximum weight gain of 11%. *(14)* The larger weight gains observed for our bacterial cellulose nanoparticles may be the result of the significantly higher surface areas present with the nanoparticles relative to the whole fibers, which provides a higher initiator concentration. Currently, it is still not understood whether the large weight gain recognized in the grafted particles results from the grafting of PCL at multiple surface hydroxyl sites, or from the polymerization of high molecular weight PCL chains from fewer hydroxyl sites. NMR data discussed below, indicates that a large fraction of C6 surface hydroxyls are substituted.

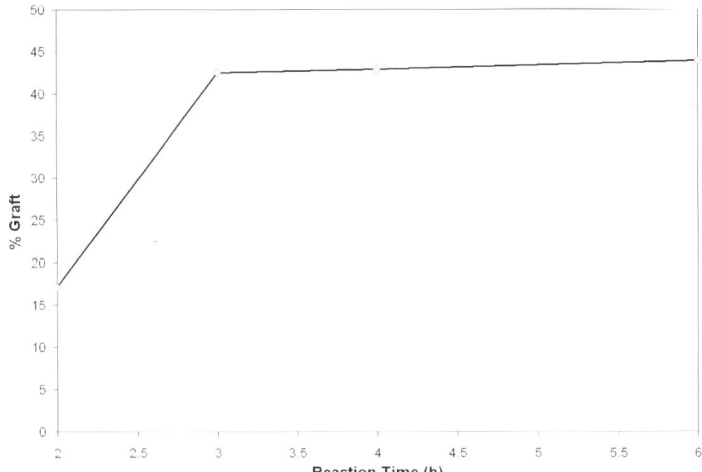

Figure 2. Graft weight % as a function of reaction time for the cellulose-graft-PCL

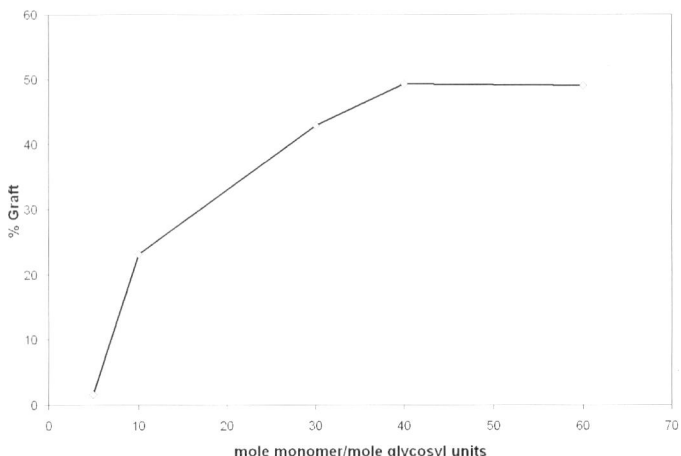

Figure 3. Graft weight % as a function of the molar ratio between the monomer and glycosyl units in the cellulose

Spectroscopic Analysis

Chemical grafting was determined by FTIR analysis (Figure 4) and ^{13}C CP-MAS NMR (Figure 5). The FTIR spectrum of the chemically grafted BCNC shows the most notable feature, a strong band at 1740 cm^{-1}, corresponding to the ester carbonyl groups in the PCL. This signal is absent in the BCNC prior to modification. To distinguish between the chemical attachment of the PCL to cellulose and the physical association of PCL with cellulose, a physical mixture of the un-reacted BCNC and free PCL extracted from the reaction system was prepared as a reference, and analyzed under the same conditions. The mixture shows an ester carbonyl signal at 1730 cm^{-1}, which is consistent with pure PCL and 10 cm^{-1} lower than that observed in the chemically grafted material. The difference in frequency of the carbonyl ester bands observed between the chemically and physically prepared materials indicates that there is a difference in the chemical environments surrounding the ester groups. This result provides a strong indication that the PCL is covalently attached to the BCNC.

^{13}C-CPMAS NMR analyses also support the grafting of the PCL (Figure 5). In the BCNC spectrum, a distinct shoulder at 63 ppm is observed on the main C6 peak at 65 ppm. This shoulder arises due to a separate chemical shift for the hydroxymethyl groups on the surface and in the amorphous regions of the particles from those present in the crystal interior. *(24, 25)* The difference in chemical shift at the surface or in the amorphous regions is related to the presence of multiple conformations of the hydroxymethyl groups in these regions. In the crystal interior the hydroxymethyl group is locked into the trans-gauche conformation, giving rise to the higher chemical shift observed at 65 ppm. *(24)* Owing to the highly crystalline nature of our cellulose preparations (~85%) *(22)* and large surface areas, the bulk of the contribution to the shoulder signal is presumed to arise from the surface C6 groups.

A similar situation is observed for the C4 carbon (87 ppm) of cellulose and is apparent from the spectra in Figure 5. As the grafting reaction proceeds, the C6 surface signal begins to gradually decay in intensity. Concomitantly, the signals in the region of 72 ppm increase in intensity; this region includes C3, C5, and the signal for grafted C6. A similar occurrence was observed for surface maleated cellulose microfibers, where the C6 surface peak disappeared after long reaction times, and the ratio of signal intensities in the C2, C3, and C5 cluster changed. *(26, 27)* After a 4 h reaction time, the C6 surface signal is significantly diminished. A slight shoulder is still observable, indicating that surface substitution is not complete, but has proceeded to a significant extent. The absence of other spectral changes surrounding the C2 and C3 signals suggests that modification occurs predominantly at the C6 position of the glycosyl residues. Substitution at C2 or C3 would be expected to yield more significant changes in the carbohydrate signal region of the NMR spectra.

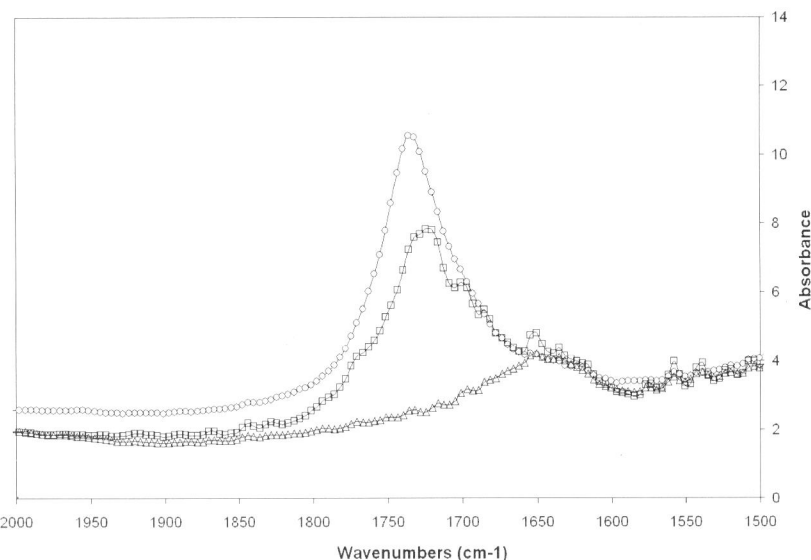

Figure 4. FTIR overlay of unmodied bacterial cellulose nanocrystals (Δ), cellulos-PCL prepared using a 4h reaction time and 40:1 ratio of monomer:glycosyl units (○), and a physical blend of bacterial cellulose nanoparticles and free PCL extracted from the reaction mixture (□).

Also, with the progression of the reaction, methylene carbon signals from the PCL appear between 20 to 35 ppm, and the carbonyl carbon signal appears near 175 ppm. Under conditions of longer reaction times and as the graft weight % increases, these signals corresponding to the PCL, sharpen and increase in intensity. The ratio of the integration values for the methylene signals from PCL to those of the anomeric carbons of cellulose provides an estimate for the graft weight %. The estimate generated from NMR analysis corresponds well with that measured gravimetrically (graft % NMR = 46%, graft % gravimetric = 50%). It should be noted that the intensity or area of the carbonyl carbon signal cannot be used to accurately evaluate the graft %, because this carbon does not have an attached proton that would allow for efficient magnetization transfer in the CP-MAS experiment, thus underestimating the relative content of side group to backbone groups. From the spectral features observed in the FTIR and the ^{13}C CP-MAS data, we can be reasonably certain that the PCL chains are covalently attached to the cellulose particles, predominantly at the C6 position.

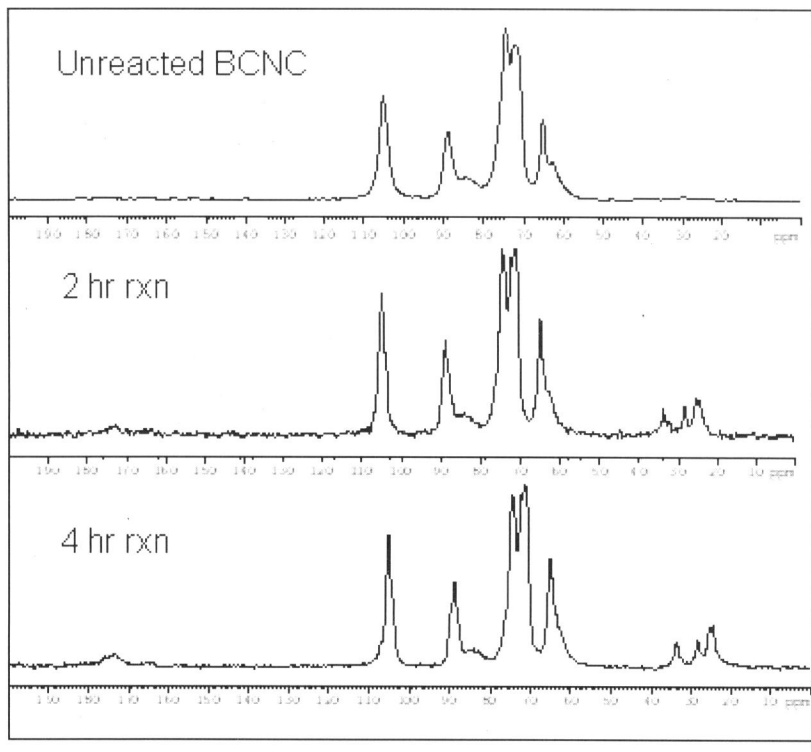

Figure 5. ^{13}C CPMAS NMR spectra of unmodified and surface grafted bacterial cellulose.

Thermal Properties of Composites

DSC was used to study the thermal properties of the cellulose-PCL / PCL nanocomposites to draw correlations between the thermal behavior and the interactions of the reinforcing phase with the matrix. Measured properties of the composites including the T_g, T_c, T_m, the enthalpy of fusion (ΔH_m), and percent crystallinity (X_c) were investigated and the results are presented in Table I. The crystallinity of the composite materials was estimated using the following equation: *(28)*

$$X_c = [\Delta H_m / \Delta H_m^{\,o}]*(100/w)$$

where w is the weight fraction of PCL, ΔH_m is the enthalpy of fusion, and $\Delta H_m^{\,o}$ is the hypothetical enthalpy of fusion for purely crystalline PCL. The estimate of $\Delta H_m^{\,o}$ was taken as 136 J g^{-1} for PCL crystals of infinite thickness. *(29)*

The thermal properties were measured and compared with those of composite samples containing bacterial cellulose fibers, unmodified cellulose nanoparticles, or montmorillonite nanoclay (MMT), to establish the effect of the surface grafting on the thermal properties of the composites, and evaluate the performance of the cellulose-PCL filler relative to more conventional fillers. For the neat PCL matrix, the T_g, T_c, T_m, and X_c were found to be –65 °C, 32 °C, 56 °C, and 39%, respectively. The thermal properties for the neat PCL and the prepared composite measured by DSC are collected in Table I. The fillers did not exhibit any thermal transitions in the temperature range investigated, thus all transitions arose from the matrix phase.

Table I. Thermal properties of PCL composites measured by DSC. Cellulose-PCL, and Cellulose-NC designations are for PCL grafted cellulose nanoparticles, and unmodified cellulose nanoparticles, respectively.

Material (wt% filler)	w_F	T_g (°C)	T_c (°C)	T_m (°C)	ΔH_c (J/g)	ΔH_m (J/g)	X_c
Neat PCL	0	-64.5	31.8	55.9	62.6	53.5	39.3
PCL/Cellulose-PCL (95/5)	0.05	-64.5	34.8	56.4	57.1	54.2	42.0
PCL/Cellulose-PCL (90/10)	0.10	-65.7	34.2	57.0	53.3	54.0	44.1
PCL/Cellulose-NC (95/5)	0.05	-67.7	35.3	55.9	53.4	54.1	41.9
PCL/Cellulose-NC (90/10)	0.10	-68.2	35.8	56.0	51.2	53.8	44.0
PCL/Montmorillonite (95/5)	0.05	-64.9	35.8	56.4	51.7	52.0	40.2
PCL/Montmorillonite (90/10)	0.10	-64.3	37.3	56.5	51.0	52.3	42.7
PCL/Cellulose Fiber (95/5)	0.05	-65.4	34.0	55.8	51.6	54.2	42.0
PCL/Cellulose Fiber (90/10)	0.10	-66.4	34.1	56.2	50.1	53.1	43.4

The T_g is an important characterisitc of composites that can control different properties of a material, and dictate end-use applications. In a DSC experiment, T_g is measured by the change of heat flow in the sample relative to an inert reference at a temperature corresponding to an onset of segmental polymer chain motion. Increases in T_g for composites over the neat material indicate restricted chain motions, which may be the result of particle-matrix interactions hindering matrix chain dynamics. The unmodified cellulose nanoparticle composites (cellulose-NC) showed a supression in T_g by 3°C to 4°C relative to the pure matrix material, representing an unfavorable particle-matrix interaction. The cellulose-PCL reinforced composites had T_g values similar to the matrix PCL indicating that these paticles have little interaction with the amorphous PCL phase. Other work involving grafted cellulose nanoparticles in PCL found no significant increases in T_g with the addition of filler. (16,17) The authors attributed this to the crystallization of PCL at the surface of the particles. Our observations appear to be consistent with these reports. (16,17)

The effects of the reinforcement phase on the T_m of the composite materials can also be a useful indicator of particle-matrix behavior. An increase in T_m can indicate favorable particle-matrix interactions, as the crystals require higher temperatures to overcome the interactions and proceed to the melt state. The T_m of the cellulose-PCL composites was only slightly greater than the pure PCL or any of the other filled composites, suggesting a minimal change in the particle-matrix interactions upon surface grafting of the particles. The heating and cooling cycle for the cellulose-PCL composites, shown in Figure 6, demonstrates the slight shift in T_m with the cellulose-PCL filled composites.

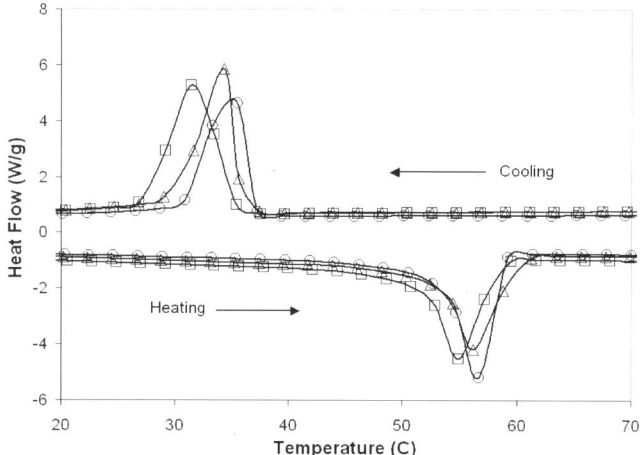

Figure 6. DSC scans of PCL nanocomposites with cellulose-PCL filler at different loadings (exotherm up). Filler loadings 0% (□), 5% (Δ), 10% (○).

The crystallization temperature (T_c) was found to increase with increasing filler loadings, for all filler types. The 10 wt% cellulose-PCL material experiences an elevation in T_c by as much as 3 °C, as shown in Figure 6. The likely explanation for the increase in T_c, is that the particles serve as nucleation sites allowing for faster crystallization and crystal growth of the PCL matrix upon cooling from the melt. This is also supported by the fact that the X_c increases with increasing filler content. The most marked increase occurs with the 10 wt% cellulose-PCL composite where a 5 % increase in crystallinity is observed. Other work reports similar crystallinity observations to our own, and goes further to describe that composite crystallinity actually declines at higher filler loadings (40 – 50 wt%) due to crystallization competing with increases in viscosity. *(16)*

Certain cellulose and chitin thermoplastic composites have been reported to exhibit a phenomenon known as transcrystallization. *(28,30)* Transcrystallization involves a high degree of matrix ordering occurring at the particle surface, and is characterized by a high density of crystallites growing perpendicularly to the particle surface. This behavior can be revealed through DSC studies on the composites, where elevated crystallization temperatures and degrees of crystallinity are observed in the presence of filler. *(28)* Based on data from thermal analysis experiments, transcrystallization appears to occur in our composite systems, particularly with the unmodified cellulose nanoparticle fillers. Contrary to our initial hypothesis, surface grafting of the cellulose nanoparticles does not appear to significantly enhance particle-matrix interactions in the composites.

Mechanical Properties

The PCL composites filled with cellulose-PCL were analyzed by dynamic mechanical analysis at different filler loadings. Comparisons of these materials to composites containing unmodified cellulose nanoparticles and nanoclay fillers were performed. For all PCL composites, the tensile storage modulus (E') remained relatively constant at 3×10^9 Pa in the glassy state. At the T_g, the modulus of the materials falls to approximately 10^8 Pa. The reduction in modulus with increasing temperature is a result of the softening of the matrix at higher temperatures.

Figure 7 shows the tensile storage moduli of cellulose-PCL composites at different filler loadings (0%, 5%, and 10%). The storage moduli at $T > T_g$ increase with increasing filler loading. In the glassy state, only the material at 10% loading shows an increase in modulus. Above Tg, enhanced E' proportional to filler loading is evident. The largest difference in E' is observed at temperatures at or above room temperature, in the typical usage range of the material. A 150% increase in E' was observed for the material with only 5% loading, and a 300% increase was observed at 10% loading of the cellulose-PCL filler at ambient temperature. The modulus values are collected in Table II, and demonstrate the improved load transfer properties of the composites over the pure matrix. The modulus values increase as a result of the rigid filler phase stiffening the bulk material. Large improvements in modulus observed at low filler loadings are highly desirable properties of these composites, as they reduce material requirements and overall product weight.

A comparison of the cellulose-PCL filler to unmodified celllose nanoparticle filler was performed to observe the effect of the surface grafted PCL on the mechanical behavior of the composites. At both 5% and 10 % filler loadings there is only a modest increase in the storage modulus at room temperature (5-10%). The largest difference in modulus is seen at -40 °C, where the 10% modified filler has a modulus value 16% greater than the unmodified material at the same loading. It was hypothesized that the surface modification would have more of an enhancing effect on the modulus. We expected that the grafted PCL would appreciably enhance the bonding of the nanoparticles with the PCL matrix at the boundary layer between the two materials, and in turn increase the modulus of the bulk material. DMA data suggests that there is only a very weak enhancing effect on the modulus imparted by surface derivatization. This could be partly related to the fact that a lower concentration of grafted particles relative to unmodified particles was added to the matrix at the same weight percent, owing to the higher mass of the grafted particles containing large side groups.

A similar, marginal increase in tensile modulus was observed for surface modified chitin whiskers in natural rubber nanocomposites, and PCL grafted cellulose composites. *(16,31)* The author of these reports hypothesizes that the mechanical enhancement of the unmodified whiskers was greater because surface modification led to the partial destruction of the 3-dimensional network of chitin and cellulose whiskers. *(16,31)* The native particles are able to form networks through hydrogen-bonding, where in hydrophobically modified particles, the potential for hydrogen-bonding and other interactions that would lead to particle-particle network formation is reduced. It is likely that such small gains in moduli for the topochemically modified cellulose composites are the result of a lack of improvement in particle-particle network formation. This would suggest that mechanical percolation may lead to greater mechanical performance gains than improved particle-matrix adhesion.

PCL composites filled with nanoclay were used as a reference material for the cellulose filled composites. The values of E' for the nanoclay reinforced composites are nearly the same as the surface modified cellulose nanoparticles. The nanoclay samples exhibit a slightly lower E' below T_g but then increased to approximately the same level as the cellulose-PCL filled composites. These results indicate that our cellulose reinforced nanocomposites have comparable mechanical performance to that of composites reinforced with nanoclay.

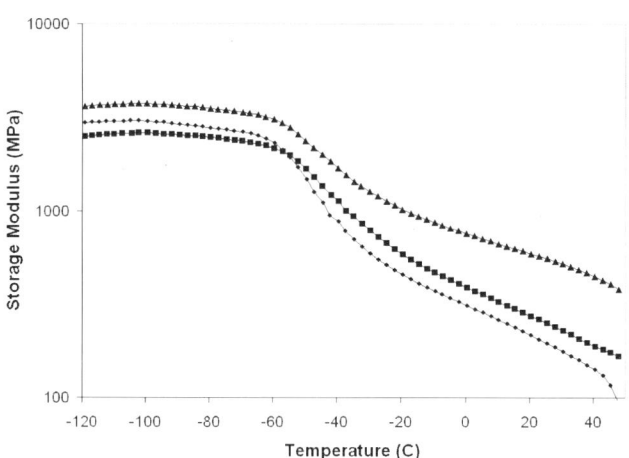

Figure 7. Tensile storage moduli of PCL with cellulose-PCL filler at different loadings. 0% Cell-CCL (♦), 5% Cell-PCL (■), 10% Cell-PCL(●).

Table II. Tensile storage modulus (E') for the PCL composites at −60 °C and 30 °C.

Material (wt% filler)	E' at -60 °C (MPa)	E' at 30 °C (MPa)
Neat PCL	2292	176
PCL/Cellulose-PCL (95/5)	2157	228
PCL/Cellulose-PCL (90/10)	3178	519
PCL/Cellulose-NC (95/5)	2392	301
PCL/Cellulose-NC (90/10)	2869	472
PCL/Montmorillonite (95/5)	2854	516
PCL/Montmorillonite (90/10)	2615	550

Conclusions

Nanoparticles of bacterial cellulose were isolated and surface derivatized with PCL. Through several characterization methods, we have confirmed the covalent attachment of PCL to the surface of the bacterial cellulose nanoparticles. The intent of surface modification was to improve the hydrophobic character and interfacial adhesion of these materials with a hydrophobic plastic material in composite applications. Nanocomposites were successfully prepared from the reinforcement of PCL, a biodegradable thermoplastic polyester, with cellulose nanoparticles using melt processing techniques. Thermal properties of the composite materials indicate only slight improvements in particle-matrix adhesion as a result of surface modification. The tensile properties of the composites were determined to be superior to the pure matrix material, and the surface grafting of the cellulose nanoparticles was shown to produce only marginal mechanical improvements over the unmodified cellulose. Our results suggest that despite sucessful functionalization of the cellulose nanoparticles, little improvement in particle-matrix interactions were recognized. As has been described in other similar composite systems, efforts focusing on enhancing particle-particle network formation may be more beneficial to improving mechanical performance of composites. The mechanical properties of the cellulose composites were also referenced against equivalently prepared nanoclay composites. The results show that 'green' nanocomposites based on cellulose can compete in mechanical performance with conventional composite fillers.

References

1. Bledzki, A. K.; Reihmane, S.; Gassan, J. *J. Appl. Polym. Sci.* **1996**, 59, 1329.
2. Glasser, W. G.; Taib, R.; Jain, R. K.; Kander, R. *J. Appl. Polym. Sci.* **1999**, 73, 1329.
3. Sakurada, I.; Nukushina, Y.; Ito, T.; *J. Polym. Sci.* **1962**, 57, 651.
4. Matsuo, M.; Sawatari, C.; Iwai, Y.; Ozaki, F. *Macromolecules.* **1990**, 23, 3266.
5. Favier, R.; Chanzy, H.; Cavaille, J.-Y. *Macromolecules.* **1995**, 28 (18), 6365.
6. Grunert, M.; Winter, W.T.; *J. Polym. Environ.* **2002**, 10 (1/2), 27.
7. Samir, M. A. S. A.; Alloin, F.; Dufresne, A. *Biomacromolecules.* **2005**, 6(2), 612.
8. Goodrich, J.D.; Winter, W.T. *Biomacromolecules.* **2007**, 8(1), 252.
9. Antonio, J.; O'Reilly, T.; Cavaille, J. Y.; Gandini, A.; *Cellulose.* **1997**, 4, 305.
10. Baiardo, M.; Frisoni, G.; Scandola, M.; Licciardello, A. *J. Appl. Polym. Sci.* **2002**, 83, 38.
11. Sipahi-Saglam, E.; Gelbrich, M.; Gruber, E. *Cellulose.* **2003**, 10, 273.
12. Hayashi, S.; Takeuchi, Y.; Eguchi, M.; Iida, T.; Tsubokowa, N. *J. Appl. Polym. Sci.* **1999**, 71, 1491.
13. Carrot, G.; Rutot-Houze, D.; Pottier, A.; Degee, P.; Hilborn, J.; Dubois, P. *Macromolecules.* **2002**, 35, 8400.
14. Hafren, J.; Cordova, A. *Macromol. Rapid. Comm.* **2005**, 26, 82.
15. Tsubokawa, N.; Yoshihara, T. *J. Polym. Sci. Part A: Polym. Chem.* **1993**, 31, 2459.
16. Habibi, Y.; Dufresne, A. *Biomacromolecules.* **2008**, 9, 1974.
17. Habibi, Y.; Goffin, A-L.; Schlitz, N.; Duquesne, E.; Dubois, P.; Dufresne, A. *J. Mater. Chem.* **2008**, 18, 5002.
18. Kowalaski, A.; Duda, A.; Penczek, S. *Macromolecules.* **2000**, 33(20), 7359.
19. Henderson, L.A.; Svirkin, Y.Y.; Gross, R.A.; Kaplan, D.L.; Swift, G. *Macromolecules.* **1996**, 29(24), 7759.
20. Lonnberg, H.; Zhou, Q.; Brummer, H. III.; Teeri, T. T.; Malmstrom, E.; Hult, A. *Biomacromolecules.* **2006**, 7, 2178.
21. Persson, P. V.; Schroder, J.; Wickholm, K.; Hedenstrom, E.; Iversen, T. *Macromolecules.* **2004**, 37, 5889.
22. Roman, M.; Winter, W.T. *Biomacromolecules.* **2004**, 5, 1671.
23. Casa, J.; Persson, P. V.; Iversen, T.; Cordova, A. *Adv. Synth. Catal.* **2004**, 346, 1087.
24. Vietor, R. J.; Newman, R. H.; Ha, M. −A.; Apperley, D. C; Jarvis, M. C. *Plant J.* **2002**, 30, 721.
25. Newman, R. H.; Davidson, T. C. *Cellulose.* **2004**, 11, 23.
26. Jandura, P.; Kotka, B. V.; Riedl, B. *J. Appl. Polym. Sci.* **2000**, 78, 1354.
27. Bhattacharya, D; Winter, W. T. Manuscript in preparation.
28. Amash, A.; Zugenmaier, P. *Polym. Bull.* **1998**, 40, 251.
29. Avella, M.; Errico, M. E.; Laurienzo, P.; Martuscelli, E.; Raimo, M.; Rimedo, R. *Polymer.* **2000**, 41, 3875.
30. Dufresne, A.; Kellerhals, M. B.; Witholt, B. *Macromolecules.* **1999**, 32 (22), 7396.
31. Nair, K. G.; Dufresne, A. *Biomacromolecules.* **2003**, 4, 1835.

Chapter 9

Versatile Concept for the Structure Design of Polysaccharide-based Nanoparticles

Thomas Heinze[*], Stephanie Hornig

Centre of Excellence for Polysaccharide Research, Friedrich Schiller University of Jena, Humboldtstraße 10, D-07743 Jena, Germany
*Member of the European Polysaccharide Network of Excellence, EPNOE
(www.epnoe.eu)

Defined functionalized derivatives of cellulose, dextran, and xylan as well as commercially produced cellulose esters self-assemble into nanoparticles by using different techniques of nanoprecipitation. By varying the polysaccharide, the type of functional group, the degree of substitution, and the preparation conditions, nanoparticles ranging in size from 60 to 600 nm can be prepared. A tailored functionalization of the polysaccharides yields nanoparticles applicable for specific requests. For example, fluorescence marked dextran derivatives yield biocompatible and tunable nanosensors that can be used for ratiometric pH measurements. Pharmacologically active substances can be covalently bond to the polymer backbone resulting in potential prodrug nanoparticles.

The structure and properties of nanomaterials differ significantly from those of atoms and molecules as well as from those of the bulk material. Synthesis, structure, properties, and applications form the theme of the emerging area of nanoscience. Each of these aspects is strongly connected with chemistry, in particular in the field of polymeric nanostructures (1). The successful synthesis of triblock copolymers offered the possibility to form nanostructure by ordered self-assembly of these highly engineered polymers. In addition, there are several "simple" techniques potentially useful for the preparation of polymeric nanoparticles like emulsion-based methods, salting out technique, spray-drying, and nanoprecipitation as recently reviewed for poly(lactic acid) usually called PLA (2). PLA is a promising polymer for biomedical and medical applications including controlled drug delivery since it is one of the most well-known bio-absorbable and non-toxic polymers.

Nanoparticles are of interest in the field of material science, *e.g.*, for the functionalization of surfaces. They create enhanced surface roughness, leading to superhydrophobic surfaces in combination with tunable interactions of the particles for the fabrication of functional coatings (3,4).

Polymeric nanoparticles are of particular interest for drug delivery devices. The controlled release of medicaments remains an important tool for drug delivery. Therefore, a wide variety of reports can be found in the open literature dealing with drug delivery systems. In this context, the use of nano- and microparticles devices has received special attention during the past two decades (2). Comparing to the use of liposomes as drug carriers, polymeric particles are preferred from the point of view of stability, and because they offer the possibility of modulation of the drug-release profile. Furthermore, these systems exhibit improved efficacy, reduced toxicity, and convenience in comparison to conventional dosage forms.

Furthermore, biologically compatible nanoparticles show great potential as sensor materials. The incorporation of environmentally sensitive fluorescent dyes into a nanoparticle matrix yields fluorescent sensors, which may help to understand physiological and pathological mechanisms on the molecular level. The combination, *e.g.*, of a pH sensitive- and a reference dye enables pH value measurements in cell culture media and cellular compartments after cellular uptake (5).

Polysaccharides and their semi-synthetic derivatives are of great importance for biological- and biomedical applications; they are biocompatible, non-toxic, and may even efficiently interact with the highly branched carbohydrates on cell surfaces, the glycocalix (6). Moreover, the multifunctionality of polysaccharides (mostly hydroxyl groups but also amino-, carboxyl- and other moieties may appear) opens up different paths for structure design by chemical modification reactions, and hence to adapt the molecular structure to the desired application.

It is very well known that polysaccharides form various naturally occurring superstructures. For example, the extensive hydrogen bond system of cellulose (β-1-4-glycosidic linked glucose), including crystalline and amorphous phases, determines the properties of this unique biopolymer (7). The interaction of cellulose and hemicelluloses in the cell wall of plants is another fascinating example for the formation of supramolecular assemblies that are still not completely understood (8). Artificial nanostructures may be formed by the Langmuir-Blodgett (LB) technique with cellulose ethers; trimethylsilyl cellulose LB films can even be regenerated to cellulose nanofilms (9).

Another fascinating polysaccharide is dextran that finds successful applications in the biomedical field, *e.g.*, as a blood plasma substitute (10). Dextran and carboxymethyl dextran are also applied as coatings of ferrite nanoparticles for hyperthermia in cancer therapy (11,12). Dextran, having an α-1,6-linked main chain of glucose monomers with varying proportions of linkages and branches depending on the bacterial source, is also an interesting starting biopolymer for the design of structures by polymer-analogous reactions, from the chemist's point of view.

The authors' studies of nanostructures, in particular nanoparticles, are focused on the use of polysaccharides and their derivatives, synthesized by advanced procedures including homogeneous phase chemistry and regiochemistry (*13,14*). Polysaccharide derivatives of various structures are valuable starting materials for the formation of well-defined spherical nanoparticles obtained by nanoprecipitation.

The method of nanoprecipitation was first described by Fessi *et al.* and is based on the interfacial deposition of polymers following displacement of a semi-polar solvent miscible with water from a lipophilic solution (*15*). It constitutes an easy and reproducible technique that has been widely used in the preparation of PLA- and PLGA based nanoparticles (*16,17*). Important advantages associated with this method are:

(1) large amounts of toxic solvents are avoided,

(2) nanoparticle sizes with narrow size distribution are obtained,

(3) no use of external energy sources is needed.

One first classification of the polymer particles can be made taking into account their size and preparation process. The term microparticle designates systems larger than 1 mm whereas nanoparticle is used to define submicron particles. The particle size is a fundamental parameter related to the way of administration of the drug. If the drug has to be directed to target tissues via systemic circulation or across the mucosal membrane (as well as in the cases of oral administration), particles of less than 500 nm are required (*18,19*)

Nanoprecipitation allows the preparation of polysaccharide derivative nanoparticles below 500 nm in size. The polysaccharide derivatives are dissolved in a water miscible organic solvent (*e.g.*, *N,N*-dimethyl-acetamide (DMAc), acetone), and the slow addition of water results in self-assembly of the polymer molecules into regular nanostructures (*20-23*). The organic solvents used can be removed either by a dialysis process or by evaporation in case of organic solvents with a low boiling point as schematically illustrated in Figure 1. Typically, DMAc is the preferred solvent for dialysis and acetone for subsequent solvent evaporation, respectively.

172

Controlled precipitation of polysaccharide derivatives

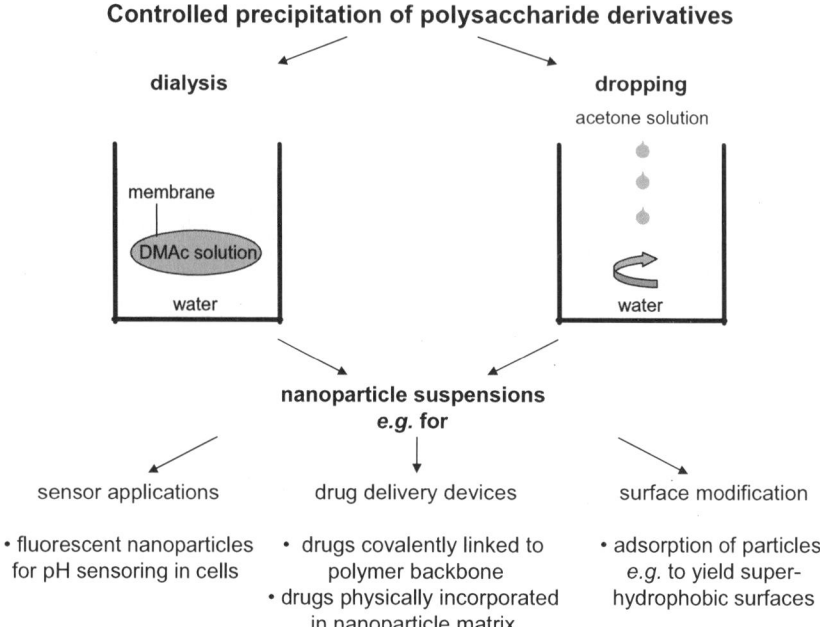

Figure 1. Schematic illustration of the preparation of nanoparticles based on polysaccharide derivatives including potential applications.

The formation of the particles depends strongly on the solvent used, the concentration of the polymer solution, the type of the polysaccharides and substituents, and the degree of substitution. The preparation technique has an important influence on the size and shape of the particles. For instance, there is a considerable difference between the dropping techniques (*22*). If the acetone solution of the polysaccharide derivative is dropped into water, small but non-uniformly distributed particles will be obtained (~100 nm). By applying the reverse procedure, particles are approximately double in size but possess very low polydispersity index (PDI). Dialysis as well as the dropping technique of dextran esters (dextran furoate pyroglutamate, DS_{Fur} 0.79, DS_{Pyr} 1.27; perpropionylated dextran pyroglutamate, DS_{Pyr} 1.99) lead to regular spherical particles as depicted in the SEM images in Figure 2 (*24*). However, dialysis of a DMAc solution of cellulose acetate (DS_{Ac} 2.46) yields almost spherical particles, whereas by dropping water into acetone solutions of the polymer, bean-shaped particles were obtained (*22*). Therefore, dialysis of DMAc solutions of cellulose acetate is recommended if regularly shaped cellulose acetate nanospheres are needed.

dialysis   **dropping**

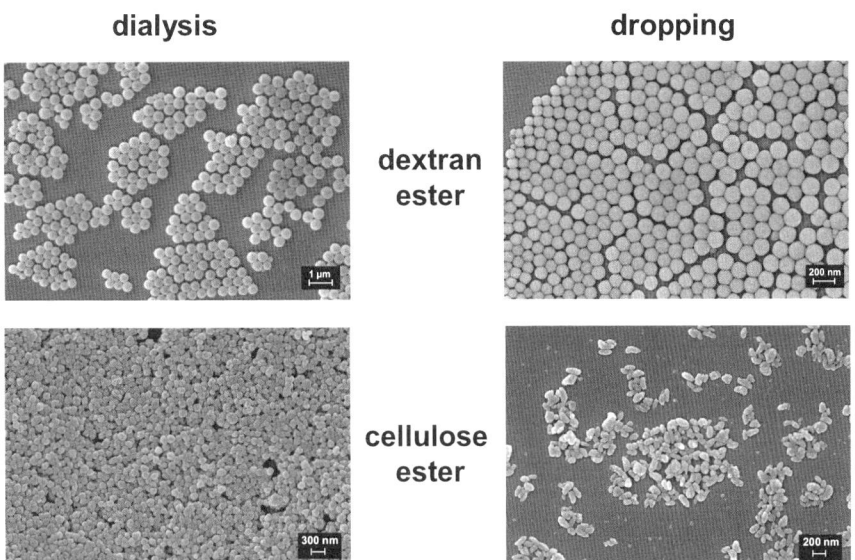

Figure 2. SEM images of cellulose acetate (DS$_{Ac}$ 2.46) nanoparticles and dextran furoate pyroglutamate (DS$_{Fur}$ 0.79, DS$_{Pyr}$ 1.27) and perpropionylated dextran pyroglutamate (DS$_{Pyr}$ 1.99) nanoparticles prepared by dialysis of DMAc solution and dropping of an acetone solution into water.

An important criterion for the formation of small and uniformly distributed nanoparticles is a polymer concentration below the critical overlap concentration. Figure 3 reveals the influence of the concentration of a cellulose acetate (DS$_{Ac}$ 2.46) dissolved in DMAc on the reduced viscosity, from which the critical overlap concentration (c*) can be calculated, and the resulting particle sizes and PDI values after dialysis of the polymer solution against water. Below c* of 14.4 mg/mL, the molecules are in a dispersed state and can separate into nanodomains after adding the non-solvent water. Particle sizes ranging from 142 to 234 nm and small PDI values can be observed from 1 to 10 mg/mL. At 15 mg/mL, the PDI and the size of the particles increase. Aggregation occurs at concentrations above 25 mg/mL.

The formation of nanoparticles via nanoprecipitation is feasible for various polysaccharide derivatives, mainly esters, including cellulose (*22*), dextran (*20,21,24*), xylan (*25*), and pullulan (*26*). Among these are commercially available products (*e.g.* cellulose acetate), copolymers (*e.g.* poly(ethylene glycol) grafted pullulan acetate), and defined functionalized polysaccharide esters (*e.g.* dextran esters bearing chiral-, photo-crosslinkable-, complexing-, biologically active-, or ion-sensitive groups). In order to prepare multiple functionalized derivatives, novel acylation systems are required, in particular for the functionalization with complex and sensitive acids.

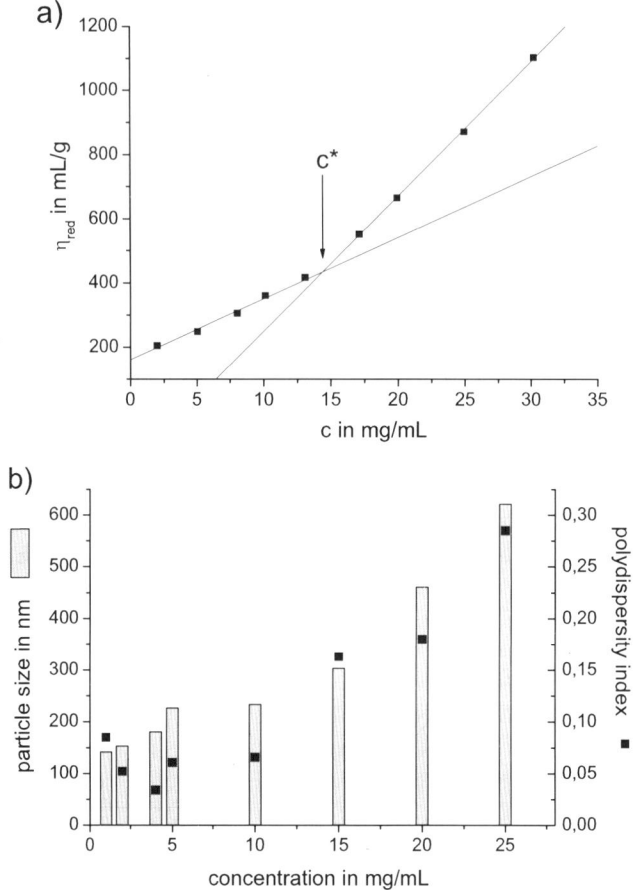

Figure 3a) Plot of the reduced viscosity versus the concentration of cellulose acetate (DS_{Ac} 2.46) dissolved in DMAc (c* = critical overlap concentration); b) dependency of the particle size and the polydispersity index of the resulting nanoparticles after dialysis of the sample dissolved in DMAc.

Table 1. Degree of substitution (DS) of functionalized polysaccharide derivatives and mean diameters of the resulting nanoparticles after dialysis (DMAc/H$_2$O) determined by dynamic light scattering

Polysaccharide	Substituent(s)	DS	Size in nm	PDI	Reference
Cellulose	Acetate	2.46	181	0.03	22
Cellulose	Acetate	0.18	399	0.10	22
	Propionate	2.40			
Cellulose	Acetate	1.19	337	0.12	22
	Phthalate	0.73			
Dextran	Pyroglutamate	1.96	87	0.08	21
	Propionate	3.00			
Dextran	Furoate	0.22	260	–	24
	Pyroglutamate	1.34			
	Propionate	2.78			
Dextran	18-Crown-6	0.22	222	–	29
	Propionate	2.78			
Dextran	15-Crown-5	0.79	154	–	29
	Propionate	2.21			
Dextran	Ibuprofen	2.08	287	0.19	30
Dextran	Naproxen	1.62	387	0.21	30
Dextran	Fluorescein	0.027	374	0.13	31
	Propionate	1.81			
Dextran	Sulforhodamine B	0.0015	524	0.23	31
	Propionate	3.00			
Xylan	Furoate	0.74	73	–	25
	Pyroglutamate	0.74			

Recent studies have shown that the *in situ* activation of carboxylic acids with *N,N'*-carbonyldiimidazole (CDI) is an efficient tool to obtain polysaccharide esters possessing defined structures and adjusted properties like dextran furoates and -pyroglutamates or crown ether esters of cellulose (*27,28*). Table 1 gives an overview of typical polysaccharide derivatives that form nanoparticles after dialysis of a DMAc solution against water.

For nanoprecipitation, it is important that a certain balance of polar, *i.e.*, hydroxyl groups or carbonyl functions, and non-polar moieties exists within the polymer backbone. A rather high amount of unmodified hydroxyl groups in the polymer derivatives prevents the formation of nanostructures due to intensive hydrogen bond formation. The hydrophilic-hydrophobic balance of the derivatives is either appropriate for nanoprecipitation or can be adjusted by subsequent functionalization with hydrophobic acyl moieties (acetate, propionate). The character of functional groups, the DS, the course of reaction, and consequently the distribution of the substituents influence the size of the resulting nanoparticles.

The dependency of the nanoparticle formation on the DS was specified for dextran benzoate (*29*). As demonstrated in Figure 4, the particles obtained by dialysis exhibit sizes in the range of 265 to 500 nm and PDI values from 0.13 to 0.26. The perpropionylated derivatives, which means that all remaining hydroxyl groups of the polysaccharide ester are subsequently functionalized with propionate, were also examined according to their self-assembly behavior. It is noteworthy that in all cases smaller particles within a close interval of between 210 to 260 nm with a more narrow size distribution result. The ratio of polar carbonyl groups and short hydrophobic alkyl chains seems to be especially appropriate for organization into small and uniform nanospheres. The concept of perpropionylation for reduction of particle sizes can be applied for other derivatives (*e.g.* dextran furoate) as well.

In general, the nanoparticle suspensions based on polysaccharides, in particular cellulose- and dextran derivatives, are stable in water under neutral conditions for at least three weeks. Particles below 200 nm do not agglomerate and sediment even up to one year after preparation. It was shown that the nanoparticle suspensions are stable under standard sterilizing procedures in an autoclave at 120 °C and 2 bar for removing germs, which is an important prerequisite for biological experiments. Furthermore, the suspensions are constant in their size and shape from pH 4 to 11. The stability of the nanoparticle suspensions can be tracked via zeta potential measurements. Cellulose acetate- (DS_{Ac} 2.46) and perpropionylated dextran pyroglutamate (DS_{Pyr} 1.99) nanoparticles exhibit a zeta potential of -22.2 mV and -27.8 mV, respectively, at pH 7.1 (Figure 5). Agglomeration of cellulose ester particles occurs below pH 3.4; dextran esters agglomerate below pH 4. At low pH values, the surface charge, which is quantified by the zeta potential, is neutralized by H^+. Consequently, the zeta potential decreases and repulsion between the particles becomes less intensive. The majority of colloidal systems dispersed in water acquire a negative surface charge, probably due to preferential adsorption of hydroxyl ions (*32*). The addition of a surfactant, such as polymeric non-ionic Tween 80 (polyoxyethylene (20) sorbitan monooleate), avoids aggregation at low pH values due to sterical repulsion of the nanospheres.

With increasing pH value, the zeta potential of the nanospheres increases. However, saponification of the polysaccharide esters proceeds above pH 11, which leads to aggregation of pure cellulose particles due to the formation of a strong intermolecular hydrogen bonding, or dissolution in case of dextran derivatives.

The stability of the nanoparticles and the presence of a hydrophobic core, which was proven by the encapsulation of the non-polar fluorescence probe pyrene, are requisites for a potential application as delivery systems for drugs and hydrophobic compounds, *e.g.* for cosmetic applications. Important objectives may be achieved by using such polymeric carriers, including stabilization of the therapeutic agent, enhancement of the drug solubility, improvement of the circulation life time and the therapeutic index, and reduction of side-effects (*33*). Pharmacologically active agents can either be covalently bound to the polymer backbone, physically incorporated into a polymeric matrix, *e.g.* polymeric micelles or porous particles, or form polyelectrolyte complexes of oppositely charged polymer/drug systems (*34*).

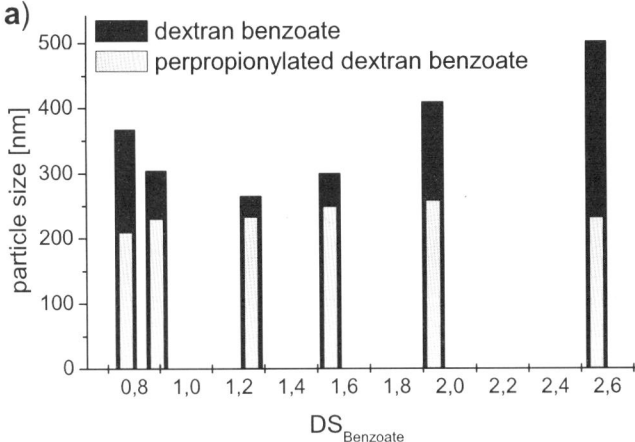

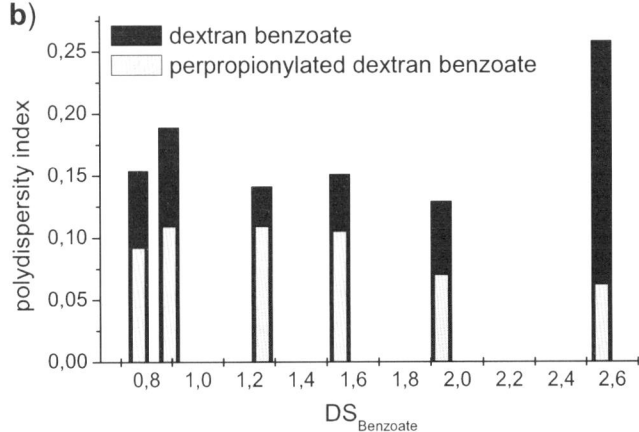

Figure 4 a) Particle size and b) PDI of nanoparticles prepared by dialysis (4 mg/mL, DMAc/H₂O) based on dextran benzoate and its perpropionylated derivatives at different DS values.

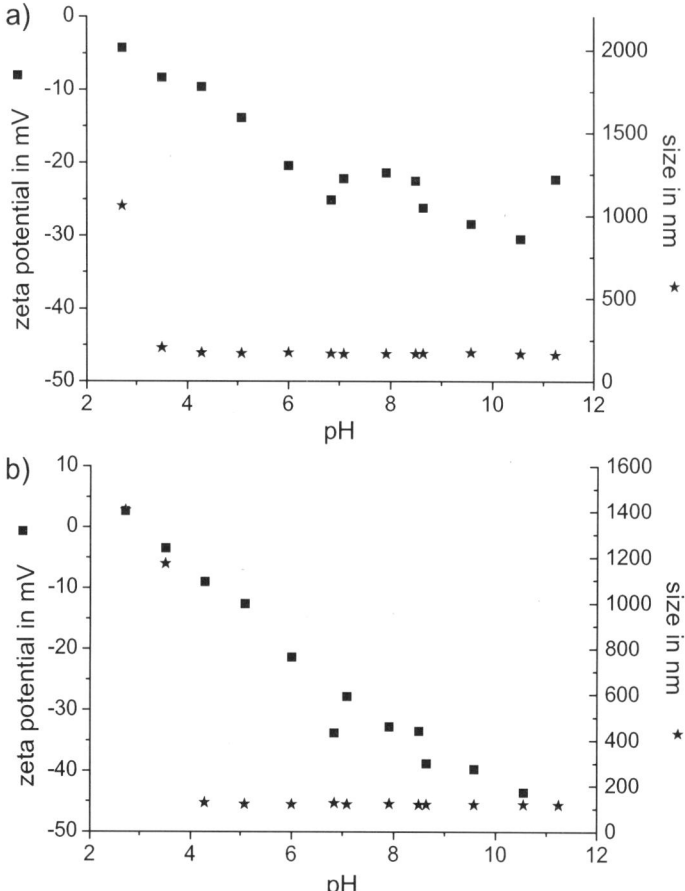

Figure 5. Zeta potential and particle size of a) cellulose acetate (DS_{Ac} 2.46) and b) perpropionylated dextran pyroglutamate nanoparticles (DS_{Pyr} 1.99) depending on pH.

Polysaccharides are promising, physiologically-evaluated carriers for physically encapsulated drugs on the one hand (*35,36*). On the other hand, the high amount of hydroxyl groups facilitates the introduction of, *e.g.*, proteins, aptamers, or drugs into the polymer backbone yielding prodrug systems. However, in case of water-insoluble drugs, the degree of substitution (DS) is limited due to requirements of water solubility. This limitation can be overcome by using the presented concept of nanoprecipitation of hydrophobic polysaccharide derivatives. In this way, prodrugs with an enhanced loading efficiency of hydrophobic drugs can be prepared while retaining the ability to transport in aqueous systems.

The non-steroidal anti-inflammatory drugs ibuprofen and naproxen were chosen as low water-soluble carboxylic acids to evaluate the potential of hydrophobic polysaccharide esters as prodrugs (*29,30*). The *in situ* activation of the carboxylic acids with CDI was used to functionalize dextran yielding derivatives with controlled degrees of substitution. The drug content of the dextran esters varies from 37 to 71 wt% and is remarkably high compared to the loading capacity of other ibuprofen release materials (*37-39*). By applying nanoprecipitation via dialysis of a DMAc solution against water, ibuprofen- and naproxen dextran ester nanoparticles in the range from 102 to 287 nm, and from 177 to 387 nm, respectively, are accessible. With increasing DS, the particles become larger and less uniformly distributed. The preparation of nanoparticles by dropping of an acetone solution into water (and the reverse path) is also feasible and yields nanoparticle suspensions. Figure 6 displays the schematic structure of an anhydroglucose unit of an ibuprofen dextran ester (DS$_{Ibu}$ 0.62) and a SEM image of resulting nanoparticles prepared by dialysis.

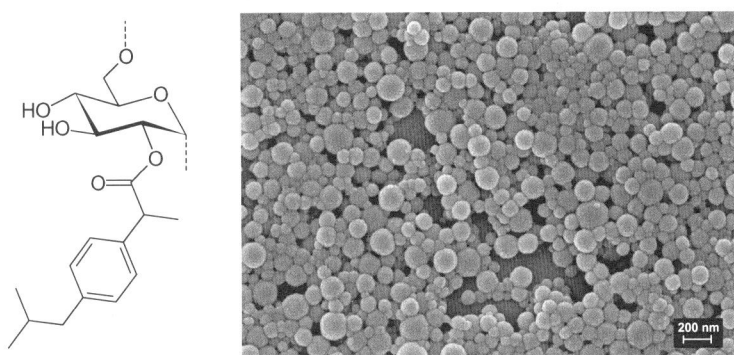

Figure 6. Schematic illustration of a repeating unit and SEM image of the nanoparticles prepared by dialysis (c = 4 mg/mL, DMAc/H$_2$O) of an ibuprofen dextran ester (DS$_{Ibu}$ 0.62)

The feasibility of ester cleavage of the dextran derivatives is a prerequisite for its use as potential prodrugs in drug delivery systems. Because of its hydrolysis at pH values above the physiological range, attempts need to be made concerning the enzymatic cleavage of the ester bond with lipases, esterases, or dextranase. However, it is known that the introduction of functional groups alters the biodegradation process (*40*). The introduction of a spacer between drug and polymer might minimize the steric hindrance of the substituents and, thus, enhance the susceptibility of ibuprofen dextran ester nanoparticles to undergo enzymatic reactions.

A benefit of the nanoparticles based on dextran esters is their compatibility with biological systems. Studies on the cellular uptake of dextran ester particles show an incorporation into cell compartments without the use of stimulating agents (*31*). The exploration of chemical microenvironments on the cellular level is of great importance in medical and biological research. To understand physiological and pathological mechanisms on the molecular level in detail, tools to measure, *e.g.*, the pH value in the extra- and intracellular space are essential prerequisites. In recent years, optical analyte detection via fluorescent sensor particles has become a very promising method (*41*). In order to obtain fluorescent nanoparticles, hydrophobic dextran derivatives (*e.g.*, dextran propionate) are functionalized with fluorescein isothiocyanate (FITC) and sulforhodamine B acid chloride (SRB) (*31*). The combination of an indicator dye (FITC-dextran propionate) and an inert reference dye (SRB-dextran propionate) allows the ratiometric detection of the pH value.

By mixing the FITC- and SRB-dextran derivatives in the desired ratio prior to particle preparation, fluorescent nanoparticles bearing both indicator- and reference dye with a desired fluorescence emission are accessible. Fluorescence spectra of various SRB/FITC-dextran propionate ratios are depicted in Figure 7a. Reasonable spectra of the respective dextran derivatives for ratiometric pH measurements are obtained in the ratios 7/1 and 3/1. The particle suspensions need to be calibrated against buffer solutions of pH 4.9-8.2 (Figure 7b). The fluorescence intensity of fluorescein decreases with decreasing pH, whereas the sulforhodamine B intensity remains stable. The small decrease of the sulforhodamine B emission peak results from the overlap with the fluorescein emission. In addition, the determination of the pH value in living cells can be carried out by confocal laser scanning microscopy. A z-stacking through the cell proves that the particles are located inside the cells and do not stick on the cell surface. The particles located in the more acidic cytosol have distinct spectral properties compared to particles in the more alkaline surrounding solution.

Due to the great variety of fluorescent dyes and indicators, various indicator-reference dye combinations can be tailored to specific applications. The covalent attachment of any fluorescent dye, if it contains at least one functional group for the reaction with hydroxyl groups or a spacer moiety, is conceivable. Another fluorescence dye, which was successfully immobilized onto dextran to form nanoparticles, is sulforhodamine Q5, a potential reference dye. By choosing sensitive dyes and varying the ratio of sensor to reference dye, tailoring of sensor nanoparticles for the detection of any requested analyte might be possible. Hence, these are basic requisites for the development of a comprehensive toolbox for single-cell analysis.

In conclusion, the authors present a versatile concept for the formation of defined nanostructures based on polysaccharides. The possibility of multiple functionalization yields tailor-made polysaccharide derivatives, which may self-assemble into nanostructured materials. By applying the nanoprecipitation technique, nanoparticles in the range of 60 to 600 nm are accessible, depending on the preparation conditions, the nature of the polysaccharide backbone, the substituents and their degree(s) of substitution. The variation possibilities of the highly engineered polysaccharide derivatives open new perspectives for advanced applications in many fields, *e.g.* in drug delivery devices or as sensor materials.

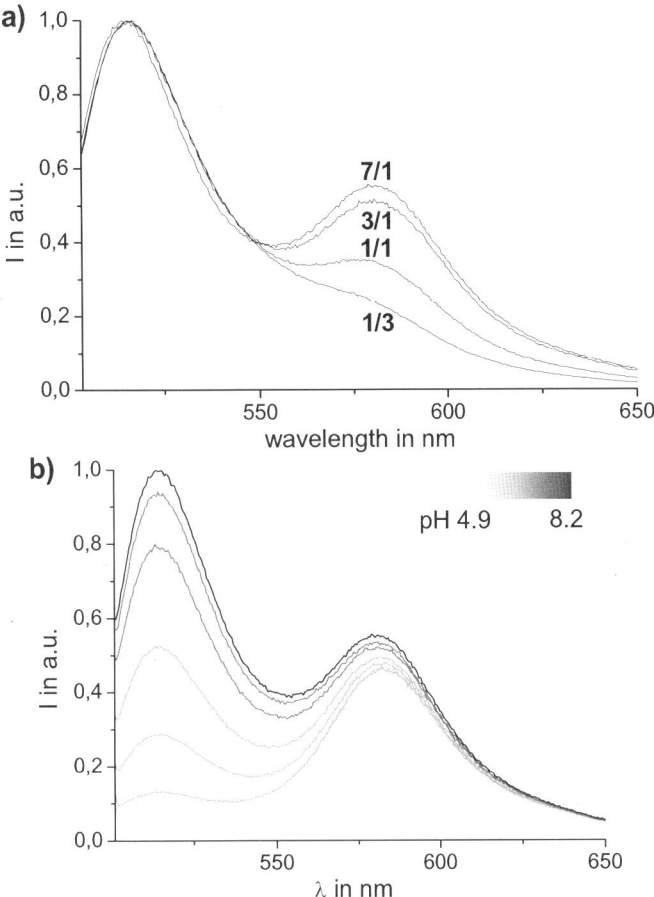

Figure 7. Fluorescence emission spectra (λ_{exc} = 488 nm) of nanoparticle suspensions a) at pH 8.1 with varying ratios (w/w) of SRB-dextran propionate (λ_{max} = 581 nm) to FITC-dextran propionate prior to the dialysis process and b) at pH 4.9 (bottom), 5.4, 6.0, 6.5, 7.0, 7.5 and 8.2 (top) of FITC-dextran propionate/SRB-dextran propionate = 1/7.

Acknowledgements

Th. H. gratefully acknowledges the general financial support of the "Fonds der Chemischen Industrie" of Germany as well as of Borregaard Chemcell, Hercules, Dow Chemicals, Wolff Cellulosics, ShinEtsu, and Rhodia.

References

1. *The chemistry of nanomaterials*; Rao, C. N. R.; Müller, A.; Cheetham, A. K., Eds.; Wiley-VCH: Weinheim, 2004; Vol. 1 and 2.
2. Lassalle, V.; Ferreira, M. L. *Macromol. Biosci.* **2007**, *7*, 767-783.
3. Motornov, M.; Sheparovych, R.; Lupitskyy, R.; MacWilliams, E.; Minko, S. *J. Colloid Interface Sci.* **2007**, *310*, 481-488.
4. Shang, H. M.; Wang, Y.; Takahashi, K.; Cao, G. Z.; Li, D.; Xia, Y. N. *J. Mater. Sci.* **2005**, *40*, 3587-3591.
5. Burns, A.; Sengupta, P.; Zedayko, T.; Baird, B.; Wiesner, U. *Small* **2006**, *2*, 723-726.
6. Marchesi, V. T.; Furthmayr, H.; Tomita, M. *Ann. Rev. Biochem.* **1976**, *45*, 667-698.
7. Krässig, H. A. *Cellulose, Structure, accessibility, and reactivity*; Gordon and Breach Science Publishers: Amsterdam, 1993.
8. Gradwell, S. E.; Renneckar, S.; Esker, A. R.; Heinze, T.; Gatenholm, P.; Vaca-Garcia, C.; Glasser, W. *Comptes Rendus Biologies* **2004**, *327*, 945-953.
9. Schulze, M.; Seufert, M.; Fakirova, C.; Tebbe, H.; Buchholz, V.; Wegner, G. In *Cellulose derivatives, modification, characterization and nanostructures*; Heinze, T.; Glasser, W. G., Eds.; ACS Symp. Series: Washington DC, 1998; 688, pp. 2-18.
10. Heinze, T.; Liebert, T.; Heublein, B.; Hornig, S. In *Polysaccharide II, Advances in Polymer Science*; Klemm, D., Ed.; Springer Verlag: Heidelberg, 2006; 205, pp. 199-291.
11. Lemarchand, C.; Gref, R.; Couvreur, P. *Eur. J. Pharm. Biopharm.* 2004, 58, 327-341.
12. Clement, J. H.; Schwalbe, M.; Buske, N.; Wagner, K.; Schnabelrauch, M.; Görnert, P.; Kliche, K. O.; Pachmann, K.; Weitschies, W.; Höffken, K. *J. Cancer Res. Clinical Oncol.* **2006**, *132*, 287-292.
13. El Seoud, O.; Heinze, T. In *Polysaccharide I, Structure, Characterization and Use, Advances in Polymer Science*; T. Heinze, Ed.; Springer Verlag: Heidelberg, 2005; 186, pp. 103-149.
14. Heinze, T. In *Polysaccharide: Structural Diversity and functional versatility*; Dumitriu, S., Ed.; Marcel Dekker: New York, 2004; 2nd edition, pp. 551-590.
15. Fessi, H.; Puisieux, F.; Devissaguet, J.; Ammoury, N.; Benita, S. *Int. J. Pharm.* **1989**, *55*, R1-R4.
16. Guterres, S.; Fessi, H.; Barrat, G.; Puisieux, F.; Devissaguet, J. *Int. J. Pharm.* **1995**, *113*, 57-63.
17. Chacon, M.; Berges, L.; Molpeceres, J.; Aberturas, M.; Guzman, M. *Int. J. Pharm.* **1996**, *141*, 81-91.
18. Avgoustakis, K. *Curr. Drug Deliv.* **2004**, *1*, 321-333. ; Jain, R. *Biomaterials* **2000**, *21*, 2475-2490.
19. Delie, F.; Blanco-Prieto, M. J. *Molecules* **2005**, *10*, 65-80.
20. Liebert, T.; Hornig, S.; Hesse, S.; Heinze, T. *J. Am. Chem. Soc.* **2005**, *127*, 10484-10485.
21. Hornig, S.; Heinze, T. *Carbohydr. Polym.* **2007**, *68*, 280-286.

22. Hornig, S.; Heinze, T. *Biomacromolecules* **2008**, *9*, 1487-1492.

23. Aumelas, A.; Serrero, A.; Durand, A.; Dellacherie, E.; Leonard, M. *Colloids Surf., B* **2007**, *59*, 74-80.

24. Hornig, S.; Heinze, T.; Hesse, S.; Liebert, T. *Macromol. Rapid Commun.* **2005**, *26*, 1908-1912.

25. Heinze, T.; Petzold, K.; Hornig, S. *Cell. Chem. Technol.* **2007**, *41*, 1-6.

26. Jung, S.-W.; Jeong Y.-I.; Kim, Y.-H.; Kim, S.-H. *Arch. Pharm. Res.* **2004**, *27*, 562-569.

27. Hornig, S.; Liebert, T.; Heinze, T. *Macromol. Biosci.* **2007**, *7*, 297-306.

28. Liebert, T.; Heinze, T. *Biomacromolecules* **2005**, *6*, 333-340.

29. Hornig, S. Ph.D. thesis, Friedrich-Schiller-University, Jena, Germany, 2008.

30. Hornig, S.; Bunjes, H.; Heinze, T. *J. Controlled Release*, to be submitted.

31. Hornig, S.; Biskup, C.; Gräfe, A.; Wotschadlo, J.; Liebert, T.; Mohr, G. J.; Heinze, T. *Soft Matter* **2008**, *4*, 1169-1172.

32. Marinova, K. G.; Alargova, R. G.; Denkov, N. D.; Velev, O. D.; Petsev, D. N.; Ivanov, I. B.; Borwankar, R. P. *Langmuir* **1996**, *12*, 2045-2051.

33. Panyam, J.; Labhasetwar, V. *Adv. Drug Delivery Rev.* **2003**, *55*, 329-347.

34. Duncan, R. *Nature Reviews* **2003**, *2*, 347-360.

35. Edgar, K. J. *Cellulose* **2007**, *14*, 49-64.

36. Felt, O.; Buri, P.; Gurny, R. *Drug Dev. Ind. Pharm.* **1998**, *24*, 979-993.

37. Vallet-Regi, M.; Ramila, A.; del Real, R. P.; Perez-Pariente, J. *Chem. Mater.* **2001**, *13*, 308-311.

38. Kuntsche, J.; Westesen, K.; Drechsler, M.; Koch, M. H. J.; Bunjes, H. *Pharm. Res.* **2004**, *21*, 1834-1843.

39. Galindo-Rodriguez, S. A.; Puel, F.; Briancon, S.; Allemann, E.; Doelker, E.; Fessi, H. *Eur. J. Pharm. Sci.* **2005**, *25*, 357-367.

40. Vercauteren, R.; Bruneel, D.; Schacht, E.; Duncan, R. *J. Bioact. Compat. Polym.* **1990**, *5*, 4-15.

41. Wang, F.; Tan, W. B.; Zhang, Y.; Fan, X.; Wang, M. *Nanotechnology* **2006**, *17*, R1-R13.

Chapter 10

Modified Galactoglucomannans from Forestry Waste-water for Films and Hydrogels

Margaretha Söderqvist Lindblad[1,3], Olof Dahlman[2], John Sjöberg[1,4], and Ann-Christine Albertsson[1]

[1] Department of Fibre and Polymer Technology, Royal Institute of Technology, SE-100 44 Stockholm, Sweden
[2] STFI-Packforsk AB, Box 5604, SE-114 86 Stockholm, Sweden
[3] Current address: Södra Cell, Research and Development, SE-430 24 Väröbacka, Sweden
[4] Current address: PRV, Box 5055, SE-102 42 Stockholm, Sweden

Hemicelluloses are among the most abundant natural polymers in the world and are consequently a potential source for sustainable materials, that has so far been underexploited. Galactoglucomannans are the principal hemicelluloses in softwoods and can be found in, for example, industrial wood processing waste-water. Currently, we are investigating the fractionation and purification of O-acetyl-galactoglucomannans from newsprint and fiberboard mill waste-waters, as well as the preparation of new barrier films with low oxygen permeation and hydrogel materials from the fractions obtained. Self-supporting films have been formed by solution-casting. Interesting oxygen barrier and mechanical strength properties were achieved for films obtained from a physical blend of O-acetyl-galactoglucomannan and either alginate or carboxymethylcellulose. To create oxygen barrier films with high resistance towards moisture, benzylated derivatives of O-acetyl-galactoglucomannan were made. A hydrogel is a polymeric material that swells in water but does not dissolve, valuable for applications including drug delivery. In order to obtain the right properties, we performed tailored cross-linking to create a flexible network structure. The chemical modification procedure involves a methacrylation reaction carried out under mild conditions. Herein we review past work and present some new data on fractionation and purification of galactoglucomannans.

Galactoglucomannans

Hemicelluloses are one of the world's most abundant renewable feedstocks for production of polymers and chemicals. Hemicellulose is the collective name for a group of several different heteropolysaccharides exhibiting a large structural diversity, comprising predominantly xylans and glucomannans (see Figure 1). The structure of the hemicellulose polysaccharide, i.e., the degree of branching, polymerization and substitution, varies depending on the natural source, but is also affected by the isolation and purification methods employed. However, the structural features of hemicelluloses present properties and possibilities which make these polysaccharides very interesting sources of value-added polymers and chemicals.

The hemicelluloses are found in substantial quantities in the majority of trees and terrestrial plants and, hence are available for example in the form of waste from forestry and agriculture. In contrast to cellulose, which is the basis of a large chemical industry sector, hemicellulose is to a great extent unexplored as a raw material for polymers and chemicals. However, in recent years interest in the production of polymers and other compounds from hemicellulose has been increasing. For this purpose, we have previously investigated steam extraction as a means for isolating water soluble softwood hemicellulose, mainly *O*-acetyl-galactoglucomannan, from spruce chips (*1* , *2* , *3* , *4*). Water soluble hemicelluloses, *O*-acetyl-(4-*O*-methyl-glucurono)-xylan and *O*-acetyl-glucomannan, extracted from hardwoods have also been characterized in similar studies (*3, 5*).

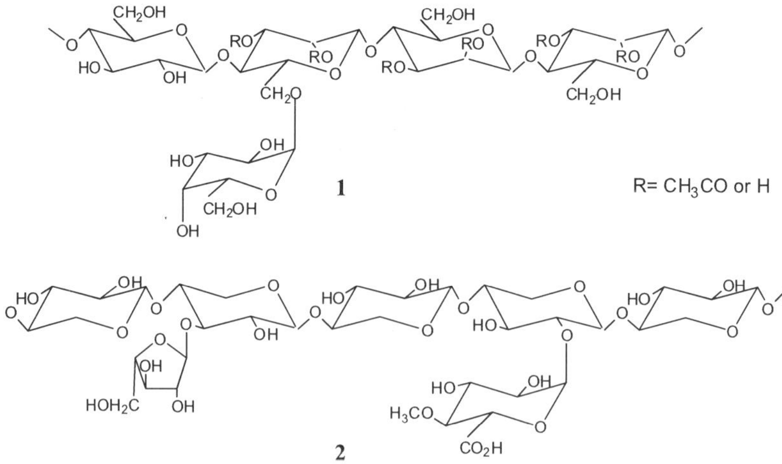

*Figure 1. Structural formulae for predominant hemicelluloses in softwood, O-acetyl-galactoglucomannan (**1**) and arabino-4-O-methyl-glucuronoxylan (**2**).*

Galactomannans from forestry waste waters

The forest products industry would seem to be a most viable source of hemicelluloses. However, O-acetyl-galactoglucomannan cannot be easily isolated economically by conventional methods in polymeric form by direct extraction of softwood wood chips due to its location in the complex matrix within the secondary layer of the fiber wall. Conversion of the wood to holocellulose by e.g., chlorite delignification, followed by removal of the arabinoglucuronoxylan by extraction with dimethylsulphoxide is usually required to make the O-acetyl-galactoglucomannan accessible to extraction with water (6, 7). In contrast, mechanical pulping causes the cleavage of the fiber wall and subsequent defibration of the wood, which renders parts of the hemicelluloses soluble in the process water without extensive degradation of the polysaccharide structures. Similarly, partly degraded hemicelluloses are dissolved into the process water during the Masonite process that utilizes steam explosion to process the wood into fibers suitable for hardboards. Thus, process waters from both mechanical pulping and hardboard (Masonite) production are potential resources for large quantities of hemicelluloses.

In the present study (8) process waters from mechanical pulping of softwood (spruce) were obtained from a Swedish newsprint mill, Stora Enso Kvarnsveden Mill AB. Samples from both stone ground wood pulping (SGW) and thermomechanical pulping (TMP) production lines were investigated. Also included in this study was process water from steam explosion of softwood (mixed pine and spruce) obtained from Masonite AB, Sweden. The chemical compositions of the mill process waters are reported in Table I. The polysaccharide content was highest in the process waters from steam explosion (Masonite) followed by the water from thermomechanical pulping (TMP) and the lowest content was for the SGW water, as evident from the data in the Table I. In each case the hemicellulose fraction was dominated by O-acetyl-galactoglucomannan (75-85 %) with minor contributions from arabinogalactan (15-20%). Only traces of arabino-4-O-methyl-glucuronoxylan were detected in two of the investigated mill process waters (SGW and Masonite).

Table I. Dry matter and polysaccharide contents, and composition of the hemicellulose in process waters from mechanical pulping of spruce (denoted SGW and TMP) and steam explosion of mixed softwoods (denoted Masonite) (8).

Substance	SGW	TMP	Masonite
Dry matter (g/l)	~1	~2.5	~5
Polysaccharides (g/l)	0.3	1	1.3
Galactoglucomannan (%)*	75	85	80
Arabinogalactan (%)*	20	15	15
Arabinoxylan (%)*	5	0	5

* Expressed as wt % of polysaccharide fraction.

Sequential ultrafiltration and diafiltration were employed in order to remove inorganic salts and low molecular size organic impurities from the process waters and to increase the hemicellulose concentration prior to drying. Following pre-treatment of the process water by centrifugation (to remove fines and other particles) ultrafiltration was performed at room temperature using a regenerated cellulose membrane (cut-off 1 kDa). Figure 2 is a schematic presentation of the ultrafiltration and diafiltration process sequence employed for the preparation of hemicelluloses from the process waters from mechanical and steam explosion pulping.

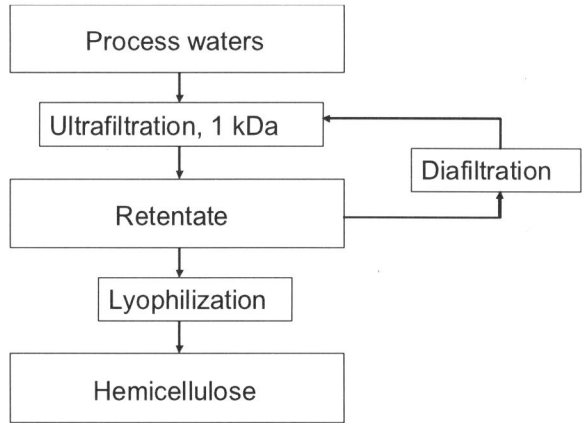

Figure 2. Schematic representation of the sequential ultrafiltration and diafiltration procedure employed for the isolation of hemicelluloses from the process waters from mechanical and steam explosion pulping.

The ultrafiltration and diafiltration treatment of the process waters yielded purified streams (retentate) exhibiting much higher concentration of hemicellulose (up to 60 g/l) than that for the corresponding untreated mill process waters. More than 80 % of the total quantity of oligo- and polysaccharides in the original mill process waters could be recovered as purified hemicelluloses with the high molar mass retentate fraction. In contrast, substantial quantities of the water soluble lignin fragments (~60 %) were removed from the product streams together with the inorganic salts with the permeate streams.

Freeze-drying of the retentate fractions yielded hemicelluloses as voluminous, light-tan powders. They were characterized with respect to molecular weight by Size-Exclusion Chromatography by employing a previously described method (3). Table II reports the weight-average molar mass and degree of substitution of acetyl groups (DS_{Ac}) for the purified *O*-acetyl-galactoglucomannan fractions prepared by ultrafiltration and subsequent diafiltration from the mill process waters from mechanical pulping (SGW and TMP) and steam explosion (Masonite). The hemicellulose isolated from the thermomechanical pulping process water demonstrated the highest weight-average molar mass followed by that from the SGW water, whereas the lowest molar mass value was measured for the steam explosion (Masonite) hemicellulose. All three hemicellulose fractions exhibited approximately the same degree of substitution of acetyl groups (DS_{Ac}), see Table II.

Table II. The weight-average molar mass and degree of substitution of acetyl groups (DS_{Ac}) for the purified _O_-acetyl-galactoglucomannan fractions from mechanical pulping waters (denoted SGW and TMP) and from steam explosion water (denoted Masonite) (8).

Parameter	SGW	TMP	Masonite
M_w	8600	9100	6600
DS_{Ac}	~0.5	~0.5	~0.5

Experimental conditions used for isolation of galactomannans

The carbohydrate compositions were determined by employing the hydrolysis conditions described in the TAPPI –standard method (T 249 cm-00, Carbohydrate Composition of Extractive-Free Wood and Wood Pulp by Gas Chromatography. Atlanta, TAPPI Press). However, the sugars obtained after the hydrolysis step were determined by ion-exchange chromatography (IC) employing a Dionex DX500 IC analyzer equipped with a gradient pump (Dionex, GP50), electrochemical detector (Dionex, ED40), Dionex, PA1 separation column, and applying a sodium hydroxide/acetate gradient buffer eluent.

The degree of acetyl substitution (DS_{Ac}) was calculated from the quantity of acetyl residues (as determined by alkaline hydrolysis followed by Ion Chromatography, IC) and the carbohydrate composition. A weighed portion of the dry hemicellulose sample was first dissolved in 2 ml 1.0 M sodium hydroxide and then treated at 80 °C for 1 h in order to cleave the sugar acetate ester linkages. After filtering through a Teflon filter (pore diameter 0.2 μm) an aliquot (0.20 ml) of the filtrate was diluted to 10 ml with pure water. The quantity of acetate ions in the dilute alkaline hydrolysate was subsequently determined by IC employing a Dionex ICS-2000 Ion Chromatography System (with an electrochemical detector), Dionex, GA15 guard column, SA15 separation column, and applying a potassium hydroxide (35 mM) buffer eluent.

The molecular weight parameters for the hemicelluloses purified by ultrafiltration were determined by employing aqueous size exclusion chromatography. Pre-filtered hemicellulose samples (containing approximately 1 mg dry matter) were injected into the SEC column system which consisted of three columns containing Ultrahydrogel 120, 250 and 500 (Waters Assoc. USA), respectively. The columns were linked in series to each other and to a refractometer (Waters Assoc. USA). The eluent system utilized was 50 mM ammonium acetate pH 7. The signal from the refractometer was processed on a standard PC using the PL Caliber SEC software and interface (Polymer Laboratories Ltd., UK).

Oxygen barrier films

Film forming ability of galactoglucomannan

Since many polysaccharides possess the ability to form films with oxygen barrier properties, it was of interest to study galactoglucomannan in this respect (9). Spruce-derived mannans have later been used for film-production by Mikkonen et al. (10).

Films were cast from aqueous solutions of O-acetyl-galactoglucomannan (AcGGM), obtained from TMP process waters according to the isolation procedure described above, and an additive (Table III). Plasticizers used were glycerol, sorbitol and xylitol and in addition the renewable polymers alginate and carboxymethylcellulose (CMC) were investigated. The films were evaluated by measuring dynamic mechanical properties under a humidity scan. For the films with the low molecular weight additives (glycerol, sorbitol and xylitol) a drop in storage modulus was observed between 35% and 50% RH as shown in Figure 3. The film composed of a mixture of AcGGM, alginate and glycerol showed a drop in storage modulus at 50% RH, see Figure 3.

Table III. Manufactured films.

Composition	Cast from 14 ml water
AcGGM:Alginate	0.26:0.14[a]
AcGGM:CMC	0.26:0.14[a]
AcGGM:Glycerol	0.30:0.10[a]
AcGGM:Sorbitol	0.26:0.14[a]
AcGGM:Xylitol	0.26:0.14[a]
AcGGM:Alginate:Glycerol	0.26:0.07:0.07[a]

[a] Values given are in grams.

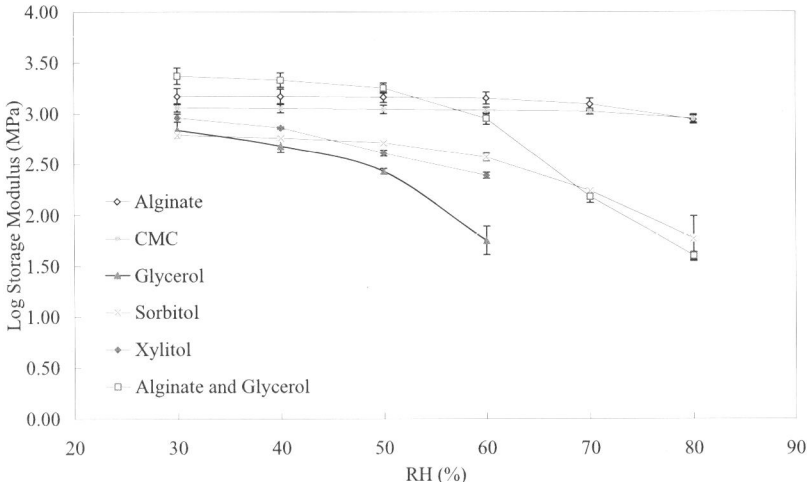

Figure 3. Storage modulus of AcGGM films with additives, as a function of varying moisture content (mean values of three samples per film).

Solutions of alginate or CMC and AcGGM formed films whose mechanical properties were comparatively insensitive to humidity. When CMC was used, almost no softening was observed up to 80% RH, whereas there was a small drop in storage modulus for the film containing alginate at 70% RH.

The oxygen permeabilities were 0.6 and 1.3 (cm^3 μm) / (m^2 24h kPa) at 50 % RH for the films containing alginate and CMC, respectively. In the case of the plasticizers, sorbitol addition gave a lower permeability, 2.0 (cm^3 μm) / (m^2 24h kPa), than glycerol (4.6 (cm^3 μm) / (m^2 24h kPa)) or xylitol (4.4 (cm^3 μm) / (m^2 24h kPa)). Ethylene vinyl alcohol, a synthetic polymer commercially used as oxygen barrier film, has oxygen permeability from 0.1 (cm^3 μm) / (m^2 24h kPa) at 0% RH to 10 (cm^3 μm) / (m^2 24h kPa) at 95% RH (*11*).

Films obtained from hydrophobized galactoglucomannan

A deficiency of oxygen barrier films made from renewable resources, especially when low molecular weight additives are used as plasticizers, is that they only are effective under rather dry conditions. As long as the films are dry, the high amount of hydrogen bonding results in slow oxygen diffusion; however, as soon water is picked up from humid air the rigid internal structure is loosened and the oxygen mobility increases dramatically. Another problem associated with the hydrogen bondings is the resulting rigidity and brittleness of the material. In most cases a compromise has to be made between oxygen barrier properties, flexibility and moisture tolerance. The next section is a summary of an attempt (*12*) to improve the moisture resistance.

A comprehensive study of ways to chemically modify hemicellulose is to be found in (*13*).

Hydrophobic modification

Benzylation of acetylated galactoglucomannan was carried out in water by a Williamson's ether synthesis, in similar fashion as reported for other comparable polysaccharides (*14, 15, 16*). In short, to AcGGM (aq) the phase-transfer agent tetrabutylammonium iodide (TBAI) and sodium hydroxide were added. The aqueous solution temperature was held at 40°C for two hours and was thereafter raised to 100°C. Due to the alkaline treatment, the AcGGM material was deacetylated and the hydroxyl groups were activated. After the activation step, benzyl chloride was added and benzylated GGM (BnGGM) was formed as a yellowish precipitate. After quenching the reaction with hydrochloric acid, the product was collected by filtration, washed, and dried under vacuum. In Scheme 1, the reaction is depicted at the theoretical maximum molar conversion.

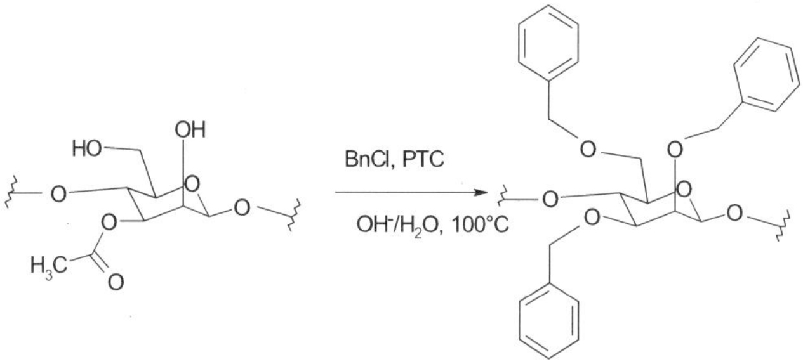

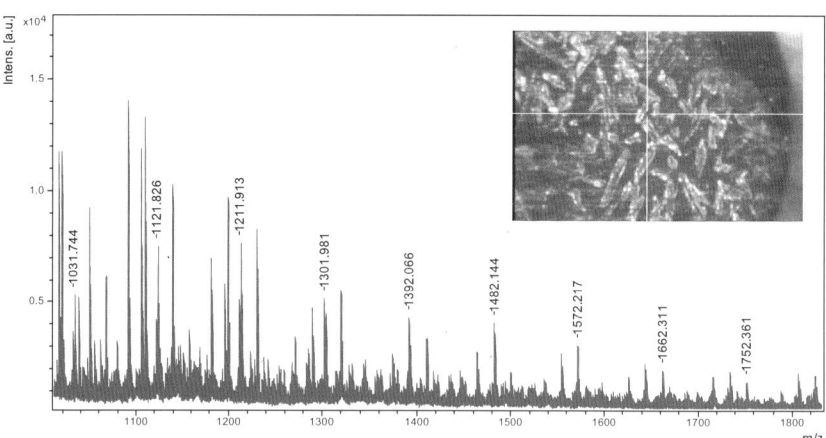

Scheme 1. Theoretical full benzylation of a hexose unit (mannose in this case) in a AcGGM chain

MALDI-TOF-MS of BnGGM showed a series of substituted GGM fragments. The spectrum (Figure 4) shows a series of peaks with DS (benzyl) of approximately 1.3, ranging from DP 3 to DP 8.

Figure 4. Segment of a MALDI-TOF spectrum showing the DP 5 mass series of sample BnGGM mixed with a DHB matrix in DMSO. A crystal of BnGGM in the matrix is also shown in the inserted sample picture.

Some segregation of the sample droplet appeared due to the low volatility of DMSO (see sample image inserted in Figure 4).

Films obtained from benzylated galactoglucomannan

Films were cast from solutions of benzylated galactoglucomannan (BnGGM) in DMF into glass Petri dishes and were left to dry in a closed hood (~ 23°C, RH < 50 %) for at least one week. The films were conditioned in a desiccator at 51 (±3) % RH and 21 (±1) °C prior to analysis.

The hydrophobic properties of the obtained films were confirmed by measuring the contact angle which was found to be around 60-70°. It was also concluded that BnGGM-films resisted contact with water very well and hardly any weight changes were observed over a period of 20 days in water. For the best films obtained, the oxygen permeability was 150-170 (cm^3 μm) / (m^2 24h kPa) at 83 % RH.

In order to improve the oxygen barrier properties and still maintain a high water tolerance, we investigated lamination of a base film of unmodified AcGGM with BnGGM. The laminated film was indeed found to be a better oxygen barrier. The oxygen permeability was 9 (cm^3 μm) / (m^2 24h kPa) at 73 % RH. Furthermore, surface modification of an unmodified base film of AcGGM using styrene was attempted using either a plasma activation (17) or a vapor-phase (18) method. The effects of those treatments were, however, small in comparison with the described lamination. The lamination approach was thus found to be a powerful route for obtaining good barrier properties even at high humidity, using predominantly renewable resources.

Hydrogels obtained from acetylated galactoglucomannan

Acetylated galactoglucomannan hemicellulose (AcGGM) can advantageously be utilized for obtaining high-quality hydrogel materials (19, 20). A hydrogel is a material that swells but does not dissolve. Hydrogels can serve as drug delivery vehicles in several different drug administration routes, where the peroral stands out as the most important one (21). A recently published review contains more general information about biomedical applications of polysaccharide hydrogels (22).

In a recently published study (23) the release of two model drugs from neutral and ionic hydrogels was examined. The preparation of neutral hydrogels followed Scheme 2. In the first step, the AcGGM obtained from TMP process waters (dissolved in DMSO) was reacted with 2-[(1-imidazolyl)formyloxy]ethyl methacrylate (HEMA-Im) and the formed modified AcGGM product (M-AcGGM - methacrylated) was precipitated by pouring the reaction mixture into ice-cold 2-propanol. The degree of modification was estimated by ^{1}H-NMR and ranged between 2 – 50% depending on the reaction time (2-100 h). In the second step, the M-AcGGM product was dissolved in a small amount of deionized H$_2$O and 2-hydroxymethacrylate (HEMA) was added as a co-monomer (60/51/42 w-% M-AcGGM). Polymerization was initiated by subsequent addition of sodium peroxodisulfate and sodium pyrosulfite as a radical initiator system, then the polymerizing solution was injected into cylindrical molds to form discs. It was also possible using related techniques to make microspheres consisting of the same novel hydrogel material (*24*).

For ionic hydrogels the HEMA-modified AcGGM was additionally modified by reaction with maleic anhydride to introduce carboxylic acid functionalities. Since the kinetics were much faster for this reaction, a fixed reaction time of one hour was set and the degree of substitution was instead regulated by the amount of added maleic anhydride. Hydrogels were synthesized in the same way as described above for neutral hydrogels using the respective equimolar amount of the double-modified AcGGM (called CM-AcGGM – carboxylated-methacrylated).

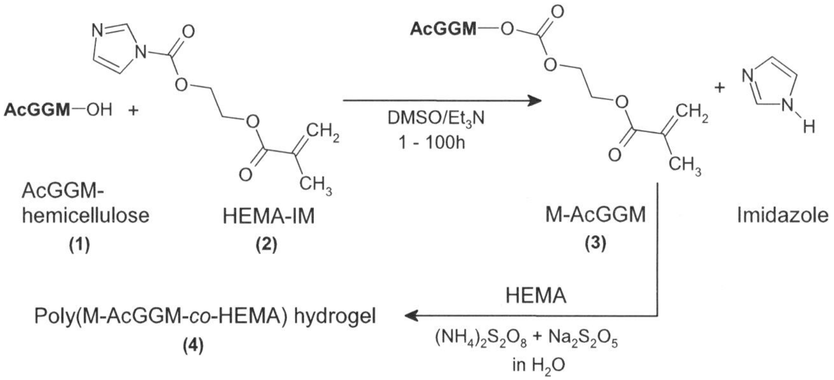

Scheme 2. Synthetical route applied for the preparation of poly(M-AcGGM-co-HEMA) hydrogels.

Release experiments

For the release experiments, drug-loaded hydrogels were synthesized by using a 23 mg/100 ml drug solution (caffeine or vitasyn blue) as polymerization medium. Drug release analysis from the hydrogels was performed after immersing the samples in a VanKel 7010 dissolution bath (Varian, Inc., Palo Alto, Calif.) using a Varian Cary 50 spectrometer (Varian, Inc., Palo Alto, Calif.) with a 12 channel multiplexer–equipped UV fiber optic probes with 10 mm probe tips (C Technologies, Bridgewater, NJ) at 37 °C and 50 rpm mechanical stirring.

The times needed for 50 wt % of the drugs to be released (T_{50}) are shown in Figure 5. The hydrogel consisting of 60 wt % M-AcGGM with a DS_M of 0.22 and 40 wt % HEMA resulted in the slowest release ($T_{50} = 35$ min for caffeine and 90 min for vitasyn blue). A further increase in the M-AcGGM content and its modification degree (DS_M) is expected to continue to prolong the release time. The equilibrium swellings (Q_{eq}) for the hydrogels in Figure 5 were in order from left to right 7.4, 6.1, 4.2, 6.0, 4.6, and 4.1 meaning that the hydrogels swell 4 – 7 times their own weight in water.

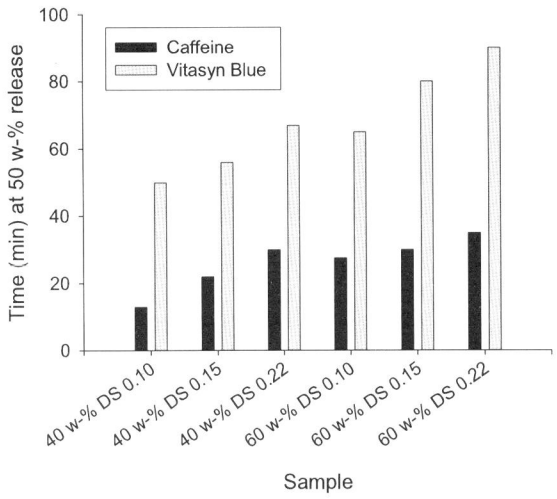

Figure 5. Time for 50 w-% release of caffeine and vitasyn blue (in 900 ml H₂O, pH 7, 37 °C, 50 rpm stirring) from initially dry hydrogels loaded during polymerization.

Drug release experiments using caffeine were performed for the acidic poly(CM-AcGGM-*co*-HEMA) hydrogels as well. The release was then tested both in neutral and acidic conditions. In Figure 6 it is shown that the release was faster at pH 7.0 compared to pH 3.0. This is due to a dissociation of the carboxylic acid functions which leads to a swelling of the matrix due to electrostatic repulsion. It can also be seen that the release from a comparable neutral hydrogel was faster than from the acidic hydrogels.

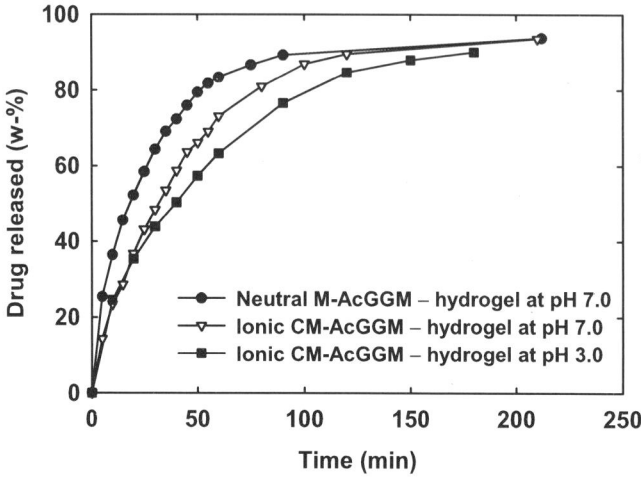

Figure 6. Caffeine release (in 900 ml H₂O, pH 7, 37°C, 50 rpm stirring) from initially dry neutral (DS$_M$ 0.10) and ionic hydrogels (DS$_M$ 0.10, DS$_C$ 0.31) loaded during polymerization. The hydrogels were all produced with 50 wt % HEMA as co-monomer.

Concluding remarks

Isolation of water-soluble hemicelluloses from steam explosion and mechanical pulping process waters by ultrafiltration and dialfiltration proved to be a useful way for obtaining substantial quantities of *O*-acetyl-galactoglucomannan fractions for further processing into barrier films and hydrogels. Those water-soluble hemicelluloses were found to be excellent candidates for making new renewable barrier materials. To obtain good oxygen barrier properties an appropriate plasticizer or a high-molecular weight renewable polymer can be used in a blend with the galactoglucomannan. However, a drawback with those films is that they are comparatively sensitive to moisture. An appropriate way to obtain good oxygen barrier properties and maintain a high water tolerance is to laminate a base film of galactomannan with benzylated galactomannan. Galactoglucomannan can advantageously be utilized for obtaining high-quality hydrogel materials. Gels crosslinked with 2-hydroxyethylmethacrylate can be utilized for drug release.

References

1   Lundqvist, J.; Teleman, A.; Junel, L.; Zacchi,G.; Dahlman, O.; Tjerneld, F.; Stålbrand H. *Carbohydr. Polym.* **2002**, *48*, 29.

2   Lundqvist, J.; Jacobs, A.; Palm, M.; Zacchi, G.; Dahlman, O.; Stålbrand, H. *Carbohydr. Polym.* **2003**, *51*, 203.

3   Jacobs, A.; Lundqvist, J.; Stålbrand, H.; Tjerneld, F.; Dahlman, O. *Carbohydr Res.* **2002**, *337*, 711.

4   Stålbrand, H.; Lundqvist, J.; Andersson, A.; Hägglund, P.; Andersson, L.; Tjerneld, F.; Jacobs, A.; Teleman, A.; Dahlman, O.; Palm, M.; Zacchi, G. In *Hemicelluloses: Science and Technology*; Gatenholm, P.; Tenkanen, M., Eds.; ACS Symposium. Series No. 864; American Chemical Society:Washington, DC, 2004; pp 66-78.

5   Teleman, A.; Antonsson, M.; Tenkanen, M.; Jacobs, A.; Dahlman, O. *Carbohydr. Res.* **2003**, *338*, 525.

6   Hägglund, E.; Lindberg, B.; McPherson, J. *Acta Chem. Scand.* **1956**, *10*, 1160.

7   Lindberg, B.; Meier, H. *Svensk papperstidning* **1957**, *60*, 785.

8   Dahlman, O.; Jacobs A. Unpublished results

9   Hartman, J.; Albertsson, A.-C.; Söderqvist Lindblad, M.; Sjöberg, J. *J. Appl. Polym. Sci.* **2006**, *100*, 2985.

10 Mikkonen, K. S.; Yadav, M. P.; Cooke, P.; Willför, S.; Hicks, K. B.; Tenkanen, M *BioResources* **2008**, *3*, 178.

11 Olabarrieta, I. Ph.D. thesis, KTH Fibre and Polymer Technology, Stockholm, Sweden, 2005.

12 Hartman, J.; Sjöberg, J.; Albertsson, A.-C. *Biomacromolecules* **2006**, *7*, 1983.

13 Söderqvist Lindblad, M.; Albertsson, A.-C. In *Polysaccharides: Structural, Diversity and Functional Versatility*, 2nd ed.; Dumitriu, S., Ed.; Marcel Dekker, Inc.:New York, NY, 2004; pp 491-508.

14 Chen, Y.; Zhang, L.; Lu, Y.; Ye, C.; Du, L. *J. Appl. Polym. Sci.* 2003, *90*, 3790.

15 Ebringerova, A.; Novotna, Z.; Kacurakova, M.; Machova, E. *J. Appl. Polym. Sci.* **1996**, *62*, 1043.

16 Vincendon, M. *J. Appl. Polym. Sci.* **1998**, *67*, 455.

17 Olander, B.; Wirsen, A.; Albertsson, A.-C. *Biomacromolecules* **2002**, *3*, 505.

18 Edlund, U.; Källrot, M.; Albertsson, A.-C. *J. Am. Chem. Soc.* **2005**, *127*, 8865.

19 Söderqvist Lindblad, M.; Ranucci, E.; Albertsson, A.-C. *Macromol. Rapid Commun.* **2001**, *22*, 962.

20 Söderqvist Lindblad, M.; Albertsson, A.-C., Ranucci, E.; Laus M.; Giani, E. *Biomacromolecules* **2005**, *6*, 684.

21 Peppas, N. A.; Bures, P.; Leobandung, W.; Ichikawa, H. *E. J. Pharm. Biopharm.* **2000**, *50*, 27.

22 Söderqvist Lindblad, M.; Sjöberg, J; Albertsson, A.-C.; Hartman, J. In *Materials, Chemicals and Energy from Forest Biomass*; Argyropoulos, D. S., Ed.; ACS Symposium Series No. 954 American Chemical Society:Washington, DC, 2006; pp 153-167.

23 Voepel, J.; Sjöberg, J.; Reif, M.; Albertsson, A.-C.; Hultin, U.-K.; Gasslander, U., Accepted for publication in *J. Appl. Polym. Sci.*

24 Edlund, U.; Albertsson, A.-C. *J. Bioact. Compat. Polym.* **2008**, *23*, 171.

Chapter 11

Synthesis of methylated cello-oligosaccharides

Synthesis strategy for blockwise methylated cello-oligosaccharides

Hiroshi Kamitakahara[1,*], Fumiaki Nakatsubo[1], Dieter Klemm[2]

[1]Graduate School of Agriculture, Kyoto University, Kitashirakawa-oiwake-cho, Sakyo-ku, Kyoto 606-8502, Japan
[2]Transfer group Polymet at Friedrich Schiller University of Jena, Technologie- und Innovationspark, Wildenbruchstrasse 15, D-07745 Jena, Germany

In this chapter, we show how cello-oligosaccharide synthesis can play an important role on understanding physical properties of industrially produced cellulose derivatives. As one example, a synthetic strategy for cello-oligosaccharide derivatives related to methylcellulose (MC) is discussed in detail. The physical properties of MC are attributed to its block-like chemical structure. Thus, blockwise methylated cello-oligosaccharides were synthesized as models of the structural units of MC. An aqueous solution of the blockwise methylated cello-oligosaccharides showed higher surface activity than that of industrially produced MC. Namely, it was confirmed that the synthesized cello-oligosaccharide derivatives have surface activity superior to that of MC due to controlling the block structure of substituents along the molecular backbone. Furthermore, blockwise methylated cello-oligosaccharides self-assembled to form ellipsoidal nano particles in water. These new findings are of importance not only for the investigation on phase separation phenomenon of MC but also for the development of new applications of cello-oligosaccharide derivatives.

Cellulose and its derivatives have been widely used for about 150 years (*1*). Cello-oligosaccharides having DP 1 to 10 and the biopolymer cellulose were recently synthesized chemically (*2-8*) and enzymatically (*9-11*). In future, renewable natural resources like cellulose will be used not only as raw materials but also as new functional materials. Thus, structure-property relationships of cello-oligosaccharide derivatives as models of cellulose derivatives were investigated in order to obtain new types of functional cellulose with high performance.

Among important cellulose derivatives, methylcellulose (MC) with DS about 1.8 is a fundamental large-scale product because of its water solubility and the rheological and gelation behavior of the solutions. It has been reported that thermo reversible gelation of aqueous solutions of MC is attributed to dehydration of hydrated MC molecules at elevated temperature (*12-16*). Due to the heterogeneous process of MC production, the industrial MC is said to be an alternate block copolymer which consists of densely substituted hydrophobic and less substituted hydrophilic block sequences (*13*). The MC is of academic interest because such a substituent pattern controls the physical properties of MC. Thus, blockwise methylated cello-oligosaccharides (*17, 18*) were synthesized as a partial chemical structure of MC and their structure-property relationships were investigated.

As synthetic methods for blockwise methylated cello-oligosaccharides, both general glycosylation methods (*19-26*) and ring-opening oligomerization of glucose orthopivalate derivative (*27-31*) were applied. Polydisperse and monodisperse blockwise methylated cello-oligosaccharides were prepared according to these methods. Relationships between their chemical structure and surface activities of their aqueous solutions were discussed. Furthermore, aggregation structure of blockwise methylated cello-oligosaccharides was studied by means of some analytical methods such as light scattering and transmission electron microscopic measurements.

In this chapter, we describe a novel methodology to develop new functional cellulose derivatives according to our synthesis strategy for blockwise methylated cello-oligosaccharides. Furthermore, their structure-property relationships are discussed.

Heterogeneous cellulose derivatization

Industrially produced cellulose derivatives have heterogeneous substituent patterns according to the heterogeneous reaction conditions. There are three kinds of heterogeneities of substituent patterns: 1) within the 8 (where there is only one type of substituent, as in MC) possible substituted anhydro glucosyl units (AGUs) (Figure 1, (1)); 2) sequence patterns of AGUs along one cellulose molecule (*e.g.*, blocky, random, and stereoregular; Figure 1, (2)); 3) variation in substitution and pattern between cellulose molecules (Figure 1, (3)). Our research focused on the sequence patterns along one cellulose chain (2).

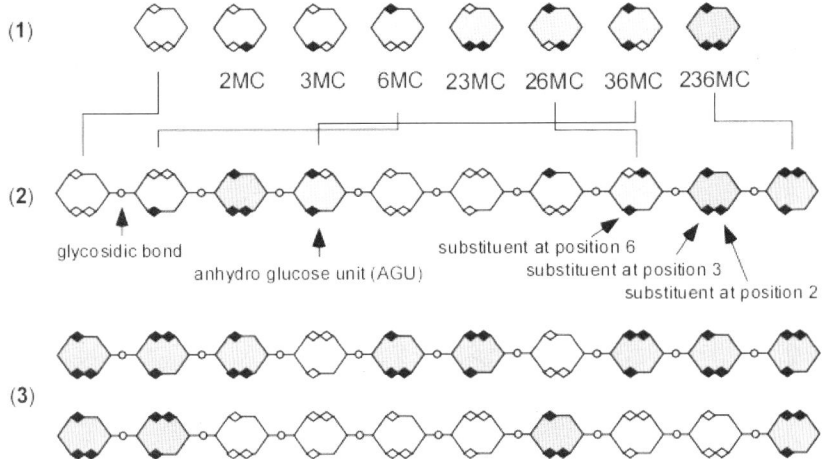

Figure 1. Three kinds of heterogeneity of substituent patterns in cellulose derivatives

Highly methylated regions (sequences of 2,3,6-tri-*O*-methyl-glucosyl residues) of the cellulose chain are said to cause micelles, that is, liquid-liquid phase separations in aqueous solution (*15*). These micelles are termed "cross linking loci' (*32*). In addition, it is well known that reversible crosslinks must exist in any reversible gel (*32*). However, the precise chemical structure of "cross linking loci" in aqueous solution of MC has not been clarified yet.

The bulk DS of industrial MC is calculated to be about 1.8. Therefore, structure-property relationships of industrially available MC must be considered statistically. For instance, 15 patterns ($_6C_2$) can be written for methylated cellohexaose consisting of 4 fully methylated and 2 unsubstituted glucoses (DS=2.0), as shown in Figure 2, neglecting the added substituents on the two end groups. Randomly substituted methylcellulose with DP=n, DS=*x* has $_{3n}C_{xn}$ substitution patterns. It is impossible to synthesize all substitution patterns of cellulose derivative having higher molecular weight. In addition, it is also impossible to fractionate and characterize all individual celluloses and cello-oligosaccharides obtained by chemical or enzymatic hydrolysis (*33-37*).

Thus, we have prepared some blockwise methylated cello-oligosaccharides individually and compared their properties in order to investigate important chemical structures of "cross linking loci".

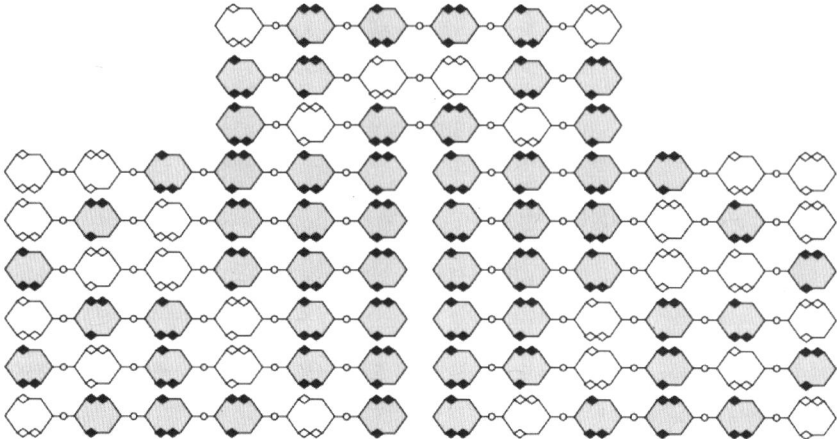

Figure 2. 15 substitution patterns of blockwise methylated cello-hexamer with DS=2.0

Basic strategy for the synthesis

Blockwise methylated cello-oligosaccharides were synthesized by glycosylation methods combined with polymerization of glucose orthopivalate derivatives. Figure 3 shows possible combinations of glycosyl donors and acceptors. Polydisperse mixtures of blockwise methylated cello-oligosaccharides were obtained using glucose orthopivalate derivatives as glycosyl acceptors. Monodisperse ones were prepared using general glycosylation methods.

Combinations of glucose and cellobiose derivatives as building blocks

Many combinations of glycosyl donors and acceptors were tried for the syntheses of blockwise methylated cello-oligosaccharides. Phenyl 2,3,6-tri-*O*-methyl-β-D-glucose-(1→4)-2,3,6-tri-*O*-methyl-1-thio-β-D-glucoside (**1**) can be a suitable glycosyl acceptor with 4-*O*-acetyl-2,3,6-tri-*O*-methyl-β-D-glucose-(1→4)-2,3,6-tri-*O*-methyl-glucosyl fluoride (**2**), 4-*O*-acetyl-2,3,6-tri-*O*-methyl-β-D-glucose-(1→4)-2,3,6-tri-*O*-methyl-glucosyl trichloroacetoimidate (**3**), and 4-*O*-acetyl-2,3,6-tri-*O*-benzyl-β-D-glucose-(1→4)-2,3,6-tri-*O*-benzyl-glucosyl trichloroacetoimidate (**4**) as glycosyl donors. Benzylated monomers are precursors of unmodified carbohydrate residues. On the other hand, 2,3,6-tri-*O*-methyl-β-D-glucose-(1→4)-2,3,6-tri-*O*-methyl-glucosyl fluoride (**5**) and 2,3,6-tri-*O*-benzyl-β-D-glucose-(1→4)-2,3,6-tri-*O*-benzyl-glucosyl fluoride (**6**) were suitable glycosyl acceptors with phenyl 2,3,4,6-tetra-*O*-methyl-β-D-glucose-(1→4)-2,3,6-tri-*O*-methyl-1-thio-β-D-glucoside (**7**) and phenyl 4-*O*-acetyl-2,3,6-tri-*O*-methyl-β-D-glucose-(1→4)-2,3,6-tri-*O*-methyl-1-thio-β-D-glucopyranoside

(**8**). Three glucosyl donors, 4-*O*-acetyl-3-*O*-benzyl-2,6-di-*O*-pivaloyl-glucosyl fluoride (**9**), trichloroacetoimidate (**10**), and phenyl 4-*O*-acetyl-3-*O*-benzyl-2,6-di-*O*-pivaloyl-1-thio-glucoside (**11**) can be good glycosyl donors to give β-glycosides based on the neighboring group participation of the pivaloyl group at position 2. 3-*O*-Benzyl-6-*O*-pivaloyl-α-D-glucopyranose 1,2,4-orthopivalate (**12**) was found to be a useful building block for the unsubstituted segments of polydisperse, blockwise methylated cello-oligosaccharides. Phenyl 2,3,4,6-tetra-*O*-methyl-1-thio-β-D-glucoside (**13**) is a building block glucosyl donor for a fully methylated nonreducing end (*38*). In addition, methylated glucosyl acceptors are also available (*38*). Other benzylated building blocks **14**, **15**, and **16** were also prepared. Pivaloylated blocks **17**, **18**, **19**, and **20** were synthesized but were not suitable for glycosylation due to their low reactivities. By selecting suitable combinations of glycosyl acceptors and donors shown in Figure 3, synthesis methods for 15 blockwise methylated cello-oligosaccharides shown in Figure 3 can be planned using retrosynthetic analysis.

Furthermore, 8 possible regioselectively methylated AGUs shown in Figure 1 (*1*) can be synthesis blocks according to the same synthesis strategy. The above-mentioned suitable combinations of glycosyl donors and acceptors will promise a library of all blockwise methylated cello-oligosaccharides theoretically. These are only the first examples in the infinite numbers of compounds. On the basis of the results obtained, suitable building blocks for cellulose and cello-oligosaccharide ethers can be selected according to the retrosynthetic analysis.

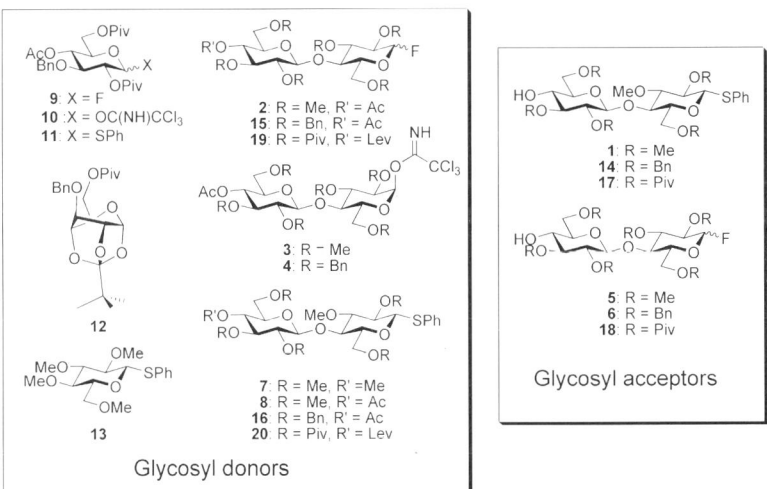

Figure 3. Possible combinations of glycosyl donors and acceptors

Syntheses of blockwise methylated cello-oligosaccharides as model compounds for methylcellulose

The blockwise methylated cello-oligosaccharides were prepared according to the combination of two synthesis methods: oligomerization of glucose orthoester derivative and glycosylation. Synthesis methods for polydisperse and monodisperse blockwise methylated cello-oligosaccharides are described below.

Polydisperse mixture of block co-oligomers of tri-*O*-methylated and unmodified cello-oligosaccharides

The polymerizations of glucopyranose 1,2,4-orthopivalate derivatives give cellulose derivatives (*7, 8*). Monodisperse cello-tetraose derivative **22** was reacted with the glucose orthoester derivative **12** to elongate the chain in the non-reducing end direction, as shown in Figure 4. The blocky oligomer **24** (DP = 4-8, average DS=2.21) was prepared according to this combined method. The mixture shows amphiphilic behavior: it is soluble in water, methanol, and chloroform.

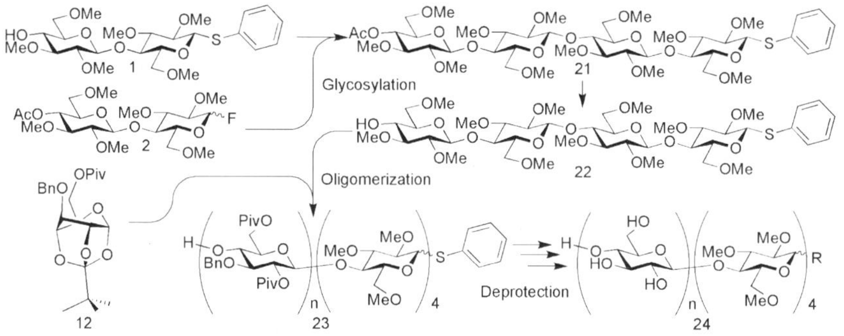

Figure 4. Synthesis scheme for the polydisperse mixture of blockwise methylated cello-oligosaccharides

Monodisperse block co-oligomers of tri-*O*-methylated and unmodified cello-oligosaccharides

Figure 5 shows schematic figure of the combinations of glycosyl donors and acceptors to yield monodisperse blockwise methylated cello-oligosaccharides. Independent blockwise methylated cello-oligosaccharides (DP=5: **25** and DP=6: **26**) included in the mixture of block co-oligomers **24** were prepared based on glycosylation method (*18*). In addition, as an analog of blockwise methylated cello-hexamer **26**, methyl β-D-glucopyranosyl-(1→4)-2,3,6-tri-*O*-methyl-β-D-glucopyranosyl-(1→4)-2,3,6-tri-*O*-methyl-D-glucopyranoside (DP=3: **27**) was synthesized according to Figure 5 as written below.   Methyl β D glucopyranosyl-(1→4)-β-D-glucopyranosyl-(1→4)-2,3,6-tri-*O*-methyl-β-D-glucopyranosyl-(1→4)-2,3,6-tri-*O*-methyl-D-glucopyranoside (DP = 4: **28**) is also available (*38*).

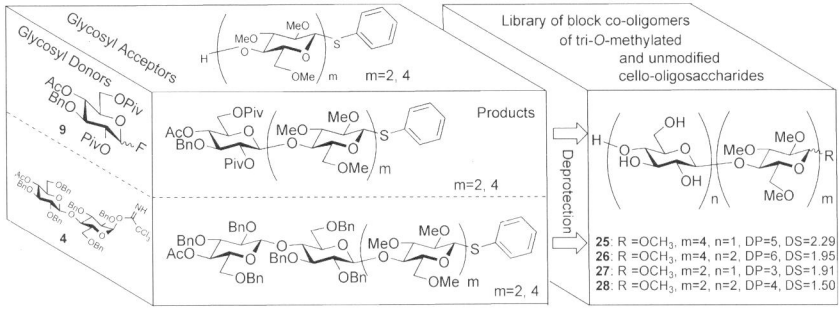

Figure 5. Schematic figure of the combinations of glycosyl donors and acceptors

The prepared series of monodisperse blockwise methylated cello-oligosaccharides are shown in Figure 6. Interestingly, all compounds were soluble in water and chloroform (*18*). Namely, blockwise methylated cello-oligosaccharides with DS about 2 were amphiphilic.

25 DS=2.29

26 DS=1.95

27 DS=1.91

28 DS=1.50

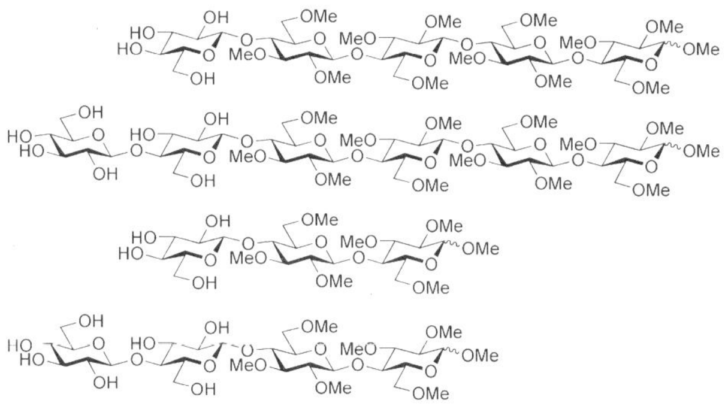

Figure 6. Monodisperse blockwise methylated cello-oligosaccharides

Surface activities of aqueous solution of blockwise methylated cello-oligosaccharides

It is well known that industrially produced methylcellulose (SM-4) acts as nonionic surfactant. The MC has a heterogeneous substitution pattern because of its heterogeneous preparation procedure. MC also has a polydisperse, statistical molecular weight distribution. By comparison of the polydisperse mixture of blockwise methylated cello-oligosaccharides **24** with monodisperse cello-oligomers of tri-*O*-methylated and unmodified cello-oligosaccharides **25**,

26 and **27**, the effect of mixing blockwise methylated cello-oligosaccharides could be studied.

The monodisperse diblock co-oligomers of tri-*O*-methylated and unmodified cello-oligosaccharides, **25** (DP=5), **26** (DP=6), and **27** (DP=3) had higher surface activities compared to the mixture of diblock co-oligomers of tri-*O*-methylated and unmodified cello-oligosaccharides (**24**) and commercially available MC SM-4.

Figure 7 shows surface tension-concentration curves of compounds **24**, **25**, **26**, **27**, and MC. The values of γ_{5000} (surface tension at the concentration of 5000 ppm (5 mg/mL or 0.5 wt. %)) of monodisperse compounds **25**, **26**, and **27** were about 30 mN/m, lower than those of polydisperse compounds **24** and MC (over 45 mN/m).

It was found experimentally that the polydispersity of a blockwise methylated cello-oligosaccharide affects the surface activity of its aqueous solution. It should be possible to improve the surface activity of aqueous MC solutions by designing the substituent pattern and molecular weight distribution for optimum surface activity.

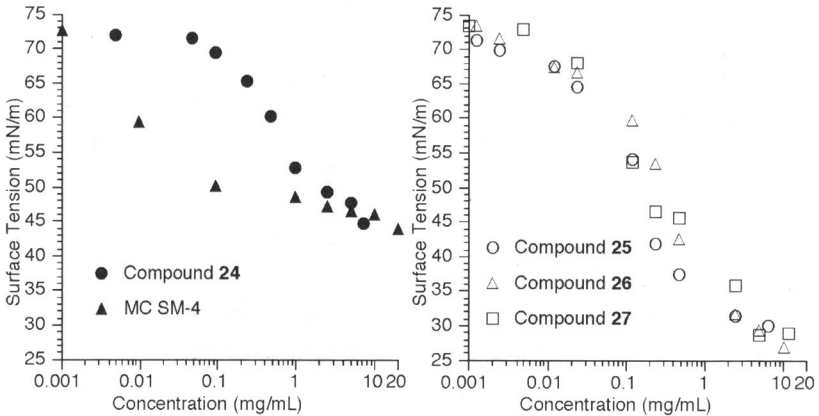

Figure 7. Surface tension concentration curves of polydisperse and monodisperse samples

Morphology

In order to understand the physical properties of aqueous MC solutions, the aggregation behavior of a polydisperse mixture of blockwise methylated cello-oligosaccharides (**24**) in aqueous solution was investigated by means of transmission electron microscope (TEM) and static and dynamic light scattering experiments. As a result, the mixture of blockwise methylated cello-oligosaccharides (**24**) was found to form ellipsoidal nano particles in its aqueous solution (*39*).

Figure 8 shows a TEM image of negatively stained blockwise methylated cello oligosaccharides (**24**) (concentration: 1 mg/mL). Ellipsoidal particles were observed in the TEM image. The measured average dimensions of one hundred of the ellipsoidal particles were about 50 nm on the semi-major axis and about 25 nm on the semi-minor axis.

Figure 8. TEM image of negatively stained blockwise methylated cello-oligosaccharides (24) (prepared at c = 1 mg/mL) bar = 200 nm.

Figure 9 shows the Zimm plot of polydisperse blockwise methylated cello-oligosaccharides **24**. The weight average M_w was 29500 g/mol and the radius of gyration (R_g) was 25 nm after extrapolation to zero concentration. The R_g increased as concentration increased, reaching 64 nm at c = 0.505 mg/mL.

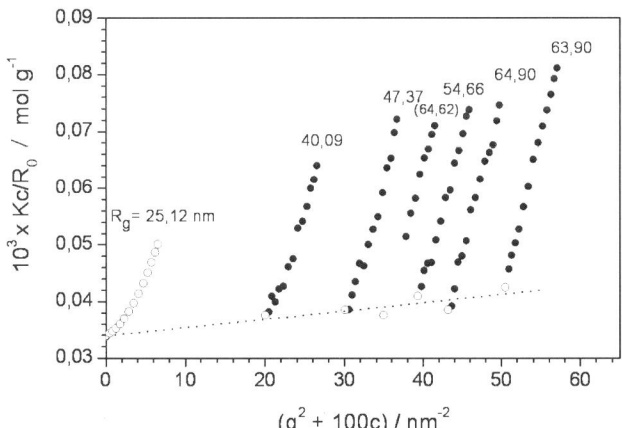

Figure 9. Zimm plot of the block-like methylated co-oligomer from concentrations c_1 = 0.200, c_2 = 0.250, c_3 = 0.301, c_4 = 0.393, c_5 = 0.432 and c_6 = 0.505 mg/mL. M_w = 29500 g/mol, R_g = 25. nm. The numbers at the different curves denote the apparent radii of gyration.

Figure 10 shows the concentration dependencies of the apparent radii of gyration ($R_{g,app}(c)$) and the hydrodynamic radii ($R_{h,app}(c)$). The conjecture of aggregation is confirmed by the ρ-parameter defined as (*40, 41*):

$$\rho = \frac{R_g}{R_h}$$

The corresponding $\rho_{app} = R_{g,app}/R_{h,app}$ remains constant ($\rho = 1$) up to concentrations of 0.35 mg/mL but increases to 1.35 at c = 0.5 mg/mL. The change in curvature and the increase of the ρ-parameter is indicative of a transition from spherical to ellipsoidal particle shapes. These clusters might be micelles or randomly branched aggregates which we could not discriminate since no critical micelle concentration could be detected.

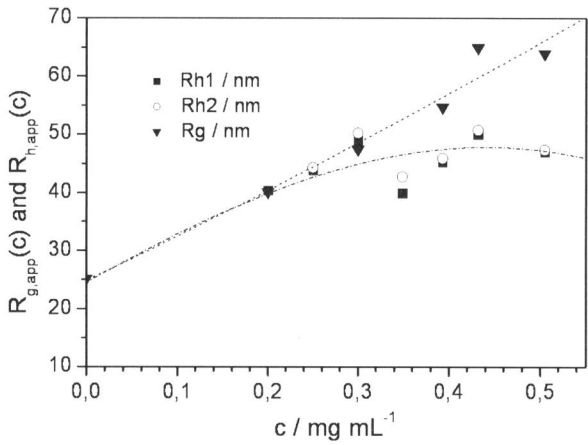

Figure 10. Concentration dependencies of the apparent radii of gyration $R_{g,app}(c)$ and hydro-dynamic radii $R_{h,app}(c)$. The open circles and the filled squares correspond to two different fitting attempts to the DLS measurements, the filled triangles represent the radius of gyration.

Aqueous solutions of industrial MC with DS about 1.8 show phase separation upon heating. Sequences of tri-*O*-methyl-glucosyl residues are believed to aggregate in the gel state of MC (*16*). The polydisperse mixture of block cooligomers of tri-*O*-methyl- and unmodified cello-oligosaccharides (**24**) self-assembled to form stable ellipsoidal nano particles at room temperature. Hydrophobic intermolecular interactions between tri-*O*-methyl glucosyl residues were experimentally confirmed in this study. The use of blockwise methylated cello-ligosaccharides as model compounds for industrial MC may clarify the structure-property relationships of MC. Morphologies of monodisperse blockwise methylated cello-oligosaccharides such as compounds **25**, **26**, and **27** in aqueous media are now under investigation.

Summary and future prospects

According to the above-mentioned synthesis strategy, all methylated cello-oligosaccharide derivatives of AB diblock and ABA triblock structures can be synthesized. As a result, the influence on properties of the patterns of distribution of substituents along the cellulose molecule and between cellulose chains can be studied. We have found that the surface activity of blockwise methylated cello-oligosaccharides can be controlled by their pattern of substituents and polydispersity. Furthermore, it was observed that polydisperse blockwise methylated cello-oligosaccharides self-assembled in water to form ellipsoidal nanoparticles. These blocky oligosaccharides should prove to be valuable tools for modeling the gelation behavior of aqueous MC solutions.

We predict that this basic strategy will prove applicable to model compounds not only for MC but also for other cellulose ethers like ethylcellulose (EC), hydroxyethylcellulose (HEC), hydroxypropylcellulose (HPC), and hydroxypropylmethylcellulose (HPMC). Basic research using cello-oligosaccharide derivatives should teach us about cellulose derivatives which possess superior properties. Development of precise synthesis methods for functional celluloses and cello-oligosaccharides will open a new field in cellulose chemistry.

Acknowledgements

The authors are sincerely grateful to Prof. A. Yoshinaga of Kyoto University for taking a TEM image, and to Dr. H. Aono, Profs. D. Tatsumi and T. Matsumoto of Kyoto University for light scattering experiments. The authors gratefully acknowledge valuable discussion with Prof. W. Burchard of Albert-Ludwig University of Freiburg on interpretation of light scattering data. We acknowledge Dr. Kazuhisa Hayakawa, Shinetsu Chemical, for providing the MC SM-4. This study is in part supported by a Grant-in-Aid for Scientific Research from the Ministry of Education, Science, and Culture of Japan (Nos. 13760132, 15780124, and 18680009). H. Kamitakahara acknowledges the Alexander von Humboldt Foundation for a research fellowship at Friedrich Schiller University of Jena.

Literature Cited

1. Klemm, D.; Heublein, B.; Fink, H. P.; Bohn, A., *Angew. Chem. Int. Ed.* **2005,** *44* (22), 3358-3393.
2. Takeo, K.; Yasato, T.; Kuge, T., *Carbohydr. Res.* **1981,** *93* (1), 148-156.
3. Takeo, K.; Okushio, K.; Fukuyama, K.; Kuge, T., *Carbohydr. Res.* **1983,** *121* (SEP), 163-173.
4. Takano, T.; Harada, Y.; Nakatsubo, F.; Murakami, K., *Mokuzai Gakkaishi* **1990,** *36* (3), 212-217.

5. Kawada, T.; Nakatsubo, F.; Murakami, K., *Cellulose Chem. Tech.* **1990**, *24* (3), 343-350.

6. Nishimura, T.; Nakatsubo, F., *Cellulose* **1997**, *4* (2), 109-130.

7. Nakatsubo, F.; Kamitakahara, H.; Hori, M., *J. Am. Chem. Soc.* **1996**, *118* (7), 1677-1681.

8. Kamitakahara, H.; Hori, M.; Nakatsubo, F., *Macromolecules* **1996**, *29* (19), 6126-6131.

9. Kobayashi, S.; Kashiwa, K.; Kawasaki, T.; Shoda, S., *J. Am. Chem. Soc.* **1991**, *113* (8), 3079-3084.

10. Okamoto, E.; Kiyosada, T.; Shoda, S. I.; Kobayashi, S., *Cellulose* **1997**, *4* (2), 161-172.

11. Egusa, S.; Kitaoka, T.; Goto, M.; Wariishi, H., *Angew. Chem., Int. Ed.* **2007**, *46* (12), 2063-2065.

12. Heymann, E., *Trans. Faraday Soc.* **1935**, *31*, 846.

13. Savage, A. B., *Ind. Eng. Chem.* **1957**, *49*, 99.

14. Kuhn, W.; Moser, P.; Majer, H., *Helv. Chim. Acta* **1961**, *44*, 770.

15. Rees, D. A., *Chem. & Ind. (London)* **1972**, *19*, 630-636.

16. Kato, T.; Kozaki, N.; Takahashi, A., *Polym. J.* **1986**, *18* (2), 189-191.

17. Kamitakahara, H.; Nakatsubo, F.; Klemm, D., *Cellulose* **2006**, *13* (4), 375-392.

18. Kamitakahara, H.; Nakatsubo, F.; Klemm, D., *Cellulose* **2007**, *14* (5), 513-528.

19. Schmidt, R. R.; Michel, J., *Angew. Chem., Int. Ed.* **1980**, *19* (9), 731-732.

20. Mukaiyama, T.; Murai, Y.; Shoda, S., *Chem. Lett.* **1981**, (3), 431-432.

21. Nicolaou, K. C.; Dolle, R. E.; Papahatjis, D. P.; Randall, J. L., *J. Am. Chem. Soc.* **1984**, *106* (15), 4189-4192.

22. Matsumoto, T.; Maeta, H.; Suzuki, K.; Tsuchihashi, G., *Tetrahedron Lett.* **1988**, *29* (29), 3567-3570.

23. Suzuki, K.; Maeta, H.; Matsumoto, T., *Tetrahedron Lett.* **1989**, *30* (36), 4853-4856.

24. Toshima, K.; Tatsuta, K., *Chem. Rev.* **1993**, *93* (4), 1503-1531.

25. Kanie, O.; Ito, Y.; Ogawa, T., *J. Am. Chem. Soc.* **1994**, *116* (26), 12073-12074.

26. Zhang, Z. Y.; Ollmann, I. R.; Ye, X. S.; Wischnat, R.; Baasov, T.; Wong, C. H., *J. Am. Chem. Soc.* **1999**, *121* (4), 734-753.

27. Hori, M.; Kamitakahara, H.; Nakatsubo, F., *Macromolecules* **1997**, *30* (10), 2891-2896.

28. Hori, M.; Nakatsubo, F., *Macromolecules* **2001**, *34* (8), 2476-2481.

29. Karakawa, M.; Mikawa, Y.; Kamitakahara, H.; Nakatsubo, F., *J. Polym. Sci. Polym. Chem.* **2002**, *40* (23), 4167-4179.

30. Karakawa, M.; Kamitakahara, H.; Takano, T.; Nakatsubo, F., *Biomacromolecules* **2002**, *3* (3), 538-546.

31. Kamitakahara, H.; Koschella, A.; Mikawa, Y.; Nakatsubo, F.; Heinze, T.; Klemm, D., *Macromolecular Bioscience* **2008**, *8* (7), 690-700.

32. Kato, T.; Yokoyama, M.; Takahashi, A., *Colloid Polym. Sci.* **1978**, *256*, 15-21.

33. Erler, U.; Mischnick, P.; Stein, A.; Klemm, D., *Polym. Bull.* **1992**, *29* (3-4), 349-356.

34. Arisz, P. W.; Kauw, H. J. J.; Boon, J. J., *Carbohydr. Res.* **1995,** *271*, 1-14.
35. Mischnick, P., *Cellulose* **2001,** *8* (4), 245-257.
36. Richardson, S.; Gorton, L., *Anal. Chim. Acta* **2003,** *497* (1-2), 27-65.
37. Heinrich, J.; Mischnick, P., *J Chromatogr A* **1996,** *749* (1-2), 41-45.
38. Kamitakahara, H., Manuscript in preparation.
39. Kamitakahara, H.; Yoshinaga, A.; Aono, H.; Nakatsubo, F.; Klemm, D.; Burchard, W., *Cellulose* **2008,** *15* (6), 797-801.
40. Burchard, W.; Schmidt, M.; Stockmayer, W. H., *Macromolecules* **1980,** *13* (5), 1265-1272.
41. Burchard, W., *Cellulose* **2003,** *10* (3), 213-225.

Chapter 12

Direct Synthesis of Partially Substituted Cellulose Esters

Kevin J. Edgar

Research Laboratories, Eastman Chemical Company, Kingsport, TN, 37662, USA
Current Address: 230 Cheatham Hall, Virginia Tech, Blacksburg, VA, 24061, USA

We describe herein a new method for direct esterification of cellulose to partially substituted cellulose esters, using insoluble sulfonated polystyrene resin beads as the catalyst. The process has been extended to the esters of cellulose with long chain acids, which are materials of significant interest that are difficult to synthesize by other means. We offer some mechanistic observations about this interesting reaction. In addition, we present here for the first time a detailed description of the complementary synthetic method, titanium-catalyzed partial esterification in amide or urea solvents, which also provides direct access to partially substituted cellulose esters, including long chain esters. Both methods are practical, efficient and broad in scope. These novel methods address a long standing problem of efficiency in cellulose ester synthesis.

Introduction

Derivatives of cellulose are among the most important biomaterials from the point of view of successful commercial application. Among cellulose derivatives, cellulose esters (CEs) have found the widest variety of end uses, resulting in over a billion pounds of CE manufacture per year. This is reflective of the fact that CE properties can be varied and finely controlled across a very wide range, due to the availability of predictably manipulated control elements.

These include most importantly degree of substitution (DS), nature and chain length of substituent, and the amount and type of plasticizer present. Great strides have been made in recent years in understanding the effects of these control elements on cellulose ester properties[1]. It is equally important to be able to manipulate these control elements, in particular those of DS and substituent type, in convenient fashion. Only in this way can one create synthetic processes for these materials, with efficiency adequate to the needs of the application areas for which CEs are well-suited by virtue of their properties.

One of the most persistent sources of inefficiency in cellulose ester synthesis and manufacture has been the inability to develop practical methods for direct esterification to the desired DS. Cellulose is a profoundly insoluble material, due for the most part to its extensive hydrogen bonding network in the crystal lattice. Most of the processes that have been developed for cellulose ester synthesis involve reactive dissolution of cellulose in a mixture of activated acyl (usually carboxylic anhydride) and catalyst (usually mineral acid). This affords a mixture of unreacted and fully reacted cellulose, until the reaction is complete and all the cellulose has been esterified to the triester stage. It is not useful to stop the reaction at an intermediate, partially substituted stage, because of the heterogeneity of reaction; the result would be a highly heterogeneous polymer that would not be amenable to solution or thermal processing. Manufacturers therefore must fully esterify cellulose and then back-hydrolyze to obtain a relatively homogeneous, partially substituted product. This of course necessitates a separate hydrolysis step.

Many investigators have attacked this problem over the last 25 years. One approach that works very well in the laboratory, and has wide flexibility with respect to DS and acyl type, is the esterification of cellulose in solution in N,N-dimethylacetamide (DMAC) and LiCl. This solvent system for cellulose was patented nearly simultaneously by McCormick[2] and Turbak[3]. It was utilized by McCormick[4] to prepare a variety of cellulose esters, and has been intensively exploited by Heinze[5], Glasser[6], Edgar[7] and many others. Cellulose when dissolved in DMAC/LiCl can be readily esterified, in the presence of mineral acid or alkaline catalysts, or thermally without a catalyst, by reaction with carboxylic anhydrides, acid chlorides, and other activated acyl reagents. Unfortunately, practical applications of this method are limited, due to the high cost of the lithium salt, the dilute solutions required, and the complex dissolution process. Similarly, Petrus[8], Heinze[9,10,11] and others have reported on DMSO/tetrabutylammonium fluoride (TBAF) solvent systems for cellulose that are suitable for etherification and/or esterification of the dissolved cellulose. These systems also suffer from issues of expensive reagents, water content of the TBAF salt hydrate, and dilute solutions.

Grishin and co-workers in Russia have reported[12,13] in the patent literature on sulfuric acid-catalyzed esterification of cellulose in DMAC, to afford cellulose acetate (CA), cellulose acetate propionate (CAP), and cellulose acetate butyrate (CAB). The authors found it necessary to incorporate a terminal hydrolysis period in order to obtain acetone-soluble cellulose acetate.

Esters of cellulose with carboxylic acids of chain length 6 or longer (often referred to as "long chain esters" of cellulose) are materials of special interest due to the fact that the hydrocarbon chains of the ester substituents may serve as internal plasticizers. This eliminates the need for admixture with small molecule plasticizers as is normally required for thermal processing of esters of cellulose with acids of 2-4 carbon atoms. Unfortunately synthesis of these long chain esters of cellulose is difficult if not impossible with conventional mineral acid catalyzed chemistry, due to the slow acylation, and competing cellulose chain cleavage catalyzed by the mineral acid[14,15]. The classical method for synthesis of these cellulose long chain esters is the so-called impeller method, in which the long chain acid is reacted with an activated anhydride, such as chloroacetic[16] or trifluoroacetic anhydride[17]. This generates a sufficient concentration of the mixed anhydride, which reacts with cellulose to afford fully substituted long chain esters. In addition to the fact that this chemistry is limited to triesters, it is further limited by the small amounts of residue of the reactive acyl groups (*e.g.*, trifluoroacetyl groups) that inevitably remain bound to cellulose hydroxyls, contributing to product thermal instability. The solution acylation of cellulose, for example in DMAC/LiCl, affords access to essentially the full range of partially and fully substituted long chain esters, but as mentioned earlier these are not practical processes for manufacture.

Experimental Section

Materials

Cellulose used in these experiments was mixed hardwood dissolving grade pulp, made by a Kraft process. In most cases it was Natchez HVX, obtained from International Paper. Amide and urea solvents were dried over 3 Å molecular sieves. Diketene was from Eastman Chemical Company. Anhydrides were reagent grade materials, used as received (nonanoic anhydride was manufactured by Eastman Chemical Company from reaction of nonanoic acid and acetic anhydride). Sulfonated polystyrene resins were purchased from Aldrich, washed thoroughly with deionized water, and then dried overnight at 40°C under vacuum prior to use. "Standard reaction flask" refers to a 500 mL, three-neck round bottom flask equipped with N_2 inlet, thermometer, overhead stirrer, and reflux condenser; all reactions were run under a N_2 atmosphere.

Measurements

DSC spectra were obtained using a Du Pont 2100 differential scanning calorimeter. In order to equalize thermal history, each sample was heated to 240°C at 20°C/min, then quenched to -78°C. The spectrum then was acquired while heating at 20°C/min from -78°C. All glass transition temperatures (T_g) are reported as the midpoint of the transition observed in the second scan. Gel permeation chromatography (GPC) data were acquired on a Waters Model 150C gel permeation chromatograph. Unless otherwise noted, the mobile phase was N-methyl pyrrolidinone (NMP), and the sample size was 20-25 mg/10 mL. All molecular weights are absolute (using viscometric detection). IV was measured using 0.25% solutions in the appropriate solvent, at 25°C. DS was determined by proton NMR in DMSO-d_6, based on the methods of Buchanan *et al.*[18] Spectra were acquired on JEOL Model GX-400 and GX-270 NMR spectrometers, using 5-mm-o.d. NMR tubes at 80°C. Chemical shifts are reported relative to the chemical shift of the solvent at 39.44 ppm.

Procedures

Cellulose Triacetate (CTA) by Sulfonated Polystyrene Resin Catalyzed Acetylation in Acetic Acid

A standard reaction flask was charged with cellulose (10.00 g) and acetic acid (150 mL). The slurry was allowed to stand for 60 min, then acetic anhydride (75.56 g, 12.00 equiv) and Amberlyst XN-1010 resin (2.50 g) were added. The slurry was heated to reflux (118°C) with stirring, then was held at reflux until it was clear except for the black catalyst beads (liquid was very pale amber at this point). The mixture was cooled to 40°C, then filtered through a glass frit to remove the catalyst and a few small residual fibers. The clear filtrate was added slowly to 1L water with vigorous stirring, then the white fibrous precipitate was collected by filtration and washed with water until the washings were neutral. It was vacuum-dried at 50°C to afford 14.41 g of product (80.1% yield). The product was CTA by ^{1}H NMR, DS 3.08, M_n 27,000 by GPC in DMF, 87 ppm sulfur by X-ray analysis.

Catalyst Separation Experiment.

Cellulose (10.00 g, 61.67 mmol) for this reaction was activated overnight in acetic acid, then the slurry was filtered and the acetic acid-damp cellulose (25.13 g) was charged to a 500 mL three neck round bottom flask equipped with N_2 inlet, thermometer, overhead stirrer, Soxhlet extractor designed for continuous return (see **Figure 2**), and reflux condenser. Then acetic acid (1.83 g), and acetic anhydride (216 g, 2.12 mol) were added to the flask. Amberlyst XN-1010 catalyst (2.50 g) was added to the frit of the Soxhlet extractor. The slurry in the flask was heated to reflux, and held at reflux for 6 h (at this point the reaction mixture had not quite fully cleared to a solution). The refluxing liquids were condensed, dripped onto the Amberlyst catalyst, and flowed back to the reaction flask. The reflux rate was such that the catalyst stayed damp, but no liquid visibly accumulated on the frit. The temperature in the flask was 134°C, and at the frit it was 100°C during the reaction. The reaction mixture was added slowly to cold water (1L) with vigorous stirring, then the product was isolated by filtration and water washing. After vacuum drying at 40°C the product was a white fibrous solid weighing 11.26 g. See third entry, **Table IV** for product analyses.

Cellulose Acetate Propionate by Ti-Catalyzed Acylation in DMAC.

To a standard reaction flask were charged cellulose (15.00 g, 92.51 mmol) and DMAC (75 mL). The slurry was heated to 120°C, then to it were added in order propionic anhydride (54.18 g, 4.5 equiv) and titanium isopropoxide (0.45 g, 0.017 equiv). The mixture was held at 140°C for 6.5 h, during which time it became a clear, reddish brown solution. The solution was cooled to 40°C, then diluted with 300 mL acetone to reduce the viscosity. It was filtered through Celite, then the filtrate as added to 2L water with vigorous stirring. The product was collected by filtration, then thoroughly washed with water. Vacuum drying at 60°C afforded 24.24 g pale tan fibrous solid (yield 84.2%). The product was soluble in DMSO, NMP, acetic acid, tetrahydrofuran (THF), $CHCl_3$, and acetone. See entry 5, **Table I** for detailed analytical results.

Cellulose Acetate Hexanoate by Ti-Catalyzed Acylation in DMAC.

To a standard reaction flask were charged cellulose (15.00 g, 92.51 mmol) and DMAC (115 mL). The slurry was heated with stirring to 135°C, then hexanoic anhydride (39.6 g, 2.00 equiv), acetic anhydride (18.9 g (2.00 equiv), and titanium isopropoxide (0.45 g, 0.017 equiv) were added. The temperature was increased to 155°C, and the reaction mixture was stirred at that temperature until a smooth, clear, brown solution was obtained (5.6 h). The solution was cooled to 40°C and diluted with acetone (250 mL), then filtered through Celite. The filtrate was added to 3L water:methanol 1:4 (v:v) with vigorous stirring. The product was collected by filtration, then thoroughly washed with methanol until there was no more hexanoic acid odor. Then the product was washed with water, and vacuum dried at 40°C to afford 23.73 g off-white fibrous solid (yield 81.6%). The product contained 1580 ppm titanium by ICP. See entry 1, **Table II** for detailed analytical results.

Cellulose Propionate by Sulfonated Polystyrene Resin Catalysis in DMAC.

A standard reaction flask was charged with cellulose (15.00 g) and DMAC (150 mL). The slurry was heated to 110°C, then catalyst (Amberlyst XN-1010, 3.75 g) and propionic anhydride (72.25 g, 6.00 equiv) were added. The slurry was heated to 140°C and held at that temperature for 72 min. At that point the reaction mixture was a clear, viscous, and pale yellow solution, with only the black catalyst beads undissolved. The mixture was cooled and filtered through a glass frit to remove the catalyst, then the filtrate was added slowly to water (2 L) with vigorous stirring. The product was collected by filtration, then thoroughly washed with water and vacuum-dried at 60°C. The product (22.62 g) was a white, fibrous solid (yield 78.9%). The product was soluble in organic solvents including acetone, THF, DMSO and NMP. The product contained 367 ppm sulfur by X-ray fluorescence. See entry 2, **Table V** for additional analytical data.

Cellulose Acetate Hexanoate by Sulfonated Polystyrene Resin Catalysis in DMAC.

A standard reaction flask was charged with cellulose (15.00 g, 92.51 mmol) and DMAC (150 mL). The slurry was heated to 120°C with stirring, then acetic anhydride (18.89 g, 2.00 equiv), hexanoic anhydride (39.65 g, 2.00 equiv), and Amberlyst XN-1010 resin (3.75 g) were added. The mixture was heated to 150°C and held at that temperature for 2h, 33 min. The mixture had become an amber colored, viscous solution. The solution was cooled and diluted with acetone (200 mL), then was filtered through a glass frit to remove the catalyst. The filtrate was added slowly to water (2L) with vigorous stirring. The product was washed with methanol and water sequentially until the hexanoic acid odor was gone, then was dried at 60°C under vacuum. The product (4.67 g) was a white, fibrous solid, soluble in organic solvents including NMP, DMSO, acetic acid, $CHCl_3$, THF, and acetone. Detailed analytical results may be found under entry 1, Table VII.

Results and Discussion

Recently we have found that DMAC and related amide and urea solvents have interesting properties for cellulose esterification chemistry, even when salts are not added to promote dissolution of the cellulose. We have patented[19,20] and reported briefly in the open literature[21] a process for the esterification of cellulose using titanium isopropoxide catalyst[22,23] in a diluent such as DMAC. We found that this combination of a moderately strong esterification catalyst, relatively high temperatures (120-160°C), and polar aprotic solvents enabled dissolution of the intermediate cellulose ester before full esterification was attained. Because the reaction rate at the point of dissolution was relatively low, it was a simple matter to stop the reaction at the point of attaining a homogeneous solution. This gave highly repeatable results with respect to product composition, including DS. **Table I** illustrates the results obtained by esterification in DMAC with acetic (Ac_2O) or propionic (Pr_2O) anhydride, using titanium isopropoxide catalyst.

Table I. Esterification of Cellulose in DMAC with Anhydrides using Ti(O-i-Pr)$_4$ Catalyst

	prod.	anhyd.[a] (equiv/ AHG)	T (°C)	DMAC/ Cell. (mL/g)	time (h)	DS (Ac)	DS (Pr)	Mn (/1000)	T$_g$ (DSC)
1	CA	Ac$_2$O (6.0)	120	10	6	2.67	---	64	188
2	CA	Ac$_2$O (6.0)	120	10	4.5	2.62	---	46	187
3	CAP	Pr$_2$O (4.5)	120	5	17.3	0.08	2.57	81	152
4	CAP	Pr$_2$O (4.5)	120	10	23.6	0.10	2.44	61	146
5	CAP	Pr$_2$O (4.5)	140	5	6.5	0.10	2.58	52	142
6	CAP	Pr$_2$O (4.5)	140	10	8.3	0.13	2.40	50	150
7	CAP	Pr$_2$O (6.0)	120	5	11.1	0.02	2.67	78	144
8	CAP	Pr$_2$O (6.0)	120	10	14.5	0.04	2.59	64	148
9	CAP	Pr$_2$O (6.0)	140	5	6.3	0.03	2.56	72	147
10	CAP	Pr$_2$O (6.0)	140	10	10.1	0.06	2.55	80	149

[a] Ac$_2$O denotes acetic anhydride, Pr$_2$O denotes propionic anhydride.

We found that acetylation proceeded smoothly to afford a clear and viscous solution. (Note that we used in these experiments, as in all esterification experiments reported in the manuscript, commercial hardwood dissolving pulp and so all results reflect cellulose esters that may contain small amounts of hemicellulose esters, such as glucoronoxylan esters). Upon reaching this clear solution stage, each reaction was halted by removing the heat. The reaction mixture was filtered through Celite to remove a trace of residual fibers, and then added to water to precipitate the product. Acetone-soluble CA of approximate DS 2.6 was obtained, which was of high molecular weight and had plastic and thermal properties characteristic of conventionally prepared cellulose diacetate. We were quite interested in this process as a potentially efficient route to partially substituted cellulose propionates. The latter entries of **Table I** show the results of a statistically designed experiment exploring this possibility. As the table indicates, propionylation went smoothly in this system and esters were obtained that were partially substituted, high molecular weight, and highly soluble. One unexpected result was the presence of significant amounts of acetyl in the product in spite of the fact that only *propionic* anhydride was used as reagent. We suspected that the source of acetyl must be slow reaction of cellulose with the DMAC solvent. In experiments (data not included) using N,N-dimethylpropionamide as solvent, no acetylation was observed, confirming that DMAC was the source of the acetyl in the earlier experiments. The designed experiment showed us that we could attain the desired DS range of 2.5-2.6 readily, with relatively short reaction time, by using 6 equiv of Pr$_2$O per AGU at 140°C and minimum DMAC solvent. High DMAC/cellulose ratio, low Pr$_2$O equiv, and high temperature all promoted acetylation, as would be expected for a slow reaction competing with the faster propionylation for the same reactive hydroxyls. Factors which increased the rate of propionylation (low DMAC concentration, high equiv Pr$_2$O) also promoted high molecular weight of the product, indicating that CAP degradation is slower than cellulose degradation[24].

We found that the DMAC/titanium isopropoxide system worked very well for the synthesis of partially substituted long chain esters of cellulose[21]. These long chain esters were indeed self-plasticized, and had an interesting and potentially useful range of properties that could be related linearly to calculated solubility parameters. Since this was the main topic of our earlier article, we will not go into detail here but only present a snapshot of the results obtained, as **Table II**.

Table II. Long Chain Esters by Cellulose Acylation in DMAC using Ti(O-i-Pr)$_4$ Catalyst

entry	anh[a] (eq/ahg)	react. time (h)	T (°C)	DS[b]	IV[c]	M_n/ 1000[d]	T_g[e]
1	Ac (2) Hex (2)	6	155	1.87 0.75	1.24	62	148
2	Ac (2) Non (2)	11	145	2.03 0.70	1.18	44	129
3	Ac (3) Non (1)	8	145	2.44 0.26	1.71	43	161
4	Ac (1) Non (3)	13	155	1.59 1.11	1.16	44	118
5	Ac (0) Non (4)	13	160	1.11 1.35	0.89	31	110
6	Ac (3.5) Laur (2)	12	140	2.40 0.20	2.12	96	165
7	Ac (0) Hex (3)	7	140	0.00 2.73	0.44	23	104

[a] Anhydride, equivalents per anhydroglucose unit; Ac - acetyl, Hex – hexanoyl, Non – nonanoyl, Laur – lauroyl. [b] by ^{1}H NMR. [c] In NMP. [d] GPC in NMP. [e] DSC.

We will highlight entry 7 since it is of particular interest. Note that in this case a fully-substituted urea, N, N-dimethylimidazolidinone (DMI, **Figure 1**) was used as solvent instead of DMAC. DMI was an excellent solvent for this reaction, also affording partially substituted esters. Note that the product contained no acetyl groups, since the solvent contained none; this is further evidence that DMAC slowly acetylates cellulose under titanium catalysis at these high temperatures. Reaction in DMI was faster than in DMAC, but afforded somewhat lower molecular weight cellulose ester. It may be possible to reduce the temperature for DMI esterifications and obtain higher molecular weight products.

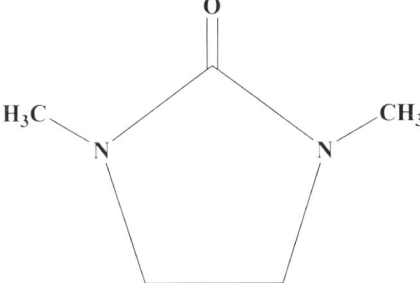

Figure 1. DMI Structure

The well-understood control elements of this long chain ester synthetic process, along with the ability to predict properties with accuracy from calculated solubility parameters, afford to the materials scientist the ability to dial in desired properties within the capabilities of this broad family of biomaterials.

We were interested in further exploring the nature and limits of this new synthetic chemistry. We wondered whether other moderately strong esterification catalysts could replace the titanium isopropoxide in the amide diluent, and still yield partially substituted cellulose esters. Initial exploration was not promising; we found that other acids including boric acid, manganese diacetate, germanium dioxide, zinc diacetate, antimony (III) ethoxide, and aluminum (III) isopropoxide all were entirely ineffective at catalyzing esterification with Ac_2O in DMAC. Sulfonated polystyrene resins were also of interest. Touey and Kiefer made the remarkable observation in a 1958 patent[25] that these sulfonated polystyrene resins could be used to catalyze cellulose esterification with carboxylic acid anhydrides, in carboxylic acid diluents. Thus, cellulose acetate could be synthesized by reaction of cellulose with acetic anhydride in acetic acid with sulfonated polystyrene catalysis (U.S. Patent 2,861,069, Example 1), affording a highly substituted cellulose acetate ("soluble in cellulose triacetate solvents"). They also reported synthesis of cellulose propionates and butyrates in this way.

This was a fascinating and strange result. We repeated Touey and Kiefer's chemistry in our laboratory, and were able to confirm smooth preparation of fully substituted cellulose esters by this sulfonated polystyrene resin catalyzed method. For example, reaction of cellulose and acetic anhydride with sulfonated polystyrene resin catalysis at 134°C gave cellulose acetate with a DS measured at 3.08 and a M_n of 74,000 (GPC in DMF). The insoluble resin could be readily removed from the final product solution by filtration and recycled at least three times with no loss of effectiveness (Table III). It was surprising that a substrate (cellulose) and a catalyst (the sulfonated resin), mutually insoluble in the acetic acid/acetic anhydride reaction mixture, could somehow come together to form a catalytic complex and achieve the observed reaction.

Table III. **Cellulose Acetate Using Recycled Amberlyst 15 Catalyst**[a]

recycle number	reaction time (min)	DS (^{1}H NMR)	IV (Phenol/TCE)	$M_n/1000$ (GPC, DMF)
0	159	3.08	0.83	74
1	76	3.08	1.20	40
2	50	3.14	1.21	108
3	79	3.05	0.94	60

[a]Runs used, per gram cellulose, 0.25g catalyst, 20 mL Ac$_2$O, 1.25 g acetic acid. Catalyst was recovered from the reaction mixture by filtration, then recycled to the next run without makeup ("Recycle 0" was the first run)

This led us to perform several experiments in an attempt to better understand the mechanism, only one of which will be described here. Sulfonation of polystyrene resins is of course a reversible reaction, and several groups have studied the stability of these resins towards desulfonation. Klein and co-workers[26] studied the rates of desulfonation of sulfonated polystyrene resins in water, and found that the rate of desulfonation was substantial at 100°C, and rather rapid at 200°C. We speculated that resin desulfonation might generate sufficient soluble catalyst (SO$_3$, CH$_3$COOSO$_3$H, or similar species[27]) to account for the observed results. We tested this theory by physically separating the catalyst from the cellulose; if acetylation still occurred at a reasonable rate, this would confirm the presence of a low molecular weight catalyst and lend strong support to the desulfonation mechanism. We placed the catalyst in a Soxhlet extractor designed for continuous return to the reaction vessel (**Figure 2**). We measured the liquid temperature both in the catalyst-containing thimble (100°C), and in the reaction vessel (134°C). Because of the temperature difference, and our lack of certainty about which step was rate-determining, we ran control experiments at both temperatures. The results (**Table IV**) supported desulfonation as the mechanism; acetylation of cellulose was indeed observed, taking approximately twice as long as the control reaction run at 100°C, that is, the temperature in the catalyst-containing thimble in our catalyst separation experiment. This should be the equivalent temperature if the rate determining step is, as we postulate, desulfonation in the thimble to generate the catalyst. The fact that the catalyst is not generated in close proximity to the cellulose may explain the small observed rate difference.

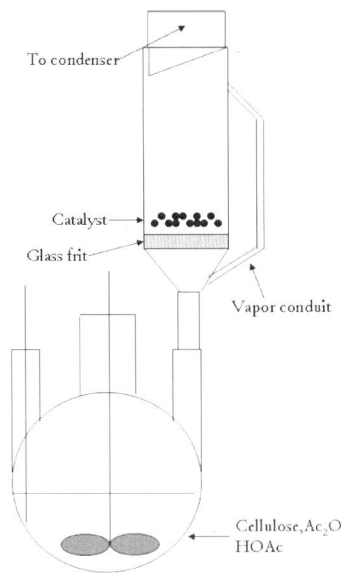

Figure 2. Apparatus for Catalyst Separation Experiment

Table IV. Cellulose Acetylation Using Physically Separated Catalyst[a]

run	T (°C)	reaction time[b] (min)	DS (^{1}H NMR)	IV (Phenol/ TCE)	Mn/1000 (GPC, DMF)
Control	134	8	3.13	1.36	114
Control	100	120	3.13	1.80	180
Soxhlet[c]	134 (vessel) 100 (thimble)	240	3.14	0.68	96

[a] Runs used, per g cellulose, 0.25g Amberlyst XN-1010, 20 mL Ac_2O, 1.70 g acetic acid.

[b] Reactions were run until all of the cellulose had dissolved.

[c] See text and **Figure 1** for description.

We reasoned that by combining the DMAC solvent with the relatively mild acid catalyst, sulfonated and crosslinked polystyrene resin, it might be possible to create a new and convenient method for direct synthesis of partially substituted cellulose esters. This proved to be the case[28,29]. **Table V** shows the results of acetylation and propionylation reactions carried out using Amberlyst XN1010 sulfonated polystyrene catalyst.

Table V. Cellulose Acylation in Amide Solvent with Resin Acid Catalysis

entry	anh.	equiv/a hg[a]	amide	T (°C)	React. time (h)[b]	DS (¹HNMR)	$M_n/1000$ (GPC, NMP)
1	Ac$_2$O	6	DMAC	120	5.25	2.88	125
2	Pr$_2$O	6	DMAC	140	1.20	2.63	52
3	Pr$_2$O	6	NMP	150	8.65	3.09	43

[a] Equivalents of anhydride per anhydroglucose unit (ahg). [b] Reactions were run until all of the cellulose had dissolved to a clear solution with suspended black catalyst particles.

Acetylation with acetic anhydride, as shown in the first entry, was smooth and the reaction mixture slowly changed from a slurry, to a viscous, fibrous gel, to a smooth and viscous solution. The reaction was halted, by removing the heat, when a clear solution was obtained. The viscous solution was diluted with solvent to reduce viscosity, then filtered through a glass frit to remove the catalyst and small amounts of residual fibers. The product was isolated by addition of the filtrate to water with rapid stirring, filtration, and water-washing. As can be seen in **Table V**, the product was a high molecular weight cellulose acetate with a DS slightly less than three, with good solubility in solvents such as NMP and DMSO. Even more interesting results were obtained with propionylation of cellulose in DMAC (entry 2, **Table V**). Again, the reaction was smooth and afforded a clear and viscous solution, from which the catalyst was separated and the product isolated in much the same way as in the previous example. The product was a partially substituted cellulose propionate, propionyl DS 2.63, with high molecular weight and good solubility in organic solvents (for example, acetone, THF, DMSO, NMP). Interestingly, this material has propionyl DS very close to that of commercial CAP used for plastic applications[1]. Entry 3 illustrates the importance of the solvent power of DMAC to a successful partial esterification; when we switched to another tertiary carboxamide solvent, NMP, we also obtained good conversion to a cellulose propionate, but the cellulose propionate was fully substituted. Since the reaction is run until the reaction mixture is clear, this simply illustrates that NMP will not dissolve all of the ester until the reaction has gone fully to completion.

The preparation of partially substituted cellulose propionate in DMAC with sulfonated polystyrene resin catalysis is repeatable and robust, as illustrated by the designed experiment described in **Table VI**.

Table VI. Designed Experiment on Cellulose Propionylation in DMAC

experiment		variables				responses		
entry	DE	equiv Pr$_2$O/ ahg	T (°C)	DMAC (mL/g cell.)	react. time (h)	DS Pr	Mn/ 1000	Mw/ 1000
1	---	4.22	135	5	4.38	2.67	36.5	130
2	-+-	4.22	150	5	4.77	2.57	34.7	134
3	++-	6.00	150	5	1.52	2.69	46.8	143
4	+--	6.00	135	5	5.58	2.73	56.5	188
5	--+	4.22	135	10	13.7	2.49	41.8	152
6	-++	4.22	150	10	2.05	2.50	32.1	161
7	+-+	6.00	135	10	2.87	2.59	52.1	211
8	+++	6.00	150	10	1.57	2.58	52.3	234

We were interested in the effects of key variables such as reaction temperature, equivalents of propionic anhydride per anhydroglucose unit (AHG), and amount of DMAC (mL per gram of cellulose) on the results. The important responses included reaction time (reactions were run until the cellulose had dissolved completely to a smooth solution), DS propionyl, and measures of molecular weight such as M$_n$ and M$_w$ by GPC. We obtained good models for the DS and molecular weight responses. It should be noted that the range of DS (2.49-2.73) obtained was relatively narrow; still, the trends observed were consistent and reasonable. DS was increased by the equivalents of anhydride, and decreased by use of more DMAC, as expected. More interesting was the observation that temperature had no effect on DS at high DMAC, and that DS actually fell with increasing temperature at low DMAC. Molecular weight-related responses all had similar dependencies. The equivalents of anhydride had the major impact in each case, with more anhydride affording higher molecular weight. This is consistent with our previous observations (*vide supra*) and supports the hypothesis that the esters of cellulose are much less prone to acid catalyzed chain cleavage than is cellulose itself. The designed experiment indicates not only that the method is a robust way to make partially substituted cellulose propionates, but also that parameter effects are sufficiently well understood to enable precise targeting of DS and molecular weight.

We were interested in whether this technique too would enable the synthesis of long-chain esters of cellulose, and very briefly explored this question (**Table VII**).

Table VII. Long-chain Cellulose Esters by Sulfonated Polystyrene Resin Catalysis in DMAC

entry	anh 1[a]	eq./ ahg[b]	anh 2	eq./ ahg	T (°C)	react. time (h)[b]	DS 1[c]	DS 2	M_n/ 1000[d]
1	Ac$_2$O	2	Hex$_2$O	2	150	2.55	1.70	0.80	45.1
2	Ac$_2$O	2	Laur$_2$O	2	160	1.12	1.95	0.61	44.4
3	Pr$_2$O	4	Diketene	1	120	11.93	2.54	0.79	39.9

[a] Ac$_2$O – acetic anhydride, Pr$_2$O – propionic anhydride, Hex$_2$O – hexanoic anhydride, Laur$_2$O – lauric anhydride. [b] equivalents of anhydride per anhydroglucose unit. [c] DS by 1H NMR in d-6 DMSO. [d] by GPC in NMP

We found that acylation with a mixture of hexanoic and acetic anhydrides (2 equiv/AHG each) gave smooth conversion to a partially substituted cellulose acetate hexanoate of high molecular weight (DS Ac 1.70, DS Hex 0.80). Similar results in terms of total DS were obtained by reaction with a mixture of acetic and lauric anhydrides (2 equiv/AHG each), affording cellulose acetate laurate (DS Ac 1.95, DS Laur 0.61). Interestingly, a single attempt to use this reaction system (sulfonated polystyrene catalyst in DMAC) to prepare a cellulose acetoacetate propionate by reaction with diketene (1 equiv/AHG) and propionic anhydride (4 equiv/AHG) was successful, giving the expected cellulose acetoacetate propionate with apparent DS propionyl of 2.54, and DS acetoacetyl of 0.79 (**Figure 3**). Thus the ester appeared to be more than fully substituted; since it is of high molecular weight, this is unlikely to be an end group effect. We know that it can be difficult to quantify acetoacetate esters by ^{1}H NMR due to keto-enol tautomerism, but do not know whether this is a contributor to the apparently excessive total DS measured for this material.

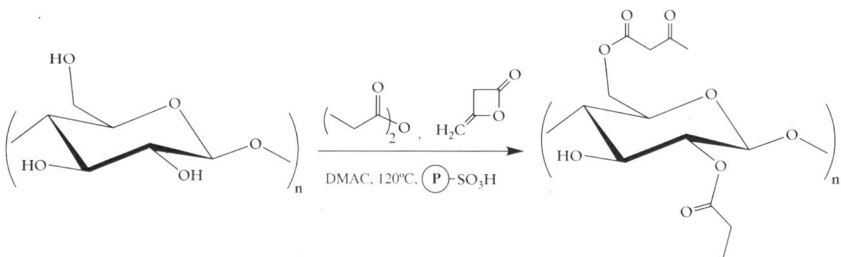

Figure 3. Cellulose acetoacetate propionate

Conclusions

We have developed two robust methods for the direct synthesis of partially substituted cellulose esters. In each case, a key role is played by the solvent, either a tertiary amide or fully substituted cyclic urea. We have shown that the sulfonated polystyrene resin works, at least for the most part, by generating low levels of sulfuric acid *via* thermal desulfonation of the resin. The mildly alkaline amide solvent serves to moderate the acidity of the released sulfuric acid at the high reaction temperature.

We note that each of these methods produces cellulose esters that contain residues of the catalyst. The titanium catalyzed esterification gives a product that contains approximately 1000 ppm Ti, as measured by ICP. We do not know whether the titanium is covalently or coordinatively attached, or is simply admixed with the cellulose ester. We do know that it has little impact on product utility; these esters give clear solutions in a broad range of organic solvents, and can be readily processed thermally. Differential scanning calorimetry of a wide variety of these esters gave results that were entirely consistent with those from well-stabilized esters made by conventional processes.

Conversely, the cellulose esters produced by the sulfonated polystyrene resin process typically contained between 200-300 ppm sulfur by X-ray analysis. These esters presumably contain covalently bound sulfate esters, as observed in conventional sulfuric acid catalyzed esterification. Since we made no effort to stabilize these sulfate esters, it is not surprising that we saw by DSC that these esters tended to decompose at temperatures lower than those observed for conventionally prepared CEs that have been well-stabilized. There is every reason to believe that these esters could easily be stabilized by conventional technology, particularly since sulfur content was surprisingly consistent across a wide variety of CEs.

It is also notable that both methods provide access to long-chain esters of cellulose. We have shown that the properties of these long-chain esters are highly predictable, even from calculated values like solubility parameters. We have also shown that, within a wide composition range, these long-chain esters are self-plasticized, requiring no added plasticizer to enable thermal extrusion. Low molecular weight plasticizers in certain applications can be extracted into the liquid contents of containers, or volatilized from the plastic in hot environments. Therefore, ready access to these long-chain esters of cellulose may open up new horizons for potential plastics applications.

We believe that these new methods have promise both for enhanced synthetic efficiency, and for broadening the material science and application possibilities.

Acknowledgments

The author thanks Thelma Watterson for her technical assistance, Tom Pecorini for the characterization and interpretation of the physical properties of the long chain esters, Peter Shang for DSC assistance, and Doug Lowman for NMR analyses.

228

References

1. Edgar, K.J.; Buchanan, C.M.; Debenham, J.S.; Rundquist, P.A.; Seiler, B.D.; Tindall, D. *Progress in Polymer Science*, **2001**, *26*, 1605-1688.
2. McCormick, C.L. Novel Cellulose Solutions. U.S. Patent 4,278,790, July 14, **1981**.
3. Turbak, A.F.; El-Kafrawy, A.; Snyder, Jr., F.W.; Auerbach, A.B. Solvent System for Cellulose. U.S. Patent 4,302,252, November 24, **1981**.
4. Dawsey, T.R.; McCormick, C.L. *J. Macromol. Sci. – Rev. Macromol. Chem. Phys.* **1990**, C30(3&4), 405-440.
5. Heinze, T.; Schaller, J. *Macromol. Chem. Phys..* **2000**, *201*, 1214-1218.
6. Davé, V.; Glasser, W.G. *J. App. Poly. Sci.* **1993**, *48*, 683-699.
7. Edgar, K.J.; Arnold, K.M; Blount, W.W.; Lawniczak, J.E.; Lowman, D.W. *Macromolecules* **1995**, *28*, 4122-4128.
8. Petrus, L.; Gray, D.G.; BeMiller, J.N. *Carb. Res.* **1995**, *268*, 319-323.
9. Heinze, T.; Dicke, R.; Koschella, A.; Kull, A.H.; Klohr, E.-A.; Koch, W. *Macromol. Chem. Phys.* **2000**, *201*, 627-631.
10. Hussain, M.A.; Liebert, T.; Heinze, T. *Macromol. Rapid Commun.* **2004**, *25*, 916-920.
11. Kohler, S.; Heinze, T. *Macromol. Biosci.* **2007**, *7*, 307-314.
12. Grishin, E.P.; Bondar, V.A.; Mironov, D.P.; Shamolin, A.I. Catalytic System for Cellulose Acylation, Process for Producing Said Catalytic System, and for its Practical Application. U.S. Patent 5,952,260, September 14, **1999**.
13. Mironov, D.P.; Shamolin, A.I.; Grishin, E.P.; Bondar, V.A. Catalytic System for Cellulose Acylation, Process for Producing Said Catalytic System, and for its. Practical Application. U.S. Patent 6,407,224, June 18, **2002**.
14. Malm, C.J.; Mench, J.W.; Kendall, D.L.; Hiatt, G.D. *Ind. Eng. Chem.* **1951**, *43*, 684-688.
15. Malm, C.J.; Mench, J.W.; Kendall, D.L.; Hiatt, G.D. *Ind. Eng. Chem.* **1951**, *43*, 688-691.
16. Clarke, M.T.; Malm, C.J. Process of Making Cellulose Esters of Carboxylic Acids. U.S. Patent 1,880,808, October 4, **1932**.
17. Morooka, T.; Norimoto, M.; Yamada, T.; Shiraishi, N. *J. Appl. Poly. Sci.* **1984**, *29*, 3981-3990.
18. Buchanan, C.M.; Edgar, K.J.; Hyatt, J.A.; Wilson, A.K. *Macromolecules* **1991**, *24*, 3050-3059.
19. Edgar, K.J.; Bogan, R.T. Direct Process for the Production of Cellulose Esters. U.S. Patent 5,750,677, May 12, **1998**.
20. Edgar, K.J.; Bogan, R.T. Direct Process for the Production of Cellulose Esters. U.S. Patent 5,929,229, July 27, **1999**.
21. Edgar, K.J.; Pecorini, T.J.; Glasser, W.G. Long-Chain Cellulose Esters: Preparation, Properties, and Perspective. In *Cellulose Derivatives – Modification, Characterization, and Nanostructures*; Heinze, T.J. and Glasser, W.G., Eds.; ACS Symposium Series 688, American Chemical Society: Washington, DC, 1998; pp 38-60.

22. Seebach, D.; Hungerbühler, E.; Naef, R.; Schnurrenberger, P.; Weidmann, B.; Züger, M. *Synthesis* **1982**, 138-141.

23. Touey, G.P.; Tamblyn, J.W. Method of Preparing Cellulose Esters. U.S. Patent 2,976,277, March 21, **1961**.

24. Malm, C.J.; Glegg, R.E.; Thompson, J.; Tanghe, L.J. *Tappi*, **1964**, *47*, 533-539.

25. Touey, G.P.; Kiefer, J.E. Method of Preparing Cellulose Esters. U.S. Patent 2,861,069, November 18, **1958**.

26. Klein, J.; Widdecke, H.; Bothe, N. *Makromol. Chem. Suppl. 6*, **1984**, pp 211-226.

27. Tanghe, L.J.; Brewer, R.J. *Anal. Chem.* **1968**, *40*, 350-353.

28. Edgar, K.J. Synthesis of long-chain esters of cellulose *via* sulfonated polystyrene catalysis. *Abstracts of Papers*, 221st National Meeting of the American Chemical Society, San Diego, CA, April 1-5, **2001;** American Chemical Society: Washington, DC, 2001.

29. Edgar, K.J. Process for Preparing Cellulose Esters Using a Sulfonic Acid Resin Catalyst. U.S. Patent 6,160,111, December 12, **2000**.

Chapter 13

Synthesis and properties of regioselectively substituted cellulose cinnamates

Tetsuo Kondo[*,†], Masanori Yamamoto[†],
Wakako Kasai[*] and Mitsuhiro Morita[†]

Bio-Architecture Center (KBAC)[*], and Graduate School of
Bioresource and Bioenvironmental Sciences[†],
Kyushu University,
6-10-1, Hakozaki,Higashi-ku, Fukuoka, Japan
Corresponding Author: T. Kondo (tekondo@agr.kyushu-u.ac.jp)

Regioselectively substituted cellulose cinnamates were synthesized in order to obtain potentially versatile functional materials. The introduced cinnamoyl groups are expected to be photosensitive and electroconductive. At the same time, regioselective substitution may strongly impact derivative properties. Thus, this paper will report the synthesis procedure, and the thermal behaviors of these regioselectively substituted cinnamoyl derivatives.

Introduction

Hydrogen bonds in cellulose and cellulose derivatives are among the most influential factors on their physical- and chemical properties. To date we have attempted to probe this structure-property relationship using regioselectively functionalized methyl celluloses[1-12] like 2,3-di-O-methyl cellulose (23MC)[13], 6-mono-O-methyl cellulose (6MC)[14], 3-mono-O-methyl cellulose (6MC)[15] and 2,3,6-tri-O-methyl cellulose (236MC)[16] as model compounds. In addition, the functionalization pattern of the methyl ether groups also had an influence on the enzymatic hydrolysis of methyl cellulose by cellulase.[17] Through those investigations, the intramolecular hydrogen bonds between the OH groups of position 3 and the ring oxygen of the adjacent anhydroglucose unit (O-5), which is supposed to determine the molecular stiffness, have also been confirmed as a key factor for properties of cellulosic materials.[18]

Cinnamic acid is well known to be capable of photo-cycloaddition, and syntheses of polymers containing the cinnamoyl substituents have been reported in the last decade[19-21]. Such polymers are expected to exhibit photoconductivity through electron accepting groups within the molecule. The preparation and the behavior under photo-irradiation of cellulose cinnamates have been extensively studied[22-28] in the 1990's, including interpenetrating polymer networks[26] and a photosensitive comb-like polymer[23]. In addition, cellulose cinnamates at a critical concentration in some organic solvents exhibited chiral nematic mesophases[25]. More recently, isopentylcellulose cinnamate as cross-linked Langmuir-Blodgett films sandwiched between a polymeric diffusion couple was investigated to provide a system for probing fundamental transport process across the ultrathin membranes[27].

Regioselectively functionalized polysaccharide cinnamates display some unique properties. The absolute helicities of acylated chitosan derivatives bearing cinnamate chromophores[20] vary depending on the degree of substitution (DS) of cinnamoyl groups, but are nearly independent of thermal stimulus. Molecular interactions including hydrogen bonding and hydrophobic interaction may be involved in this behavior. Therefore, regioselectively functionalized cellulose cinnamates would enable us to further elucidate the involvement of the inter- and intramolecular hydrogen bonds as well as photo-crosslinking behavior and other factors.

In this paper, regioselectively substituted cellulose cinnamates were synthesized in order to obtain potentially versatile functional materials. Thus, this paper will report the synthesis procedure and the thermal behaviors of these regioselectively substituted cinnamoyl derivatives in order to understand the principles of molecular design to exploit desired functional bio-based materials.

Syntheses of regioselectively substituted cellulose cinnamates

The cellulose cinnamates synthesized in this study were cellulose tricinnamate (a), cellulose-2,3-*di*-cinnamate (b), cellulose-6-*mono*-acetate-2,3-*di*-cinnamate (c), and cellulose-2,3-*di*-acetate-6-*mono*-cinnamate (d), respectively.

(i) Preparation of cellulose tricinnamate (1)

Figure 1 shows the synthesis scheme of cellulose tricinnamate. Basically, the process followed previous reports[22,24]. The starting material was commercially available microcrystalline cellulose powder (Funacell, Funakoshi Co., Japan) with a nominal degree of polymerization (DP) of *ca.* 220. The sample was dried at 105 °C for 3 hours. Pyridine was dehydrated over 3Å molecular sieves. Reagent-grade solvents and cinnamoyl chloride (Aldrich Gold Label) were used without further purification. The procedure was as follows: After 1.5 g cellulose powder was dispersed in 60 mL pyridine at 100°C for 1 hour with constant stirring, 1.5 moles of cinnamoyl chloride per mole of hydroxyl groups were added to the solution. Then, the mixture was kept at 50 °C with stirring for 5 hours.

The reaction mixture was cooled to room temperature and poured into methanol. The precipitate was dissolved in chloroform, the chloroform layer was extracted with water three times, and then evaporated under reduced pressure at 60°C to a syrup. The product was precipitated from the syrup by the addition of methanol, filtered off, and dried under vacuum at 60°C for 5 hours. The yield of product **1** was 80 %.

Cellulose 2,3-dicinnamate (**4**), cellulose 6-monoacetate-2,3-dicinnamate (**5**), and cellulose 2,3-diacetate-6-monocinnamate (**8**) were prepared starting from 6-*O*-triphenylmethyl ("trityl") cellulose in which the C-6 hydroxyl is regioselectively substituted. Tritylcellulose (**2**) was used as a starting material, since the trityl group can be easily removed by exposing the cellulose derivatives to acidic conditions[28]. This allows the preparation of cellulose derivatives that are regioselectively substituted at C-2 and C-3. Tritylcellulose was prepared according to our previous papers[13,14]: Regenerated cellulose (32.5 g) was added to 300 mL of dry pyridine, which was then heated for 3 hours at 80°C and filtered off, in order to remove water. This procedure was repeated three times.

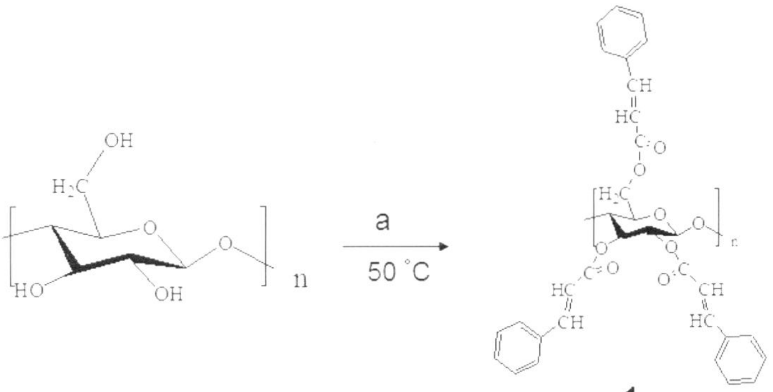

Figure 1. Synthesis of tricinnamate 1.
a – 1.5 equiv/OH cinnamoyl chloride

Then the cellulose was added to 500 mL of anhydrous pyridine and triphenylmethyl chloride (142.0 g, 2.5 moles per mol of hydroxyl group) in a 1000 mL three-necked round-bottom flask equipped with a condenser, stirrer and drying tube. The mixture was heated for 26 hours at 95°C, then cooled to room temperature and poured into methanol. The precipitate was isolated and washed in methanol for 12 hours. After a second washing, the precipitate was filtered off and dried at 100°C under vacuum. The yield of the product was 95% and the degree of substitution of trityl groups was 1.07 by 1H-NMR measurements.

(ii) Preparation of cellulose-2,3-dicinnamate (4) and cellulose-6-mono-acetate-2,3-dicinnamate (5)

The method for the preparation of cellulose tricinnamate was applied to the preparation of cellulose 2,3-dicinnamate (**4**) as shown in Figure 2. Thus, tritylcellulose (**2**: 1.0 g) was dissolved in pyridine (60 mL). After 1 h stirring under nitrogen, cinnamoyl chloride (5.2 g, 6.0 moles per mole of hydroxyl groups) was added to the solution. Then, still under a nitrogen atmosphere, the temperature was raised to 90°C and kept there for 5 hours. The mixture was cooled to room temperature and the product was isolated and purified as described for the preparation of cellulose tricinnamate. Following cinnamoylation the product (**3**) was detritylated and the sample was isolated as follows: the sample was detritylated by treating 1.0 g of the polymer (**3**) in 30 mL of dichloromethane with hydrogen chloride gas at 0°C for 4 minutes[13,14]. After the reaction, the detritylated mixture was poured into 50 mL of methanol. The product was isolated by centrifugation, washed with methanol, filtered and dried under vacuum at 60°C. The yield of the detritylated product **4** was approximately 65 %.

For acetylating compound **4**, acetic anhydride (5.8 mL, 25 moles per moles of hydroxyl group) was added to the pyridine solution containing 1.0 g of compound **4** . Then, the mixture was kept at 90 °C with constant stirring for 48 hours. After the reaction, the mixture was poured into 50 mL of methanol to obtain a precipitate.

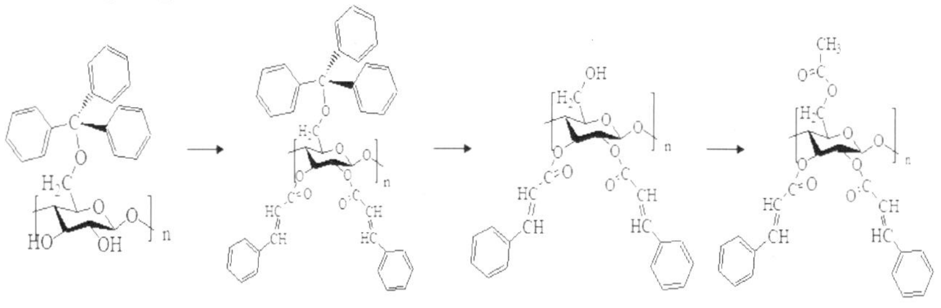

2   **3**   **4**   **5**

Figure 2. Synthesis of cellulose-2,3-dicinnamate **4** and cellulose-6-acetate-2,3-dicinnamate **5**.

The precipitate was dissolved in dichloromethane, the CH_2Cl_2 layer was extracted with water twice, and then the combined extracts were evaporated under reduced pressure at 60°C to a syrup. The product (**5**, Fig. 2) was precipitated from the syrup by the addition of methanol, filtered off, and dried under vacuum for 5 hours at 60°C. The yield of product **5** was 74 %.

(iii) Preparation of cellulose 2,3-diacetate-6-mono-cinnamate (8)

Figure 3 shows the synthesis of cellulose-2,3-diacetate-6-monocinnamate (**8**). 6-Tritylcellulose (**2**: 1.0 g) was dissolved in 40 mL of pyridine with constant stirring at 60°C for 1 h. Then, acetic anhydride (12.0 mL, 25 moles per moles of hydroxyl group) was added to the pyridine solution. The mixture was kept at 90°C with constant stirring for 48 h. Following the reaction, the mixture was poured into 150 mL of methanol to yield a precipitate. The precipitate was dissolved in dichloromethane/methanol (3/1 v/v) mixture, and filtered off. The product was reprecipitated by addition of methanol. The reprecipitation procedure was repeated twice, and then the product (**6**) was dried under vacuum at 60°C for 5 h. The products (**6**) were detritylated in 30 mL of dichloromethane according to the same procedure described above to provide cellulose 2,3-diacetate (**7**).

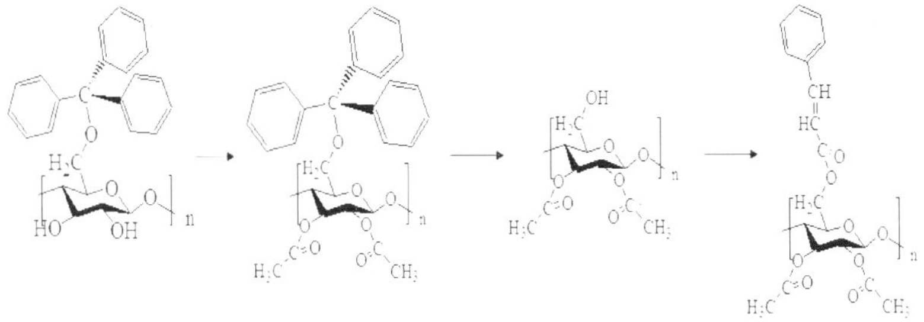

Figure 3. Synthesis of cellulose-2,3-diacetate-6-monocinnamate 8.

Cinnamoyl chloride (17.1g, 12 moles per moles of hydroxyl group) was added to a solution of cellulose 2,3-diacetate (**7**, 1.0 g) in pyridine. Then, the temperature was raised to 95°C and kept there for 5 h. The mixture was cooled to room temperature and poured into 95% (v/v) aq. methanol to precipitate the product. The isolation and purification was followed as described above for the case of 6-*O*-trityl-cellulose 2,3-diacetate. The yield of product (**8**) was 88 %.

Characterization of regioselectively substituted cellulose cinnamates

The obtained products, cellulose tricinnamate (**1**), cellulose 2,3-dicinnamate (**4**), cellulose-6-monoacetate-2,3-dicinnamate (**5**), and cellulose-2,3-diacetate-6-monocinnamate (**8**) were characterized using FTIR and ^{13}C-NMR spectroscopy. In particular, since compounds 1, 5, and 8 lacked hydroxyl groups, IR spectra showing no absorption bands due to stretching vibration of OH groups indicated complete substitution (data not shown).

Figure 4 shows ^{13}C NMR spectra of compounds **1**, **4**, **5** and **8**. As shown in each spectrum, signals were all clearly assigned to the individual chemical shifts of the component carbons. Thus, ^{13}C NMR confirmed the expected structures. Further, using ^{1}H NMR spectroscopy, the degrees of substitution (DS) of cinnamoyl group in the 4 products were found to be 3.0 for **1**, 2.1 for both **4 and 5**, and 0.86 for **8**, respectively.

236

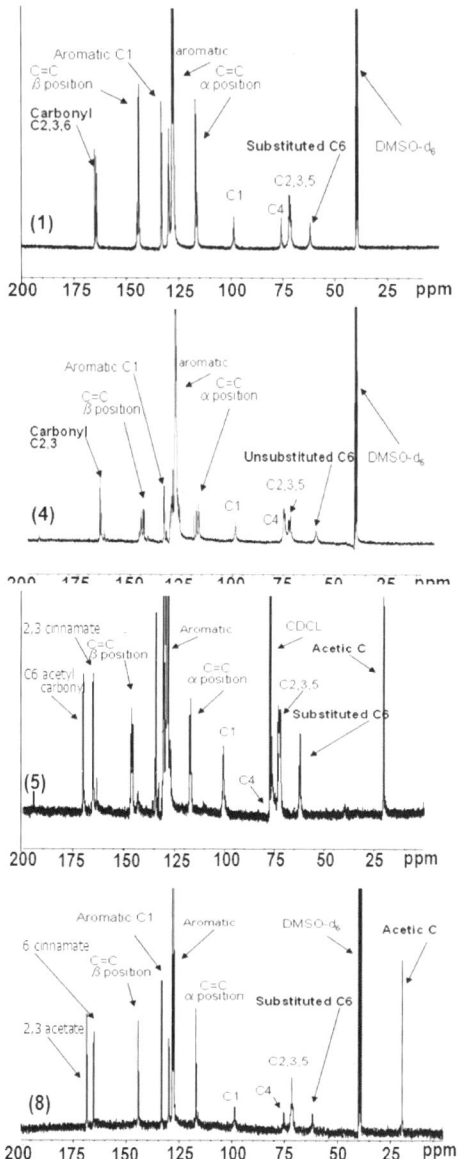

Figure 4. ^{13}C-NMR spectra of cellulose-tricinnamate (**1**), cellulose-2,3-dicinnamate (**4**), cellulose-6-monoacetate-2,3-dicinnamate (**5**) and cellulose-2,3-diacetate-6-monocinnamate (**8**).

The thermal behavior of those products exhibiting melting transition by DSC, compounds 1, 5, and 8, were investigated by obtaining equilibrium melting points (Tm_{eq}) from Hoffman-Weeks plots[29]. The Tm_{eq} was determined from DSC measurements as follows: differential scanning calorimetry (DSC) was performed under a nitrogen atmosphere using a Perkin-Elmer DSC-7. The instrument was calibrated with an indium standard. Film specimens weighing from 5 to 17 mg each were placed in aluminum sample pans which were then heated to 200°C and maintained at this temperature for 7 min to eliminate any residual crystallinity. The samples were then quenched to the selected isothermal crystallization temperature, Tic, and held at this temperature for 5 h. The samples were then cooled to 25 °C and the melting point Tm was then measured using a heating rate of 10 °C/min. Subsequently by using Hoffman-Weeks plots[30] of the Tm' points thus obtained, an equilibrium melting point (Tm_{eq}) was determined for each film as shown in Figure 5.

The measured values of Tm_{eq} for cellulose tricinnamate (1), cellulose 6-monoacetate-2,3-dicinnamate (5), and cellulose 2,3-diacetate-6-monocinnamate (8) were 146 °C, 156 °C and 160 °C, respectively. This data indicates that as the DS (cinnamate) increased, Tm_{eq} decreased.

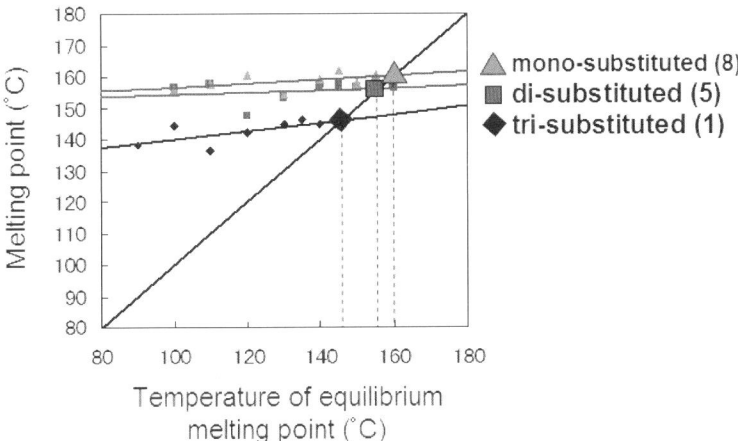

Figure 5. Equilibrium melting points of cellulose derivatives by Hoffman Weeks Plots[29]. Samples 1, 5 and 8 correspond to the compounds shown in Figs 1-3.

In order to examine the changes in morphology of products 1, 5 and 8 with increasing temperature, we observed these samples using polarized microscopy together with DSC. Observations were carried out for cast films of 1, 5, and 8 after isothermal crystallization at various temperatures, at the heating rate of 1 °C/min, with an Olympus polarized microscope BHA equipped with a hot stage (FP90, Mettler Toledo Inc.).

The T_m's determined for compounds **1**, **5** and **8** by DSC following isothermal crystallization are listed in Table 1.

Table 1. Temperature of endothermal peaks of
3 cellulose derivatives (1,5,8) in the DSC scans

cellulose tricinnamate (1)
equilibrium melting point: 146 °C

Isothermal crystallization	1ST peak($\triangle$H)	2ND peak($\triangle$H)
80 ° C 5h		
90 ° C 5h	123 ° C (0.46J/g)	138 ° C (0.72J/g)
100 ° C 5h	130 ° C (0.23J/g)	141 ° C (0.11J/g)
110 ° C 5h	137 ° C (0.82J/g)	
120 ° C 5h	143 ° C (0.97J/g)	
130 ° C 5h	146 ° C (1.15J/g)	
135 ° C 5h	147 ° C (1.26J/g)	
140 ° C 5h	145 ° C (0.58J/g)	

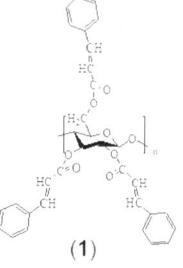

(1)

cellulose-6-monoacetate-2,3-dicinnamate (5)
equilibrium melting point: 156 °C

Isothermal crystallization	1ST peak($\triangle$H)	2ND peak($\triangle$H)
100°C 5h	136°C (0.11J/g)	157°C (1.85J/g)
110°C 5h	141°C (0.40J/g)	159°C (0.96J/g)
120°C 5h	147°C (0.48J/g)	
130°C 5h	154°C (2.49J/g)	
140°C 5h	158°C (2.16J/g)	
145°C 5h	158°C (1.95J/g)	
150°C 5h	156°C (0.53J/g)	
155°C 5h	156°C (1.18J/g)	
160°C 5h	156°C (1.58J/g)	

(5)

cellulose 2,3-diacetate-6-monocinnamate (8)
equilibrium melting point: 160 °C

Isothermal crystallization	1ST peak($\triangle$H)	2ND peak($\triangle$H)
100°C 5h	158°C (0.94J/g)	
110°C 5h	142°C (0.38J/g)	155°C (0.44J/g)
120°C 5h	147°C (0.76J/g)	162°C (0.10J/g)
130°C 5h	154°C (1.43J/g)	
140°C 5h	160°C (1.48J/g)	
145°C 5h	162°C (1.87J/g)	
150°C 5h	158°C (1.10J/g)	
155°C 5h	160°C (0.37J/g)	
160°C 5h	164°C (1.03J/g)	

(8)

Each product displayed one main endothermal peak, and depending upon the annealing temperature, a second peak at a lower temperature. The images from polarized microscopy indicated that the lower endotherm was not melting of the bulk sample. On the other hand, after traversing the higher endotherm, the images disappeared in the polarized microscopic observation. This indicates that the higher endotherm is melting of the polymer.

In cellulose tricinnamate (**1**), there appeared two endothermal peaks, one at 130°C or less and a 2nd in the range of 140-150°C. Cellulose-6-monoacetate-2,3-dicinnamate (**5**) and cellulose 2,3-diacetate-6-monocinnamate (**8**) exhibited two endothermal peaks at 140 °C and at 160 °C. Taking the results from the polarized microscopy into account, the peak at the lower temperature could be due to two possibilities; crystallization of side chain substituents independent of the bulk polymer, which is not a common event, and melting of poorly formed cellulose ester crystals as polymorphs, which is also a function of the annealing (isothermal crystallization) temperature. This commonly occurs when the annealing temperature is close to the Tg. Since there is not sufficient data at present, it would be reported in the future paper. On the other hand, the higher temperature peak could be due to melting of the crystals of cellulose main chains. Occurrence of the two kinds of crystallization growth may depend on the T_{ic} (Isothermal crystallization temperature).

In addition, cellulose tricinnamate (**1**) displayed thermotropic liquid crystalline behavior. The film subjected to isothermal crystallization at 100°C for 36 hours exhibited polarized images again at higher temperature (190°C) after melting. Compounds **5** and **8** did not display such behavior with increasing temperature.

Summary

Regioselectively substituted cellulose cinnamates were synthesized in hopes of obtaining versatile, functional cellulose derivatives. The thermal behavior of the regioselectively substituted cellulose esters depended on the DS of cinnamate. Interestingly, two melting peaks appeared depending on the thermal history at the isothermal crystallization temperature. The lower temperature peaks could be due either to crystals of poorly formed cellulose ester polymorphs or side chain crystals of cinnamoyl groups, whereas the higher temperature peaks were due to melting of the crystals of cellulose main chains. Accordingly, the basic properties of cellulose cinnamate derivatives are strongly affected by the distribution of the substituents. The information concerning the thermal behaviors in this study could provide fundamental understanding to those wishing to modify physicochemical processes, leading to desired functional bio-based materials such as photo-sensitive materials and interpenetrating polymeric materials.

References

[1]Kondo, T. *J. Polym. Sci., B: Polym. Phys.* **1997**, *35*, 717-723.

[2] Kondo, T. *J. Polym. Sci., B: Polym. Phys.* **1994**, *32*, 1229-1236.

[3] Kondo, T.; Sawatari, C.; Manley, R. St-J.; Gray, D. G. *Macromolecules* **1994**, *27*, 210-215.

[4] Itagaki, H.; Takahashi, I.; Natsume, M.; Kondo, T. *Polym. Bull.* **1994**, *32*, 77-81.

[5] Kondo, T.; Sawatari, C. *Polymer* **1994**, *35*, 4423-4428.

[6] Kondo, T., Sawatari, C. *Polymer* **1996**, *37*, 393-399.

[7] Kondo, T. *Cellulose* **1997**, *4*, 281-292.

[8] Kondo, T.; Miyamoto, T. *Polymer* **1998**, *39*, 1123-1127.

[9] Kondo, T.; Sawatari, C. Interchain Hydrogen Bonds in Cellulose-Poly(vinyl alcohol) Characterized by Differential Scanning Calorimetry and Solid-State NMR Analyses Using Cellulose Model Compounds. In: Cellulose Derivatives: Modification, Characterization, and Nanostructures, Heinze, Th., Glasser, W., Eds.; American Chemical Society: Washington, DC, ACS Symp. Ser.; Vol. 688, 1998, pp 296-305.

[10] Shin, J.-H.; Kondo, T. *Polymer* **1998**, *39*, 6899-6904.

[11] Itagaki, H.; Tokai, M.; Kondo, T. *Polymer* **1997**, *38*, 4201-4205.

[12] Kondo, T. Hydrogen bonds in cellulose and cellulose derivatives. In: *Polysaccharides: Structural diversity and functional versatility* Dumitriu, S.; Ed.; Marcel Dekker: New York, 1998, pp 131-172

[13] Kondo, T.; Gray, D. G. *Carbohydr. Res.* **1991**, *220*, 173-183.

[14] Kondo, T. *Carbohydr. Res.* **1993**, *238*, 231-240.

[15] Kondo, T.; Koschella, A.; Brigitte Heublein, B.; Klemm, D.; Heinze, T. *Carbohydr. Res.* **2008**, 343, 2600-2604.

[16] Kondo, T.; Gray, D. G. *J. Appl. Polym. Sci.* **1992**, *45*, 417-423.

[17] Kondo, T.; Nojiri, M. *Chem. Lett.* **1994**, 1003-1006.

[18] Tashiro, K.; Kobayashi, M. *Polymer*, **1991**, *32*, 1516 –1522.

[19] Suh, M-C; Suh, S-C, M; Shim, S-C; Jeong, **B-M.** *Synthetic Metals*, **1998**, *96*, 195 –198.

[20] Wu, Y.; Seo, T.; Maeda, S; Sasaki, T.; Irie, S.; Sakurai, K**.** *J. Polym. Sci., B: Polym. Phys.* **2005**, *43*, 1354-1364.

[21] Luadthong, C.; Tachaprutinun, A.; Wanichwecharungruang, S. P.; *Eur. Polym. J.* **2008**, *44*, 1285-1295.

[22] Minsk, L. M.; Smith, J. G.; van Deusen, W. J.; Wright, J. F. *J. Appl. Polym. Sci.* **1959**, *2*, 302-307.

[23] Klemm, D.; Schnabelrauch, M.; Stein, A. *Makromol. Chem.* **1990**, *191*, 2985-2991.

[24] Ishizu, A.; Isogai, A.; Tomikawa, M.; Nakano, J.; *Mokuzai Gakkaishi* **1991**, *37, 829-833.*

[25] Arai, K.; Satoh, H. *J. Appl. Polym. Sci.* **1992**, *45*, 387-390.

[26] Kamath, M.; Kincaid, J.; Mandal, B. K. *J. Appl. Polym. Sci.* **1996**, *59*, 45-50.

[27] Esker, A. R.,; Grüll, H.; Wegner, G.; Satija, S. K.; Han, C. C.; Langmuir, 2001, 17, 4688-4692.

[28] Sairam, M.; Sreedhar, B.; Mohan Rao, D. V.; *Palaniappan, S. Polym.Adv. Technol.* **2003**, *14*, 477-485.

[29] D. Horton,E. K. Just, *Carbohydr. Res.*, **1973,** 30, 349.

[30] Hoffman, J. D.; Weeks, J. J. *J. Res. Natl. Bur. Stand., Sect. A* **1962**, *66*, 13-18

Chapter 14

Self-Assembling Bolaforms from Biorefinery Polysaccharides

Joseph J. Bozell[*,a,1], **Nathan Tice**[a,2], **Nibedita Sanyal**[a], **Sunkyu Park**[a,3], **Thomas Elder**[b]

[a]**Forest Products Center, Biomass Chemistry Laboratories, University of Tennessee, Knoxville, TN 37996;** [b]**USDA-Forest Service, Southern Research Station, 2500 Shreveport Highway, Pineville, LA 71360**

Glycal-based bolaforms are synthetically flexible compounds whose structural diversity and variety of functional groups suggest that they could be useful probes of the non-covalent forces controlling molecular self-assembly at the nanoscale. The compounds are prepared using a Ferrier reaction between a glycal and a long chain α, ω-diol, followed by deacetylation under Zemplén conditions. A variety of nanostructures are observed upon self-assembly of the bolaforms in solution, but the solid state structures adopt shapes similar to those of bolaforms possessing more extensive hydrogen bonding networks, indicating that multiple hydrogen bonds are important to formation of stable, discrete nanostructures in solution, but that only a few key intermolecular interactions between bolaform headgroups may be necessary to determine the structure in the solid state. An overview of the self-assembly behavior of these bolaforms in solution and the solid state is presented, as well as preliminary results about their interaction with dissolved cellulose as a template for crystallization.

Introduction

Polysaccharides in the form of starch, cellulose or hemicellulose, or as converted into their constituent monosaccharides, comprise as much as 75% of the process streams within a biorefinery, offering a readily available supply of renewable carbon for the production of biobased chemicals and fuels. Yet our knowledge of the technology most effective for transforming these carbohydrates into marketplace chemicals remains limited when compared to the technology available for nonrenewable, petrochemical building blocks.

Narrowing this technology gap by developing conversion processes tailored to oxygen-rich carbohydrate feedstocks would greatly improve our ability to tap a vast and sustainable source of renewable carbon. A recent report from the Oak Ridge National Laboratory identified a sustainable biomass supply in the U. S. of 1.3×10^9 tons/year without upsetting normal supplies of food, feed and fiber, and without requiring extensive changes in infrastructure or agricultural practices.[4] Agriculture in the U. S. excels at the production of carbohydrates through growing of annual domestic crops and forest products. In 2007, the corn industry produced 367×10^6 tons (13.1×10^9 bushels) of corn, containing 210×10^6 tons of polysaccharides as starch, and equivalent to over 500×10^6 barrels of crude oil.[5,6] The pulp and paper industry converts over 240×10^6 tons/yr of wood, which is about 50% polysaccharide as cellulose, for the production of paper products.[7] More broadly, cellulose is the most abundant organic chemical on earth, with an annual production of about $1 \times 10^{10} - 1 \times 10^{11}$ tons.[8] Carbohydrate production from lignocellulosic feedstocks also provides a parallel supply of lignin, up to 25% by weight of the biomass and a promising source of aromatic chemicals.[9] When measured in energy terms, the amount of carbon synthesized by plants is equivalent to about ten times the world's annual consumption for energy and chemical needs.[10,11]

But despite this potentially large supply of domestic raw material, production of chemicals in the U. S. is almost exclusively based on petrochemicals - only about 2% of all chemicals are manufactured from renewables.[12] Our efforts in biorefinery development are directed at increasing this percentage by developing new conversion methodology and identifying product opportunities best suited for the structures present in biomass. This paper summarizes our current work for converting biorefinery carbohydrates and their derivatives into new nanostructural materials. Production of nanoscale materials in the biorefinery could offer a significant economic return. A driver for this research is the anticipated impact of nanotechnology on the nation's economy, with market values of over \$1 trillion within 20 years projected by NSF.[13]

Molecular self-assembly of glycal-based bolaforms

Molecular self-assembly provides access to a wide range of nanoscale materials in chemical and biological processes, and offers a way to build complex structures in a controlled manner from smaller particles.[14,15] Within the context of organic synthesis, interest in self-assembling systems represents an important transition from stepwise formation of covalent bonds to processes that create stable, non-covalent networks between individual molecules.[16-19] A number of reports describe the use of carbohydrates as key structural components in self-assembling molecular systems such as multivalent carbohydrate-protein complexes,[20] nanostructured carbon,[21] or self-assembling block copolymers.[22]

We are investigating carbohydrates as components of *bolaforms*, self-assembling molecules characterized by a long hydrophobic spacer connecting two hydrophilic headgroups.[23] Bolaforms undergo self-assembly into nanoscale micelles, monolayer membranes, tubes or vesicles. Control of self-assembly in these systems is a delicate balancing act between an array of non-covalent interactions: hydrogen and van der Waals bonding, π–π interactions, the strength and directionality of the hydrogen bonding network, chain composition, chain length, headgroup stereochemistry, and the relative influence of the hydrophobic and hydrophilic portions of the molecule. This large group of potential influences highlights the lack of understanding in prediction of shape/structure relationships during self-assembly. Relatively subtle changes in structure can lead to significant changes in the shape of self-assembled arrays, suggesting that the stereochemical diversity available within simple carbohydrates would make carbohydrate-based bolaforms useful probes of the interplay between these forces.

Background – Synthesis of Bolaforms Based on Simple Carbohydrates.
Several approaches have been developed for the synthesis of symmetrical or unsymmetrical carbohydrate based bolaforms (**Figure 1**). A conventional Lewis acid catalyzed glycosylation leads to the synthesis of a disaccharide-containing bolaform, but the overall yield of the process is less than 10%.[24]

Linkages through nitrogen (amides, carbamates, urethanes) have been widely exploited to connect carbohydrate headgroups. Sugar-based bolaforms with gluconamide and lactobionamide headgroups have been synthesized by reaction of C_6 - C_{12} diamines with gluconolactone or lactobionic acid.[25] Alternatively, reaction of galactose or lactose with long chain diisocyanates results in the formation of carbamate-linked bolaforms while reaction with isothiocyanates afforded the corresponding thiocarbamates.[26] The authors report that as the balance between the hydrophobic and hydrophilic properties of the bolaform changed, the resulting nanoscale structures observed also changed. Increasing length in the linking chain gave vesicles or tubules with varying amounts of aggregation, and ultimately, fusion of the initially formed structures.

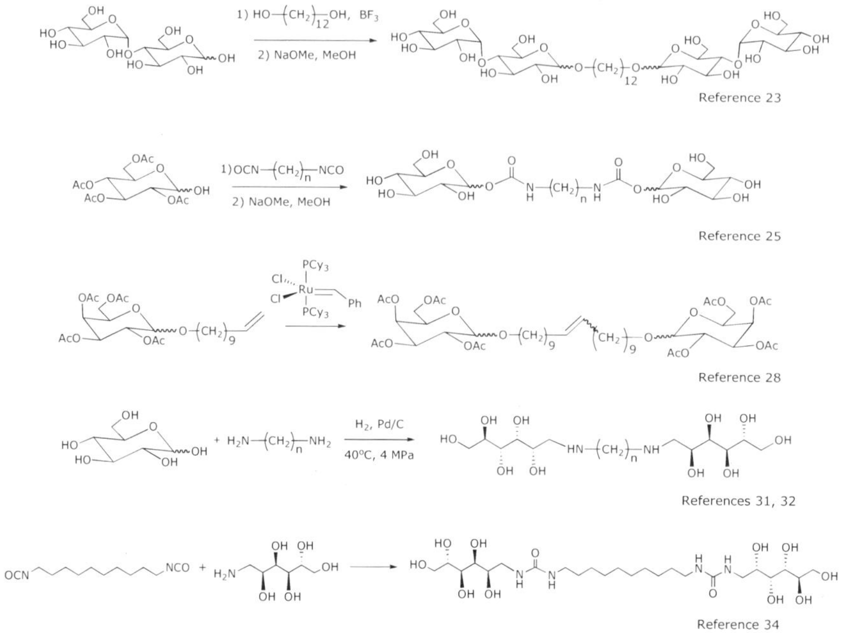

Figure 1 - Selected approaches for the synthesis of carbohydrate-based bolaforms

A family of carbohydrate-based bolaforms with varying chain length was synthesized using a multistep approach starting with etherification of isopropylidene protected sugars with long chain α, ω-diols, and following with deprotection of the alcohols with H_2SO_4. Compounds in this family with chain lengths greater than C_{12} exhibited a significant drop in the critical micelle concentration when compared to bolaforms with C_8 linking chains, indicating the increased impact on aggregation as the relative amount of nonpolar interactions increased.[27] However, the structure of the bolaform can also have an impact on the equilibrium between monomeric bolaforms, micelles and monolayer or multilamellar vesicles or tubes. In some cases, micelle formation is not observed, which has been attributed to an inability of the molecules to adopt the conformation necessary to arrange the nonpolar parts of the bolaform within a micelle structure,[26] as suggested by Gokel in his investigation of bolaforms capped with crown ether headgroups.[28]

Olefin metathesis has been used to synthesize bolaforms with unsaturated linking chains.[29,30] The critical micellar concentration of systems bearing unsaturated chains is lower than that for saturated chains, perhaps indicating improved ability for the linking chains to agglomerate.[31]

Reductive amination of glucose and galactose with diamines affords symmetrical and unsymmetrical bolaforms with open chain headgroups.[32,33] In general, cyclic carbohydrates have the advantage of possessing more directional, and hence, stronger hydrogen bonds between molecules.[34] However, increased directionality in open chain analogs results if urea linkages are used between the headgroups, because of their greater hydrogen bonding ability than either amides or urethanes.[35] Urea-linked bolaforms result from reaction of a family of structurally diverse diisocyanates with aminosugar polyols. The materials show the ability to gel water, which occurs when self-assembly entraps solvent molecules during formation of a three dimensional network.

Masuda and Shimizu have carried out extensive investigation of glycosamide bolaforms, and have developed useful models of the hydrogen-bonding network formed during the crystallization and self-assembly of both symmetrical and unsymmetrical bolaforms.[36-38] Unsymmetrical systems with glucosamide and carboxylic acid headgroups exhibit an interdigitated structure in the solid state, similar to carbohydrate based amphiphiles, and suggesting an influence of headgroup size on the mechanism of self-assembly. Symmetrically substituted glucosamide and galactosamide bolaforms possess two- and three-dimensional hydrogen bonding networks respectively. The conformation of the hydrocarbon chain linking the headgroups changes from a conventional all-*trans* arrangement to one exhibiting a distinct bend, in order to accommodate this network.[39] Molecular dynamics calculations indicate that the all-*trans* conformation is also of higher energy than the bent conformation for the assembled systems, and that a *trans* conformation will relax to the bent form.[40] Bolaforms that adopt a three dimensional hydrogen bonding network are characterized by markedly lower solubility and the formation of different self-assembled structures.[41] Benvegnu *et al* investigated the assembly of several carbohydrate-based bolaforms using freeze/fracture TEM, and found that a variety of structures resulted. The FFTEM evidence was consistent with the formation of monolayer membranes, as no structures resulting from existence of a membrane fracture plane (i. e., analogous to a bilayer) were observed.[42,43]

Modification of the linking chain between headgroups can modify the influence of hydrogen bonding. Carbohydrate based bolaforms linked by azobenzene groups exhibit chiral smectic phases and liquid crystal behavior important for potential use as display or electro-optic devices. Such properties are strongly linked to the nature of the headgroup and $\pi-\pi$ interactions between the azobenzene units. Moreover, the fine balance between the hydrophobic interactions of the linking chains and the hydrophilic interactions of the headgroups appears to play a major role in the liquid crystalline behavior exhibited by these molecules.[44] In general, the strong hydrogen bonding between headgroups precludes transmission of molecular chirality to macroscopic chirality by enhancing formation of lamellar phases. However, this lamellar self-assembly is disrupted through the introduction of azobenzene or diphenylbutadiene[45] structures within the linking chain, leading to the observation of liquid crystalline behavior, and the formation of chiral smectic C* phases. Extended azobenzene structures between the carbohydrate groups result in the formation of "super gelators" for water, forming stable gels at less than 0.1 wt%.[46,47] Solvent gelation is also observed when single aromatic rings are included as part of a longer hydrocarbon chain between headgroups.[48]

Glycals as alternative bolaform headgroups. The utility of simple carbohydrate headgroups as probes of the features controlling self-assembly is limited by multistep syntheses,[27,44,49,50] low yields,[24] or a lack of simple approaches for regiospecific modification of the sugar's secondary C-2, C-3 or C-4 hydroxyl groups. Glycals (1,2-unsaturated glycopyranosides) offer an alternative as headgroups because they are subject to a much wider range of synthetic transformations, exhibit clear differentiation of each reactive site, and offer the potential of enhanced self-assembly via π–π stacking of the double bond (**Figure 2**). Moreover, new catalytic deoxygenation technology may be able to simplify current conversions of readily available biorefinery monosaccharides to glycals.[51] We have examined the synthesis of several glycal-based bolaforms using a simple Ferrier reaction and studied how the structural differences between glycal-based bolaforms and those derived from conventional carbohydrates are manifested during self-assembly.[52]

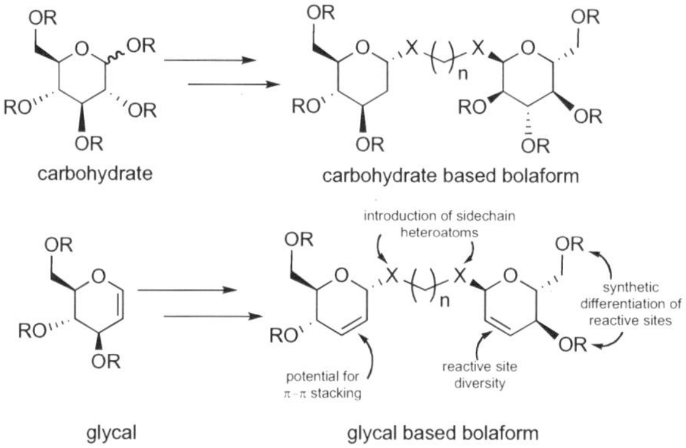

Figure 2 - Comparison of carbohydrate and glycal based bolaforms.

Bolaform Synthesis. Our synthetic approach and results are summarized in **Scheme 1**. Glycal-based bolaforms are synthesized using an iodine catalyzed Ferrier reaction.[53,54] Treatment of the glycal with a long chain α, ω-diol in the presence of 3-5% I_2 in THF gives the corresponding bolaform as the acetate. Deacetylation under Zemplén conditions (5% NaOMe in MeOH) affords the hydroxylated bolaform. The process is general, and gives fair to excellent yields of product from triacetylglucal **1a** (compounds **3a–d**) and –galactal **1b**[55,56] (compound **3e**) and diacetylxylal **1c**[57,58] (compound **3f**), via the intermediate acetates **2a-f**. The process can be carried out with linking chains containing 4 to 12 carbon atoms or in a stepwise manner to give unsymmetrical bolaforms via alcohol **4** to give unsymmetrical bolaform **5** in 76 and 50% yield for the Ferrier and Zemplén reactions, respectively. Triacetylglucal undergoes a thermodynamically controlled addition of diol to give a roughly 70:30 (NMR) mixture of α,α to α,β adducts.[59-63] In contrast, triacetylgalactal gives a stereospecific addition of diol to afford only the α,α product due to steric crowding on the β-face, and the influence of the axial acetoxy group at C-4.[64] Conversion of diacetylxylal to bolaform **3f** exhibits no stereocontrol at the anomeric center, as it lacks both the electronic influence of an axial C-4 acetoxy group and the steric influence of the CH_2OAc group at C-5.

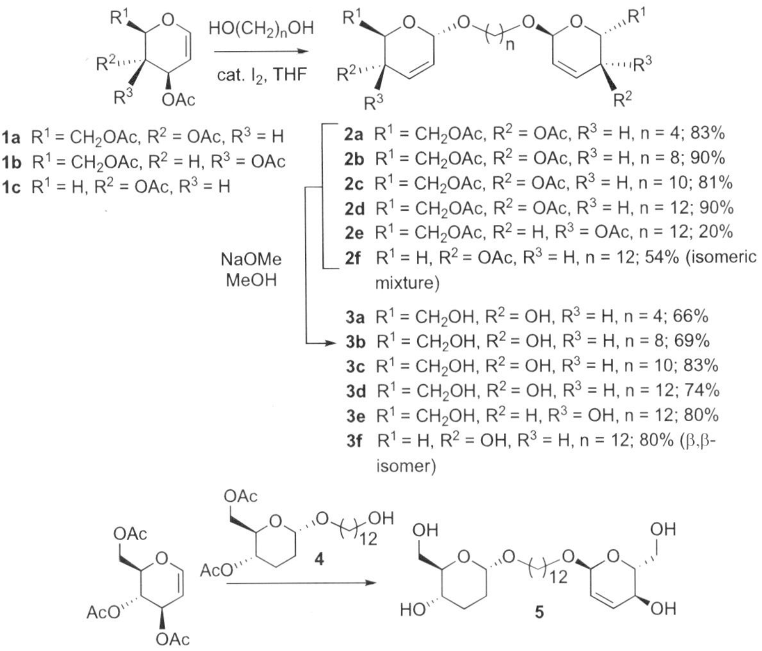

Scheme 1 - Synthesis of glycal-based bolaforms

Bolaform Self-Assembly. Carbohydrate-based bolaforms and related systems form noncovalently bound nanoscale polymers linked through an extensive hydrogen-bonding network established between the carbohydrate headgroups during self-assembly.[65] Although we expected the formation of similar networks during self-assembly of our glycal systems, replacing a carbohydrate with a glycal (**Figure 2**) results in a marked reduction of the hydrogen bonding opportunities and lowers the polarity of the headgroup.

These changes increase the bolaform's sensitivity to other intermolecular forces, such as van der Waals interactions between the hydrophobic chains linking the headgroups. For example, we observe marked solubility differences in water between compounds **3b-3d** upon increasing the chain length from C_8 to C_{12}, with compound **3b** (C_8) showing full solubility in water at concentrations of 20 mg/mL upon gentle warming, but compound **3d** (C_{12}) rapidly precipitating from hot water when cooled. Upon dispersion in aqueous solution, **3b** forms well-shaped, isolated vesicles with diameters between 500 nm and 4 microns (**Figures 3A** and **3B**). Bolaforms of longer chain length show higher aggregation, possibly driven by increased influence of the hydrophobic linking chains of the bolaform. In 2:1 water/1,4-dioxane, bolaform **3c** forms groups of vesicles of smaller diameter and somewhat more irregularly shaped than C_8 bolaform **3b** (**Figures 4A** and **4B**). Sonication appears to have little effect on **3b**, but can disrupt the initially formed aggregates in **3c**.

Bolaform **3d** exhibits the greatest diversity of structures upon self-assembly. In 1:1 water/1,4-dioxane, many samples of **3d** show the presence of large plates, sometimes interspersed with tubular structures. Nanoscale tubules with widths of a few hundred nm and lengths varying between 500 and several thousand nm appear either incorporated with microcrystalline material, or as discrete structures (**Figures 5A-5C**). Other samples show complex assemblies incorporating both tubes and spherical structures (**Figure 5D**). Tube formation is reported during self-assembly of carbohydrate-based bolaforms, and is shown to be the result of stepwise assembly of individual molecules into vesicles, helical sheets, and ultimately, stable tubes.[66] Further, these structures have been shown to coexist in self-assembling systems.[67] However, the low solubility of **3d** appears to offer a favorable parallel route to solid formation, overwhelming the ability of **3d** to exist as stable, isolated nanostructures.

The differences in self-assembly behavior between compounds **3b-3d** highlight the delicate balance between headgroup shape, length of the chain linking the headgroups, and the hydrophobic and hydrophilic properties of the molecules (**Figure 6**). When initially dispersed in solution, the bolaforms will adopt a fairly random orientation. If hydrogen bonding is the dominant noncovalent force, such as cyclic carbohydrate derivatives,[34] research has shown that a fairly rigid and ordered arrangement of the individual molecules occurs, forming a strong, intermolecular network, clearly observed in the solid state, and able to control the conformation of the linking chain.[39] In contrast, systems with fewer hydrogen bonds will be subject to greater influence of hydrophobic interactions between the linking chains. The reduced number of hydrogen bonding opportunities in **3d** may result in more transient nanostructures in solution. Other intermolecular forces in **3d** disrupt and overwhelm hydrogen bonding, entangling the chains in a less ordered manner, causing transition to larger aggregates, the formation of the much wider range of structures for **3d**, and ultimately, precipitation before establishment of an ordered hydrogen bond network occurs in solution. Our ongoing work is examining how the hydrophobic/hydrophilic balance of the bolaforms can be controlled to minimize uncontrolled aggregation and formation of well defined, discrete nanostructures.

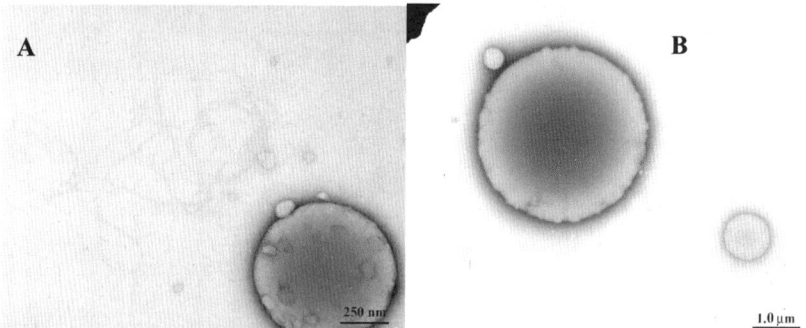

*Figure 3 - TEM Images of Bolaform **3b** (A – no sonication; B – sonicated)*

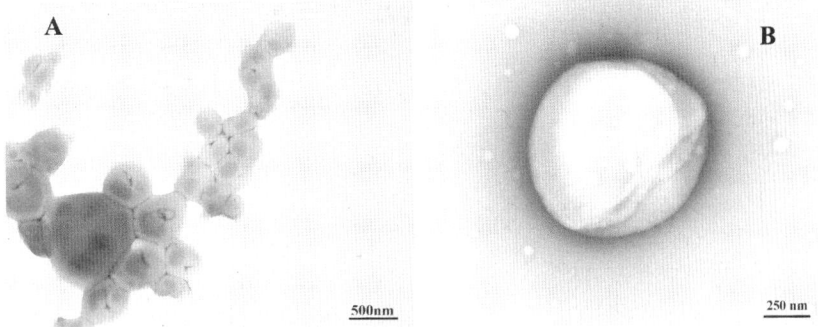

*Figure 4 - TEM images of bolaform **3c** (A – no sonication; B –sonicated)*

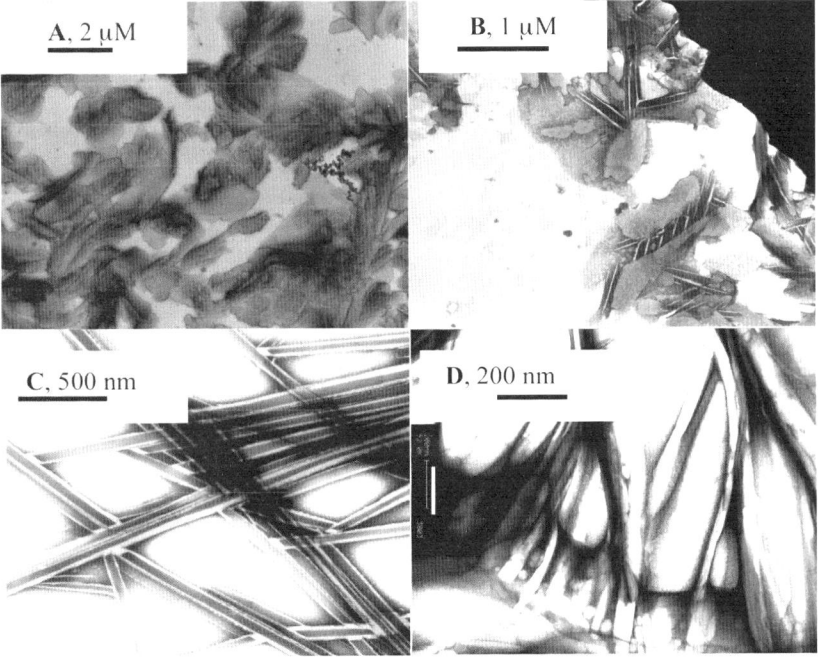

*Figure 5 - TEM images of self-assembled systems from compound **3d***

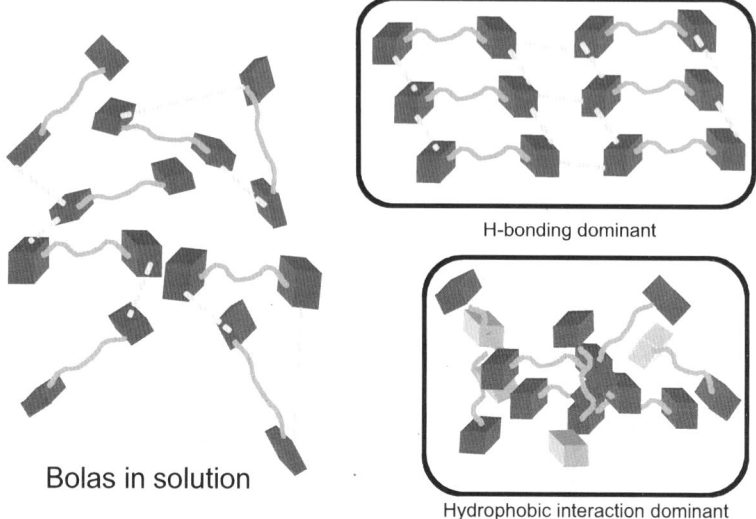

Figure 6 - Schematic of bolaform self assembly in solution
(see page 2 of color insert)

Bolaform solid state structures

To better understand the intermolecular forces in these systems, we have carried out X-ray diffraction of glucal derivative **3d** (both the major α,α- and minor α,β-diastereomers) and galactal derivative **3e**. Hydrogen bonding between headgroups induces considerable order in the solid state for each compound (**Figure 7**).[68] Evaluation of the close contacts in the solid state structure of **3d** and **3e** shows a significant decrease in the number of hydrogen bonds and a modification of the intermolecular network from that reported for similar bolaforms bearing gluco- and galacto- headgroups, respectively. The galacto analog of **3e** has been reported to have as many as 26 unique hydrogen bonding interactions in the headgroup network.[39] Close contact analysis of compounds **3d** and **3e** reveals no more than 6-8 hydrogen bonds between headgroups.

Despite the modification of the intermolecular network, the solid state packing of **3d** and **3e** is very similar to analogous bolaforms bearing simple carbohydrate headgroups. The hydrogen bonding network established by headgroup stereochemistry affords a pleated sheet structure in the solid state for glucal derivative **3d** (**Figure 8A**), and a parallel plane orientation for **3e** (**Figure 8C**), analogous to similar compounds bearing carbohydrate headgroups of identical stereochemistry at C-4 and C-5 (**Figures 8B** and **8D**). Moreover, this similarity even after loss of a significant portion of the hydrogen bonding network suggests that a relatively low number of intermolecular interactions may be necessary to induce the observed ordering. Further, preliminary semi-empirical calculations on a small array of six molecules of **3d** *in vacuo* shows a stabilization energy of approximately 42 kcal/mol consistent with an energetically favorable process upon aggregation. Further modeling is underway to better understand the energetics and mechanism of the assembly process in solution.

For those compounds that form stable nanostructures (e. g., **3b** or **3c**), we anticipate that the hydrogen bonding network is similar to that observed in the solid state. It is also reasonable that **3d** forms an analogous network. Similar retention of structure has been observed for related carbohydrate based bolaforms upon comparison of cast films with aqueous dispersions.[69] The results of the solution and solid-state studies indicate that multiple hydrogen bonds in solution are important to formation of stable, discrete nanostructures, but that only a few key intermolecular interactions between bolaform headgroups may be necessary to determine the structure in the solid state. The synthetic flexibility of glycal-based bolaforms will enhance our ability to study these effects.

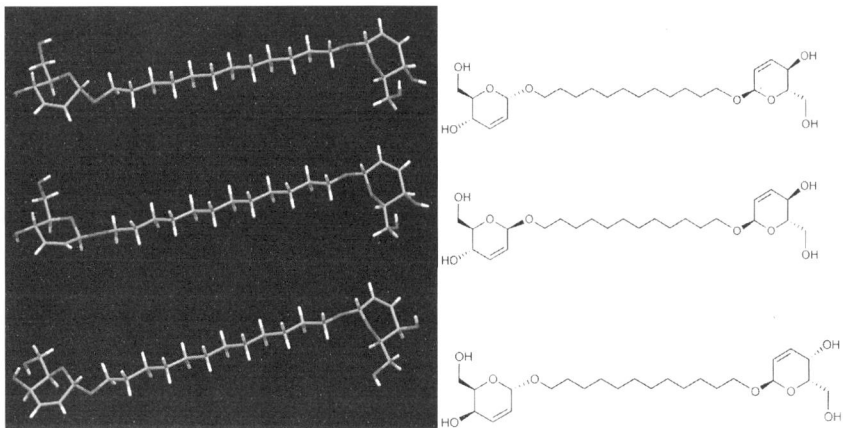

*Figure 7 - X-ray crystal structures of compounds **3d** (major and minor diastereomers) and **3e**.*
(see page 2 of color insert)

254

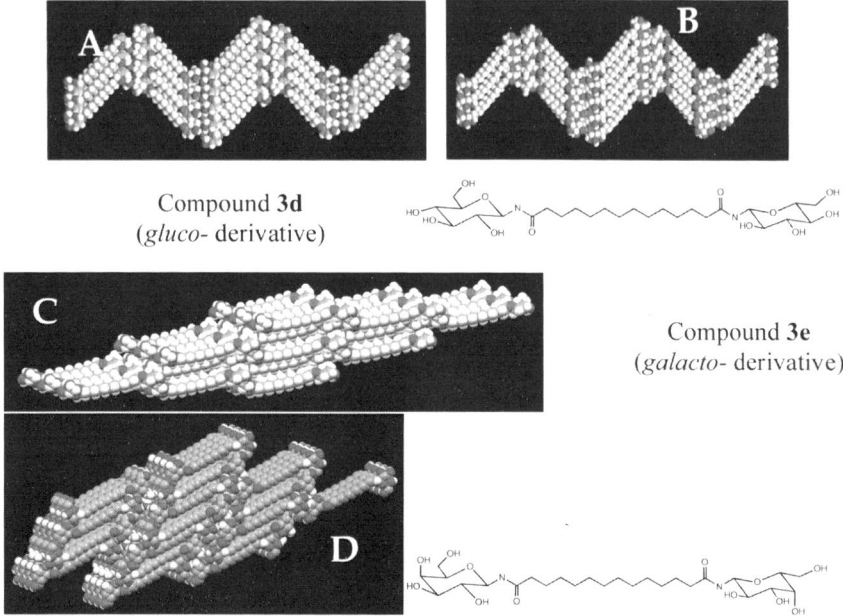

Compound **3d**
(*gluco*- derivative)

Compound **3e**
(*galacto*- derivative)

*Figure 8 - Comparative X-ray structures of glycal bolaforms **3d** and **3e** with
similar carbohydrate bolaforms
(see page 3 of color insert)*

Bolaform Self-Assembly in the Presence of Dissolved Cellulose

Systems that exhibit the ability to self-assemble may show enhanced
organization in the presence of ordered templates. Natural polymers can serve as
templates for the production of a wide range of new materials.[70] Biotemplates
such as fibers, wood, paper or cloth have been treated with ceramic precursors.
Upon removal of the template, usually through heating at high temperatures, a
ceramic material that retains the pattern of the template is produced.[71-74]
Accordingly, we have carried out preliminary investigation of the interaction of
bolaform **3d** with dissolved cellulose to see if organizational information can be
transferred from the cellulose polymer to the bolaform during assembly.

Bolaform **3d** was dissolved in a solution of 8% LiCl in dimethylacetamide,
and a drop of the solution was placed on a glass slide. Polarized light
microscopy reveals the formation of discrete, isolated crystallites (**Figure 9A**).
However, when the same crystallization is carried out in solution containing 2%
dissolved cellulose and 5-10% bolaform (based on the amount of dissolved
cellulose), a higher level of ordering of the bolaform is observed, suggesting
interaction between the polymeric cellulose aggregates in solution and the
bolaform molecules (**Figures 9B** and **9C**). When the hydrogen bonding
opportunities are reduced by carrying out the same process with dissolved
cellulose acetate, no ordering is observed. Cellulose films containing small
amounts of dissolved bolaform **3d** were also prepared and examined by AFM.

Increased order is again observed, as portions of the cellulose surface show ordered strata as opposed to cellulose films prepared in the absence of bolaform (**Figure 10**).

The factors controlling these observations are not yet understood. For example, while increased ordering is observed for the bolaform in the presence of dissolved cellulose, a wide range of crystalline patterns results among the samples. This observation may be due to the bolaform interacting with cellulose aggregates of different morphology in solution, and suggests that cellulose may only be acting as a patterned nucleating site for bolaform crystallization. In addition, the ordering on the surface of cellulose films is not uniform, again suggesting different levels of interaction as a function of cellulose morphology. The nature of these processes is currently under investigation.

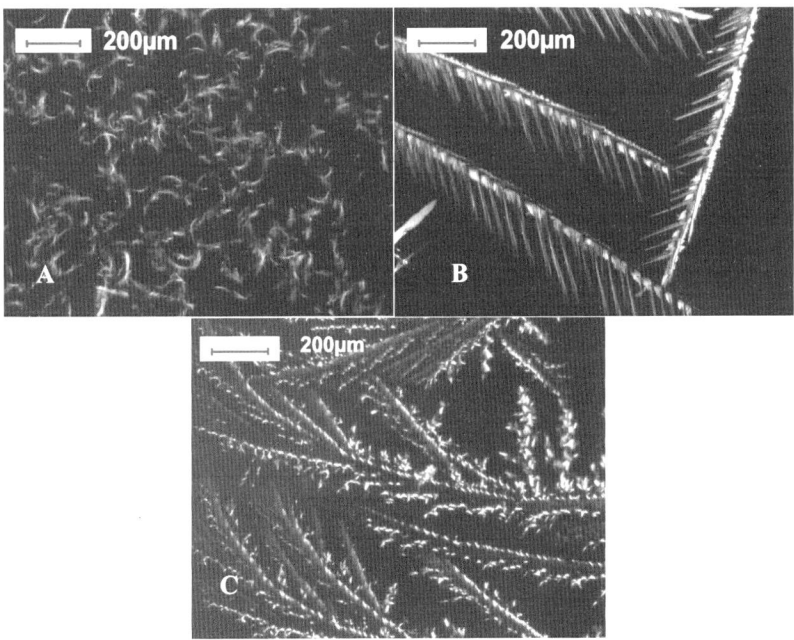

*Figure 9 - Crystallization of bolaform **3d** in the absence and presence of dissolved cellulose (**A** – no cellulose; **B**,**C** – cellulose added)*

256

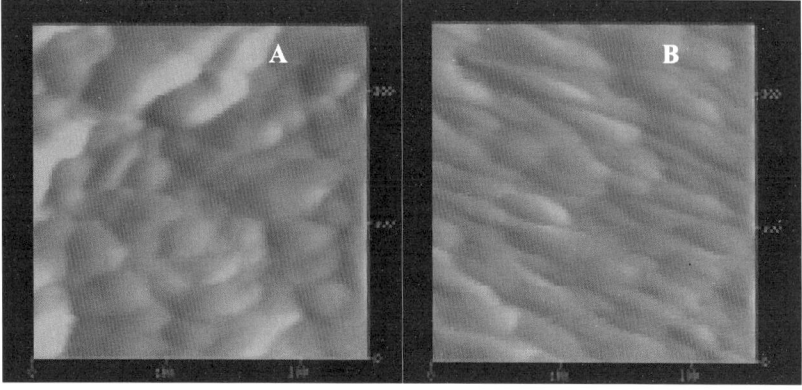

*Figure 10 – AFM of a cellulose film containing bolaform **3d** (A – no bolaform;*
B – bolaform added)

Conclusions

Biorefinery polysaccharides are a rich source of renewable and sustainable carbon for the production of new chemicals and fuels. The challenge for research lies in successfully developing those technologies best suited for polysaccharide conversion, and identifying product opportunities best able to take advantage of new technology. While attempts to identify the most promising, specific chemical structures for biorefinery research generally offers low odds of success, identification of broad chemical sectors is a valid and necessary activity.[75] Nanotechnology could be an important product area for the biorefinery to consider in this context because of its potential large economic impact. We have identified glycal-based bolaforms as possible derivatives of biorefinery process streams because of their synthetic flexibility, ease of preparation, and potential use as probes of the forces controlling self-assembly. Based on these initial results, current work is looking at the utility of glycal-based bolaforms as ordered catalysts supports for sensors, photovoltaic, or electronic devices, as multivalent biodetectors for bioactive compounds, and as compatibilizers for wood/plastic composites.

Acknowledgements

This work is supported by the USDA, the University of Tennessee Institute of Agriculture, and the University of Tennessee Forest Products Center. TEM images were collected by Dr. John Dunlap of the UT Microscopy Center.

References

(1) *jbozell@utk.edu.*
(2) *Current address: Department of Chemistry, Eastern Kentucky University, 521 Lancaster Ave., Richmond, KY 40475.*
(3) *Current address: National Renewable Energy Laboratory, 1617 Cole Boulevard, Golden, CO 80401.*
(4) Perlack, R. D., Wright, L. L., Turhollow, A. F., Graham, R. L., Stokes, B. J., Erbach, D. C. *Biomass as Feedstock for a Bioenergy and Bioproducts Industry. The Technical Feasibility of a Billion-Ton Annual Supply,* U. S. Department of Energy, DOE/GO-102995-2135, ORNL/TM-2005/66, 2005,
(5) Varadarajan, S.; Miller, D. J. *Biotechnol. Progr.* **1999**, *15*, 845-854.
(6) *The World of Corn*; National Corn Growers Association, 2008, http://www.ncga.com.
(7) *Pulp and Paper Factbook- North American Factbook 2001*; W. Mies, J. P., D. Miller, J. Kenny, ed.; Paperloop Publications: San Francisco, 2002,
(8) Hutchens, S. A.; Benson, R. S.; Evans, B. R.; O'Neill, H. M.; Rawn, C. J. *Biomaterials* **2006**, *27*, 4661-4670.
(9) Bozell, J. J., Holladay, J. E., Johnson, D., White, J. F. *Top Value Added Chemicals from Biomass. Volume II – Results of Screening for Potential Candidates from Biorefinery Lignin,* U. S. Department of Energy, PNNL-16983, 2008, www.pnl.gov/main/publications/external/technical_reports/PNNL-16983.pdf.
(10) Indergaard, M.; Johansson, A.; Crawford, B. *Chimia* **1989**, *43*, 230-232.
(11) Dale, B. E. *J. Chem. Technol. Biotechnol.* **2003**, *78*, 1093-1103.
(12) McLaren, J. S. *J. Chem. Technol. Biotechnol.* **2000**, *75*, 927-932.
(13) Roco, M. C.; Bainbridge, W. S. *Societal Implications of Nanoscience and Nanotechnology - NSET Workshop Report,* National Science Foundation, March 2001,
(14) Zhang, S. G. *Nat. Biotechnol.* **2003**, *21*, 1171-1178.
(15) Whitesides, G. M.; Grzybowski, B. *Science* **2002**, *295*, 2418-2421.
(16) Leininger, S.; Olenyuk, B.; Stang, P. J. *Chem. Rev.* **2000**, *100*, 853-907.
(17) Philp, D.; Stoddart, J. F. *Angew. Chem. Int. Ed. Eng.* **1996**, *35*, 1155-1196.
(18) Whitesides, G. M.; Mathias, J. P.; Seto, C. T. *Science* **1991**, *254*, 1312-1319.
(19) Hosseini, M. W. *Acc. Chem. Res.* **2005**, *38*, 313-323.
(20) Gour, N.; Verma, S. *Tetrahedron* **2008**, *64*, 7331-7337.
(21) Rodriguez, A. T.; Li, X. F.; Wang, J.; Steen, W. A.; Fan, H. Y. *Adv. Funct. Mater.* **2007**, *17*, 2710-2716.
(22) Suriano, F.; Coulembier, O.; Degee, P.; Dubois, P. *Journal of Polymer Science Part a-Polymer Chemistry* **2008**, *46*, 3662-3672.
(23) Fuhrhop, A. H.; Wang, T. Y. *Chem. Rev.* **2004**, *104*, 2901-2937.
(24) Gerber, S.; Garamus, V. M.; Milkereit, G.; Vill, V. *Langmuir* **2005**, *21*, 6707-6711.
(25) Garellicalvet, R.; Brisset, F.; Rico, I.; Lattes, A. *Synth. Commun.* **1993**, *23*, 35-44.

258

(26) Prata, C.; Mora, N.; Polidori, A.; Lacombe, J. M.; Pucci, B. *Carbohydr. Res.* **1999**, *321*, 15-23.

(27) Goueth, P.; Ramiz, A.; Ronco, G.; Mackenzie, G.; Villa, P. *Carbohydr. Res.* **1995**, *266*, 171-189.

(28) Munoz, S.; Mallen, J.; Nakano, A.; Chen, Z. H.; Gay, I.; Echegoyen, L.; Gokel, G. W. *J. Am. Chem. Soc.* **1993**, *115*, 1705-1711.

(29) Ramza, J.; Descotes, G.; Basset, J. M.; Mutch, A. *J. Carbohydr. Chem.* **1996**, *15*, 125-136.

(30) Descotes, G.; Ramza, J.; Basset, J. M.; Pagano, S. *Tetrahedron Lett.* **1994**, *35*, 7379-7382.

(31) Satge, C.; Granet, R.; Verneuil, B.; Champavier, Y.; Krausz, P. *Carbohydr. Res.* **2004**, *339*, 1243-1254.

(32) Schuur, B.; Wagenaar, A.; Heeres, A.; Heeres, E. H. J. *Carbohydr. Res.* **2004**, *339*, 1147-1153.

(33) Wagenaar, A.; Engberts, J. *Tetrahedron* **2007**, *63*, 10622-10629.

(34) Masuda, M.; Shimizu, T. *Chem. Commun.* **1996**, 1057-1058.

(35) Avalos, M.; Babiano, R.; Cintas, P.; Gomez-Carretero, A.; Jimenez, J. L.; Lozano, M.; Ortiz, A. L.; Palacios, J. C.; Pinazo, A. *Chem. Eur. J.* **2008**, *14*, 5656-5669.

(36) Masuda, M.; Shimizu, T. *Carbohydr. Res.* **1997**, *302*, 139-147.

(37) Masuda, M.; Shimizu, T. *J. Carbohydr. Chem.* **1998**, *17*, 405-416.

(38) Masuda, M.; Yoza, K.; Shimizu, T. *Carbohydr. Res.* **2005**, *340*, 2502-2509.

(39) Masuda, M.; Shimizu, T. *Carbohydr. Res.* **2000**, *326*, 56-66.

(40) Mikami, M.; Matsuzaki, T.; Masuda, M.; Shimizu, T.; Tanabe, K. *Comp. Mat. Sci.* **1999**, *14*, 267-276.

(41) Nakazawa, I.; Masuda, M.; Okada, Y.; Hanada, T.; Yase, K.; Asai, M.; Shimizu, T. *Langmuir* **1999**, *15*, 4757-4764.

(42) Guilbot, J.; Benvegnu, T.; Legros, N.; Plusquellec, D.; Dedieu, J. C.; Gulik, A. *Langmuir* **2001**, *17*, 613-618.

(43) Benvegnu, T.; Lecollinet, G.; Guilbot, J.; Roussel, M.; Brard, M.; Plusquellec, D. *Polym. Int.* **2003**, *52*, 500-506.

(44) Abraham, S.; Paul, S.; Narayan, G.; Prasad, S. K.; Rao, D. S. S.; Jayaraman, N.; Das, S. *Adv. Funct. Mater.* **2005**, *15*, 1579-1584.

(45) Das, S.; Gopinathan, N.; Abraham, S.; Jayaraman, N.; Singh, M. K.; Prasad, S. K.; Rao, D. S. S. *Adv. Funct. Mater.* **2008**, *18*, 1632-1640.

(46) Kobayashi, H.; Friggeri, A.; Koumoto, K.; Amaike, M.; Shinkai, S.; Reinhoudt, D. N. *Org. Lett.* **2002**, *4*, 1423-1426.

(47) Kobayashi, H.; Koumoto, K.; Jung, J. H.; Shinkai, S. *J. Chem. Soc. Perkin Trans. 2* **2002**, 1930-1936.

(48) Jung, J. H.; Shinkai, S.; Shimizu, T. *Chem. Eur. J.* **2002**, *8*, 2684-2690.

(49) Bertho, J. N.; Coue, A.; Ewing, D. F.; Goodby, J. W.; Letellier, P.; Mackenzie, G.; Plusquellec, D. *Carbohydr. Res.* **1997**, *300*, 341-346.

(50) Denoyelle, S.; Polidori, A.; Brunelle, M.; Vuillaume, P. Y.; Laurent, S.; ElAzhary, Y.; Pucci, B. *New J. Chem.* **2006**, *30*, 629-646.

(51) Gable, K. P.; Ross, B. *ACS Symp. Ser.* **2006**, *921*, 143-155.

(52) Bozell, J. J.; Tice, N. C.; Sanyal, N.; Thompson, D.; Kim, J. M.; Vidal, S. *J. Org. Chem.* **2008**, *73*, 8763-8771.

(53) Ferrier, R. J., Zubkov, O. *Organic Reactions* **2003**, *62*, 569.

(54) Koreeda, M.; Houston, T. A.; Shull, B. K.; Klemke, E.; Tuinman, R. J. *Synlett* **1995**, 90-92.

(55) Litjens, R.; den Heeten, R.; Timmer, M. S. M.; Overkleeft, H. S.; van der Marel, G. A. *Chem. Eur. J.* **2005**, *11*, 1010-1016.

(56) Bukowski, R.; Morris, L. M.; Woods, R. J.; Weimar, T. *Eur. J. Org. Chem.* **2001**, 2697-2705.

(57) Bhaskar, P. M.; Loganathan, D. *Tetrahedron Lett.* **1998**, *39*, 2215-2218.

(58) Mechaly, A.; Belakhov, V.; Shoham, Y.; Baasov, T. *Carbohydr. Res.* **1997**, *304*, 111-115.

(59) Achmatowicz, J., O.; Bukowski, P.; Szechner, B.; Zwierzch.Z; Zamojski, A. *Tetrahedron* **1971**, *27*, 1973-&.

(60) de Freitas, J. R.; Srivastava, R. M.; da Silva, W. J. P.; Cottier, L.; Sinou, D. *Carbohydr. Res.* **2003**, *338*, 673-680.

(61) Fava, C.; Galeazzi, R.; Mobbili, G.; Orena, M. *Tetrahedron-Asymmetry* **2001**, *12*, 2731-2741.

(62) Liu, Z. J.; Zhou, M.; Min, J. M.; Zhang, L. H. *Tetrahedron-Asymmetry* **1999**, *10*, 2119-2127.

(63) Ponticelli, F.; Trendafilova, A.; Valoti, M.; Saponara, S.; Sgaragli, G. P. *Carbohydr. Res.* **2001**, *330*, 459-468.

(64) Demchenko, A. V.; Rousson, E.; Boons, G. J. *Tetrahedron Lett.* **1999**, *40*, 6523-6526.

(65) Shimizu, T.; Masuda, M.; Minamikawa, H. *Chem. Rev.* **2005**, *105*, 1401-1443.

(66) Shimizu, T. *Macromol. Rapid Commun.* **2002**, *23*, 311-331.

(67) Nakashima, N.; Asakuma, S.; Kunitake, T. *J. Am. Chem. Soc.* **1985**, *107*, 509-510.

(68) Tice, N. C.; Parkin, S.; Bozell, J. J. *Carbohydr. Res.* **2008**, *343*, 374-382.

(69) Kameta, N.; Masuda, M.; Minamikawa, H.; Shimizu, T. *Langmuir* **2007**, *23*, 4634-4641.

(70) Sieber, H. *Materials Science and Engineering a-Structural Materials Properties Microstructure and Processing* **2005**, *412*, 43-47.

(71) Caruso, R. A.; Schattka, J. H. *Adv. Mater.* **2000**, *12*, 1921-+.

(72) Greil, P. *J. Eur. Ceram. Soc.* **2001**, *21*, 105-118.

(73) Huang, J. G.; Kunitake, T. *J. Am. Chem. Soc.* **2003**, *125*, 11834-11835.

(74) Walsh, D.; Arcelli, L.; Ikoma, T.; Tanaka, J.; Mann, S. *Nature Materials* **2003**, *2*, 386-U5.

(75) Bozell, J. J. *Clean* **2008**, *36*, 641-647.

Chapter 15

HR-MAS: The Other NMR Approach to Polysaccharide Solids

William T. Winter and DeAnn Barnhart

Cellulose Research Institute and Department of Chemistry, SUNY-Environmental Science and Forestry, Syracuse, NY 13210 (www.esf.edu/cellulose)

Polysaccharides often exist in forms such as gels, gums, blends, solids, or highly viscous liquids that are intractable to conventional solution high-resolution NMR methods. Crossed Polarization Magic-Angle Spinning (CP-MAS) NMR has proven useful in many cases, particularly when the system is comparatively rigid or even crystalline. High Resolution Magic-Angle Spinning (HR-MAS) NMR offers an alternative which may be useful, particularly with respect to the most intransigent systems such as gels, blends, and even fresh never-dried plant or animal tissue. For example, it can provide a facile method for identifying chemical differences between native and mutant organisms or between normal and diseased tissues. Unlike CP-MAS, this method utilizes ^{1}H detection and therefore permits most of the classical one and two dimensional homonuclear and heteronuclear experiments. This chapter provides an introduction to this method and some specific examples of applications to polysaccharides.

Introduction

Polysaccharides, *e.g.*, cellulose, starch, and pectin, are often sparingly soluble and, due to their high molecular weight and hydrophilic nature, tend to exist as highly viscous solutions, gels, dispersions, or semi-crystalline fibers and granules. *In situ*, these polymers frequently occur in association with other polysaccharides and/or proteins. Typically, such systems provide insufficient diffraction data for conformational studies, leaving NMR as the preferred approach. Both molecular weight, commonly in the $10^2 - 10^3$ kD range, and restricted molecular mobility favor shorter transverse relaxation times (T_2) and, concomitantly, larger natural line widths. Despite these complications, it remains highly desirable to develop NMR methods for investigating these systems. Enhanced sensitivity and resolution in solid-state NMR of rare nuclei, e.g., ^{13}C, in polymers were first obtained using a combination of high-power dipolar decoupling and "magic-angle" spinning to narrow line widths and "cross polarization" to eliminate the need for long recycling delays required by the longitudinal relaxation times (T_1) by Schaeffer and Stejskal in 1976 (1). By 1981, Earl and van der Hart had demonstrated that the method could be applied successfully to cellulose (2). Additionally, subsequent variations and enhancements of this method have proven considerably successful for many polysaccharides (3). With regards to cellulose, all of the resonances can now be assigned (4). Specific resonances in cellulose, such as that of C6, have proven to provide insight as to conformation (5), while details in the anomeric region (100-110 ppm) have led to the recognition that native cellulose exists as two different allomorphs, Iα and Iβ (6). Morphological differences, e.g., crystalline versus amorphous domains and accessible versus inaccessible regions, can be evaluated from the multiple C4 peaks in the 80-90 ppm region. Newman et al. have approached the analysis of these structures and their relative importance in a given material through the analysis of T_{1C} (7-12). Iversen has provided an analogous approach to the assignment of C4 chemical shifts in cellulose based upon principal component analysis (PCA) (13,14).

Experiments utilizing hydrogen detection were slower to develop largely due to the strength of the ^1H-^1H dipolar interactions, but Gil et al. demonstrated the possibility of ^1H MAS NMR at very high (>30 kHz) spinning rates by acquiring ^1H NMR spectra from cork (15). The application of high-speed spinning has been further investigated in several laboratories and is widely employed in the estimation of interatomic distance in supramolecular systems including cyclodextrins, as noted in a recent review by Chierotti and Gobetto (16).

HR-MAS Applications

High Resolution Magic-Angle Spinning (HR-MAS) NMR is a complementary tool to CP-MAS NMR that allows one to perform standard NMR experiments for the identification and quantitation of polymers as well as their interactions and mobilities. Probes for this experiment are available from Bruker, Varian (Nanoprobe), and Doty. The design of the HR-MAS probe makes it particularly well suited for experiments involving samples of very small quantities (< 1 mg) and/or restricted or anisotropic mobilities (17,18). This behavior is typical of gels, biological tissues, organisms, e.g., bacteria and algae (19), composites, soils (20), combinatorial chemistry particles (21), and biopsy tissues (22). The magic angle spinning involves rotating the sample, at speeds of 18 kHz in a 4 mm rotor and higher in narrower rotors, about an axis inclined to the magnetic field by 54.74°. This has the effect of minimizing off-diagonal contributions to the chemical shift and dipolar interaction components of the Hamiltonian so that the line widths approach those obtained from small organic molecules in solution, hence sharpening the observed signals otherwise broadened by anisotropy in conventional solid-state NMR. The HR-MAS probe permits experiments using ^{1}H and ^{13}C as well as, in some probes, many other nuclei to acquire the full range of standard two-dimensional NMR experiments, *e.g.*, COSY, TOCSY, HMQC, HMBC, and NOESY, utilized to define molecular structure, shape, and interactions in conventional solution-state NMR.

Conformation in a Starch Gel

Depicted in Figure 1 are the NOESY (A), 200 ms delay, and COSY (B) HR-MAS NMR spectra of a 10% starch gel, the assignments having been determined from previous solution-state NMR publications (*23*). The COSY spectrum illustrates the expected H1, H2 correlation (5.3, 3.6) ppm. In the NOESY spectrum, the additional peaks correlated to H1 at 3.7 and 3.9 ppm in F2 (x-axis) and 5.3 in F1 (y-axis) arise from short-range "through space" interactions involving the anomeric proton (H1) of a given starch glucose residue with H3 of the same glucose residue and H4' of the glucose residue adjacent to the glycosidic bond connecting C1 and C4'. Spectra such as these can be useful in studies of molecular conformation in gels and are readily attainable with spinning rates as low as 4 kHz. In this particular case, the (H1, H4') NOESY peak establishes that the distance between these protons in the gel is less than 0.45 nm. Since that distance correlates strongly to the values of the linkage torsion angles, the distance can be utilized as a restraint in molecular modeling of the gel-state molecular conformation and provides severe limitations to the range of the ϕ and ψ conformation angles defining the local polymer conformation.

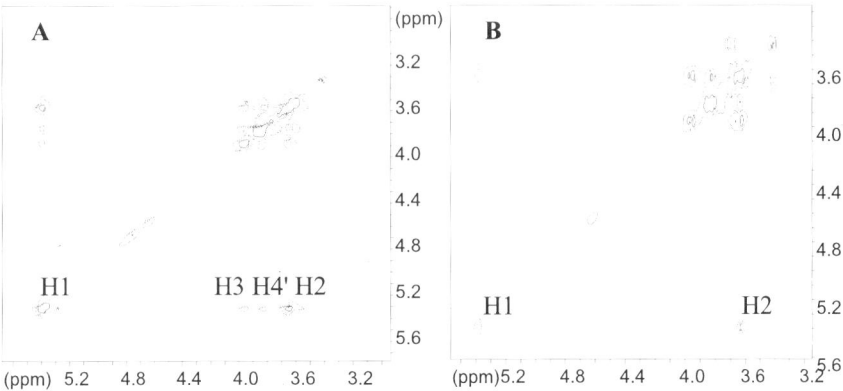

Figure 1. 600 MHz ^{1}H-^{1}H NOESY (A) with 200 ms mixing time and COSY (B) HR-MAS NMR spectra of a 10% starch paste in D_2O (24).

264

Interactions with Cellulose

Interactions between polymers represent another field of study for which NOESY / COSY comparisons may provide considerable insight. The work outlined here is preliminary and previously unpublished but is indicative of modes in which HR-MAS spectra may provide some understanding of the interactions between different molecular species in real or model systems in respect to plant tissue.

Previously, Kabel *et al.* have reported on xylan structural differences and their effects upon interactions with cellulose (*25*). The 600 MHz ^{1}H-^{1}H TOCSY HR-MAS NMR spectrum, illustrated in Figure 2, was obtained by slurrying Avicel in a birch xylan, courtesy of T. E. Timell. By comparing this spectrum with previous data from other plant xylans (*26,27*), it became clear that i) both the strong and weak row crosspeaks correlate with signals from the arabinofuranose residues, which have their anomeric proton peak at 5.3 ppm and that ii) the peaks at 2.4 and 1.6 ppm indicate the presence of acetate and 4-O-methyl glucuronic acid, respectively. Although no visual evidence of cellulose is present, it is suspected that the use of a less rigid material such as never-dried bacterial cellulose would produce different results. Currently, further research in this area is proceeding with particular emphasis upon the acquisition of NOESY spectra for the purpose of demonstrating short range interactions between different parts of the hemicellulose molecules and/or interactions within the cellulose chain itself.

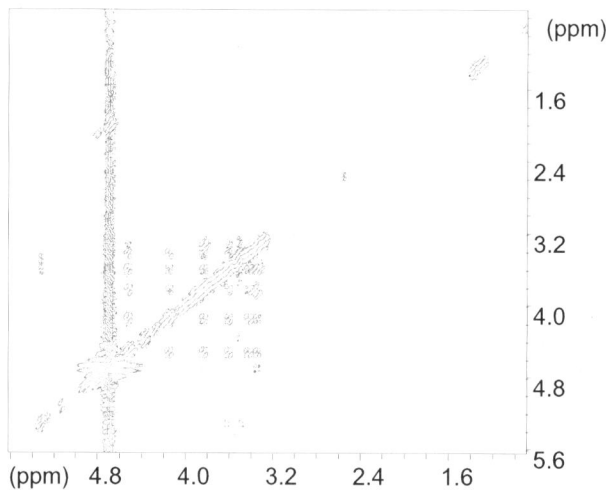

Figure 2. 600 MHz ^{1}H-^{1}H TOCSY HR-MAS NMR spectrum of birch arabinoxylan slurried onto cellulose

Applications of HR-MAS to Glycomics

The first plant to have had its genome sequenced was *Arabidopsis thaliana*, a flowering plant from the mustard family Brassicaceae. *Arabidopsis thaliana* was a good candidate for study due to its short growth cycle, small size, large seed proliferation, and small genome. This sequencing project, incorporating five pairs of chromosomes, 25,498 genes, and 115,409,949 base pairs, was completed in December 2000 (*28*). With the genome recorded, the project then turned from one of sequencing to one of annotation. The *Arabidopsis* genome was compared to those of *Escherichia coli*, *Synechocystis* sp., *Saccharomyces cerevisiae*, *Caenorhabditis elegans*, *Drosophila*, and *Homo sapiens* to detect similarities. By comparing sequences of known purpose to the *Arabidopsis* genome, it was possible to group 69% of the *Arabidopsis* genes into functional categories. Experimental methods will be needed to classify the remainder, as well as to determine the specific functions of previously classified genes (*28*).

The best way to determine the function of a particular gene is to "knock out" that gene and note the changes in the resulting mutant organism. In *Arabidopsis,* some popular targets for these induced gene mutations are the genes associated with endo-1,4-β-D-glucanases (EGases, cellulases), a large family of hydrolytic enzymes that have been shown to affect such processes as leaf abscission, fruit ripening, and cell expansion. In plants, EGases act by cleaving 1,4-β-glucosidic bonds between unsubstituted glucose residues on xyloglucan, glucomannan, and non-crystalline cellulose substrates in the cell wall (*29*). Unlike microbial EGases, most plant EGases are incapable of degrading crystalline cellulose due to the absence of a cellulose binding module (CBM), which allows attachment of the enzyme to the crystalline surface (*30*). However, *A. thaliana* is exceptional in that it has a structural subclass C that demonstrates CBM expression (*29*). Though currently the function of this CBM remains unknown, *in vitro* studies have verified the module's ability to bind to bacterial microcrystalline cellulose and modified (hydrolyzed) plant microcrystalline cellulose (*30*). Three *Arabidopsis* mutants that contain a point mutation in the genomic region affiliated with endo-1,4-β-D-glucanase so that they no longer produce a complete CBM are depicted in Table I (*29*).

Table I. Designations for three mutant *A. thaliana* plants (*29*).

Designation of mutants	Annotation	Phenotype
GH9C1-1	endo-1,4- β-glucanase 5	CBM knockout
GH9C2-2	endo-1,4- β-glucanase 6	CBM knockout
GH9C3-1	endo-1,4- β-glucanase 19	CBM knockout

266

Wild Type, GH9C1-1, GH9C2-2, and GH9C3-1 *Arabidopsis thaliana* were grown aseptically on agar plates by the research group of Dr. Jocelyn Rose at Cornell University. After three weeks of growth, the plants were transferred to SUNY-ESF for sample preparation. Aseptically, the root tissues were severed from the shoots and removed from the agar. The tissues were subjected to *in situ* 600 MHz ^{1}H HR-MAS NMR analysis ($T = 30°C$) as illustrated in Figure 3. All samples consisted of ~6 mg of root tissue, with DSS as an internal standard. The differences in peak intensities at 2.48 ppm, 2.14 ppm, and 1.96 ppm, between WT and mutant *Arabidopsis* root tissue spectra suggest changes in acetate groups, most likely due to deacetylation or deletion of certain polysaccharide residues.

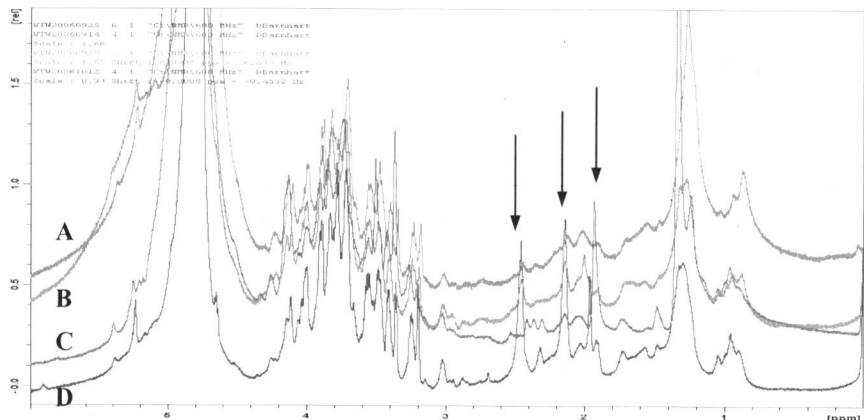

Figure 3. 600 MHz ^{1}H HR-MAS NMR spectra of WT (D), GH9C1-1 (C), GH9C3-1 (B), and GH9C2-2 (A) Arabidopsis thaliana root tissues scaled by carbohydrate region intensities. Arrows indicate significant differences in peak intensities in the acetate region of the spectra.
(see page 4 of color insert)

The WT and GH9C2-2 *Arabidopsis* root tissues were then subjected to *in situ* 600 MHz 2D HR-MAS NMR analysis ($T = 30°C$). The resulting 2D NMR spectra, shown in Figures 4-6, clearly demonstrate the absences of peaks from mutant tissue spectra in both the acetate (2.0-2.5 ppm) and the methyl (0.5-2.0 ppm) regions. Also observed in Figures 4-6 is a shift of the peaks in the carbohydrate (3.0-4.5 ppm) region, a phenomenon indicative of deacetylation. Of particular interest is the disappearance of the ~1.7 ppm peak from the GH9C2-2 spectra. Given the cell wall polysaccharides that are likely to be present in the sample, this peak in the methyl region is most likely to originate from fucose and/or rhamnose residues since both of these monosaccharides are methylated at the C6 position. Changes in the fucose and rhamnose residues would indicate alterations to the xyloglucan and pectin conformations, respectively.

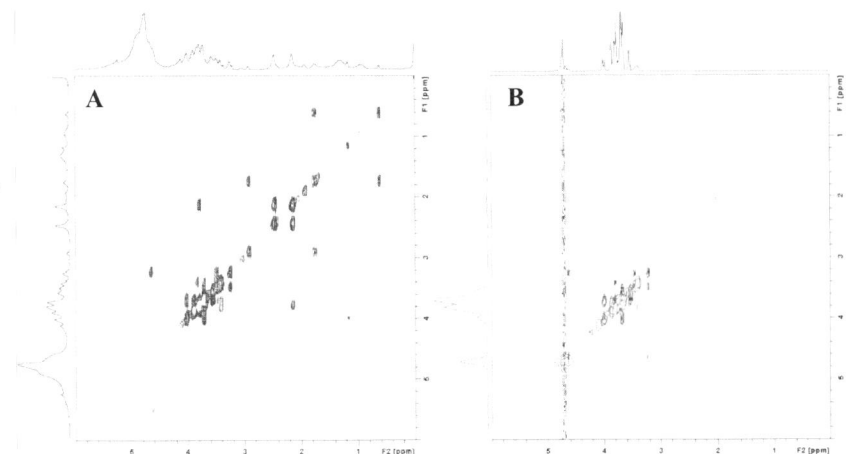

Figure 4. 600 MHz ^1H-^1H COSY HR-MAS NMR spectra of WT (A) and GH9C2-2 (B) root tissues. The absences of peaks in the acetate and methyl regions of the GH9C2-2 spectrum indicate deacetylation and/or deletion of certain residues.

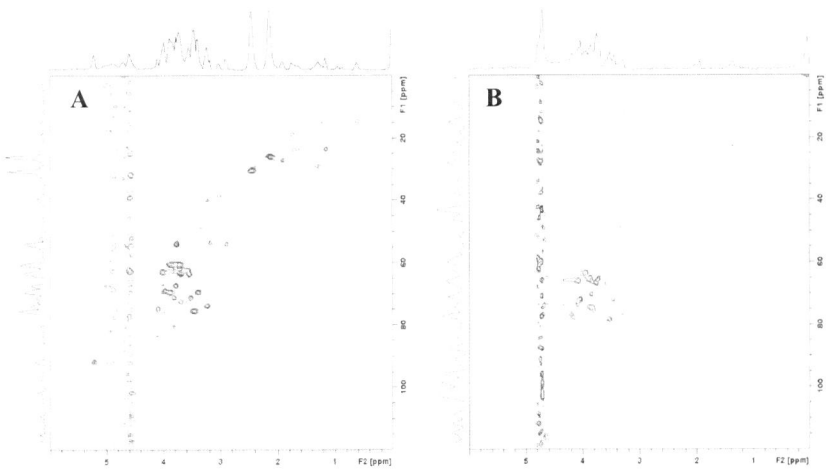

Figure 5. 600 MHz ^{13}C-^1H HMQC HR-MAS NMR spectra of WT (A) and GH9C2-2 (B) root tissues. The absences of peaks in the acetate and methyl regions of the GH9C2-2 spectrum indicate deacetylation and/or deletion of certain residues.

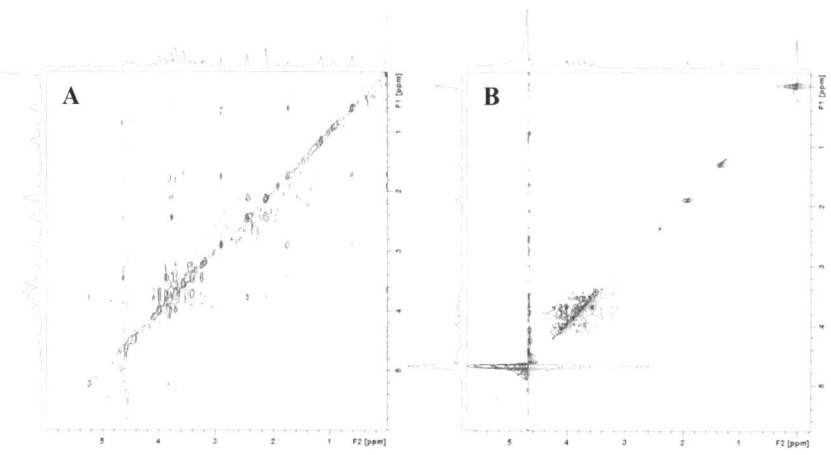

Figure 6. 600 MHz ^1H-^1H TOCSY HR-MAS NMR spectra of WT (A) and GH9C2-2 (B) root tissues. The absences of peaks in the acetate and methyl regions of the GH9C2-2 spectrum indicate deacetylation and/or deletion of certain residues.

Alterations in the pectin content were verified through FT-IR spectroscopic analysis of wild type and mutant *Arabidopsis* leaf tissues, performed by Dr. Rose's group at Cornell University. This is in agreement with the absences of peaks from the methyl regions of the GH9C2-2 NMR spectra. Since rhamnose and fucose residues comprise 3% and 1% of the monosaccharide *Arabidopsis* cell wall components, respectively (*31,32*), the C6 methyl rhamnose peak would be the more prominently visible peak in the NMR spectra. Given the highly complex and variable structure of pectin, additional research will be necessary in order to better determine the conformational changes which have been induced.

Conclusions

The intent of this article is to introduce the reader to the potential of HR-MAS NMR to supplement existing methods utilizing CP-MAS NMR and dipolar decoupled (DD) NMR. This technique offers unique advantages in that it is particularly well suited for the analysis of softer tissues and mobile surfaces of more rigid domains while still offering most of the standard organic chemistry retinue of hydrogen detected two-dimensional NMR techniques. Although not discussed in this account, numerous methods are available for filtering out either the less mobile components using Carr Purcell Meiboom Gill pulse sequences or the more mobile components utilizing diffusion correlated spectroscopies. As a final word of caution for the reader, the conventions for presenting solid-state NMR data have, until recently, been far less well-defined than those for solution and liquid-state NMR. It is perhaps a measure of the growing acceptance of the techniques described here that IUPAC is beginning to provide recommended conventions for presenting chemical shift and shielding data using solid-state experiments (*33*).

References

1. Schaefer, J.; Stejskal, E.O. *J. Am. Chem. Soc.* **1976**, *98*, 1031.
2. Earl, W. L.; vanderHart, D. L. *Macromolecules* **1981**, *14*, 570.
3. Saito, H. In *Polysaccharides (2nd Edition)*; Dimitriu, S., Ed.; Marcel Dekker: New York, 2005; pp. 253-266.
4. Kono. H.; Yunoki, S.; Shikano, T.; Fujiwara, M.; Erata, T.; Takai, M. *J. Am. Chem. Soc.* **2002**, *124*, 7506-11.
5. Horii F.; Hirai A.; Kitamaru R. In *The Structures of Cellulose: Characterization of the Solid States*; Atalla, R.H., Ed.; ACS Symp. Series: Washington DC, 1987; 304, pp. 119-134.
6. Atalla, R.H.; VanderHart, D.L. *Science* **1984**, *23*, 283–285.
7. Newman, R.H.; Hemmingson, J.A. *Holzforschung* **1990**, *44*, 351-355.
8. Newman, R.H. *J. Wood Chem. Technol.* **1994**, *14*, 451-466.
9. Newman, R.H.; Ha, M.A.; Melton, L.D. *J. Agric.Food Chem.* **1994**, *42*, 1402-1406.
10. Newman, R.H.; Hemmingson, J.A. *Cellulose* **1995**, *2*, 95-110.
11. Newman, R.H.; Davies, L.M.; Harris, P.J.; *Plant Physiol.* **1996**, *111*, 474-485.
12. Newman, R.H. *Cellulose* **1997**, *4*, 269-279.
13. Lennholm, H.; Larsson, T.; Iversen, T. *Carbohhydr. Res.* **1994**, *261*, 119-131.
14. Lennholm, H.; Westermark, U.; Iversen, T. *Carbohhydr. Res.* **1997**, *278*, 339-343.
15. Gil, A.M.; Lopes, M.H.; Pascoal-Neto, C.; Rocha, *J. Solid State Nuclear Magnetic Resonance* **1999**, *15*, 59-67.
16. Chierotti, M.R.; Gobetto, R. *Chem. Commun.* **2008**, 1621–1634.
17. *Applications of high resolution magic angle spinning spectroscopy*; Bruker; Bruker-Biospin, Inc.: Billerica, MA, 1997, Version 1.0.
18. Schnell, I.; Spiess, H. W. *J. Magnetic Resonance* **2001**, *151*, 153-227.
19. Broberg, A.; Kenne, L. *Anal. Bioch.* **2000**, *284*, 367-374.
20. Simpson, A.J.; Kingery, W.L.; Shaw, D.R.; Spraul, M.; Humpfer, E.; Dvortsak, P. *Anal. Bioch.* **2001**, *35*, 3321-3325.
21. Shapiro, M.J.; Gournarides, J.S. *Bioeng. (Comb. Chem.)* **2001**, *71*, 130-148.
22. Griffin, J.L. *Current Opinion in Chemical Biology* 2003, *7*, 648-654.

23. Falk, H.; Stanek, M. *Monatshefte für Chemie / Chemical Monthly* **2004**, *128*, 777-784.
24. Winter, W.T. *Abstracts of Papers*, 223rd National Meeting of the American Chemical Society, Orlando, FL, April 7-11, 2002; American Chemical Society: Washington, D.C. 2002.
25. Kabel, M.A.; van den Borne, H.; Vincken, J.P.; Voragen, A.G.J.; Schols, H.A. *Carbohydrate Polymers* **2007**, *69*, 94–105.
26. Sun, J.X; Sun, R.C.; Sun, X.F.; Su, Y.Q. *Carbohydrate Research* **2004**, *339*, 291–300.
27. Vignon, M.R.; Gey, C. *Carbohydr. Res.* **1998**, *307*, 107-111.
28. The *Arabidopsis* Genome Initiative. *Nature* **2002**, *408*, 796-815.
29. Urbanowicz, B.R.; Bennett, A.B.; del Campillo, E.; Catala, C.; Hayashi, T.; Henrissat, B.; Hofte, H.; McQueen-Mason, S.J.; Patterson, S.E.; Shoseyov, O.; Teeri, T.T.; Rose, J.K.C. *Plant Physiol.* **2007**, *144*, 1693-1696.
30. Urbanowicz, B.R.; Catala, C.; Irwin, D.; Wilson, D.B.; Ripoll, D.R.; Rose, J.K.C. *J. of Biol. Chem.* **2007**, *282*(16), 12066-12074.
31. Sato, S.; Kato, T.; Kakegawa, K.; Ishii, T.; Liu, Y.; Awano, T.; Takabe, K.; Nishiyama, Y.; Kuga, S.; Sato, S.; Nakamura, Y.; Tabata, S.; Shibata, D. *Plant Cell Physiol.* **2001**, *42*(3), 251-263.
32. Hu, Y.; Zhong, R.; Morrison, W.H.; Ye, Z. *Planta* **2003**, *217*, 912-921.
33. Harris, R.K.; Becker, E.D.; Cabral de Menezes, S.M.; Granger, P.; Hoffman, R.E.; Zilm, K.W. *Solid State Nuclear Magnetic Resonance* **2008**, *33*, 41-56.

Indexes

Author Index

Subject Index